"This second edition of *EU Environmental Governance* keeps the best insights of the first edition while insisting on crucial questions about the growing challenges faced by the European Union: can the EU remain a 'laboratory' for environmental politics and deal with the environment as a 'human right' whereas the Green Deal seems weakened? Can it reconcile its trade and environmental policies? Is the EU Arctic policy effective? How to integrate youth in environmental policy? A must-read collective book for all students and scholars interested in environmental governance and EU studies".

**Franck Petiteville,** *Professor of Political Science, University of Grenoble Alps, France*

"The second edition of this interdisciplinary book offers a timely and insightful reconsideration of the EU governance system and its capacity to tackle the major environmental challenges arising from the 'triple planetary crisis' at both regional and global levels. The expert input provided in this book by both academics and practitioners will meaningfully contribute to developing much-needed informed policies and to specialized teaching and research activities. It represents an indispensable source of knowledge for anybody interested in environmental matters (researchers, students, NGOs, law and policy-makers, professionals, etc.) and a valuable addition to university libraries all over the world."

**Stefania Negri,** *Associate Professor of International Law, University of Salerno, Italy*

# EU ENVIRONMENTAL GOVERNANCE

This book presents an overview of the field of environmental law and policies within the European Union (EU) and externally, from theoretical foundations to major issues and applied governance solutions.

Drawing on expertise from renowned academics and practitioners from different disciplines, *EU Environmental Governance: Current and Future Challenges* helps readers to understand the main legal, political and economic issues of environmental protection since the adoption of the European Green Deal in 2019. This new edition presents both current insights and future challenges as they seem to be emerging in the new geopolitical era in the EU after major events such as the COVID-19 pandemic and the Russian invasion in Ukraine. The authors examine a broad range of sensitive and topical environmental challenges including climate change, air and environmental pollution, waste management, biodiversity protection, environmental and human health, marine biodiversity, renewable energy, nuclear energy and sustainable Arctic governance. Overall, this volume exposes the reader to a vast array of empirical case studies, which will bolster their training and help tackle the environmental challenges faced by Europe today.

This book is a valuable resource for students, researchers and policymakers across a broad range of fields, including environmental law and policies, environmental economics, climate science and environmental sociology.

**Amandine Orsini** is Professor of International Relations at the UCLouvain Saint-Louis – Bruxelles (USL-B). She is Jean Monnet Chair on "the European Union and Global Green Politics: the Anthropocene Challenge" (EUGlobal-Green) and is the academic Director of the advanced Master on EU environmental governance (POLLEN Master) at USL-B. She has published

*Essential Concepts of Global Environmental Governance* (2020) and *Global Environmental Politics. Understanding the Governance of the Earth* (2020).

**Eleni Kavvatha** is a researcher in Arctic and EU Relations at UCLouvain Saint-Louis – Bruxelles, where she works for CReSPo (Centre of Research in Political Science) and the ISPOLE (Institut de Sciences Politique) under the supervision of Amandine Orsini. Her research is focused on the impact of EU environmental policies on the Sámi Indigenous People in the European Arctic region. She has worked in Belgium, Norway, Iceland, Svalbard, Greenland and the Finish Lapland, on subjects related to EU External Relations in the High North, non-state actors in the Arctic, Indigenous Peoples Organizations, Arctic communities and Arctic Governance. She holds a Master's degree in Environmental Governance and Sustainable Development from Panteion University of Political and Social Studies in Athens, followed by a Master's degree in Governance and Entrepreneurship in Northern and Indigenous Areas from UiT, The Arctic University of Norway. She has co-edited the previous edition of this book for Routledge publishing house under the title *EU Environmental Governance: Current and Future Challenges* and has several publications related to the Arctic.

## Routledge Studies in Environmental Policy

**Deliberative Governance for Sustainable Development**
An Innovative Solution for Environment, Economy and Society
*Franz Lehner*

**Stopping Climate Change**
Policies for Real Zero
*Paul Ekins*

**Deploying the European Green Deal**
Protecting the Environment Beyond the EU Borders
*Edited by Mar Campins Eritja and Xavier Fernández-Pons*

**The European Green Deal in Education**
*Edited by Sarah Anne McDonagh, Alessandro Caforio and Alessandro Pollini*

**EU Environmental Governance, second edition**
Current and Future Challenges
*Edited by Amandine Orsini and Eleni Kavvatha*

For more information about this series, please visit: www.routledge.com/Routledge-Studies-in-Environmental-Policy/book-series/RSEP

# EU ENVIRONMENTAL GOVERNANCE

## Current and Future Challenges

Second Edition

*Edited by Amandine Orsini and Eleni Kavvatha*

Routledge
Taylor & Francis Group
LONDON AND NEW YORK

Designed cover image: Getty Images

Second edition published 2026
by Routledge
4 Park Square, Milton Park, Abingdon, Oxon OX14 4RN

and by Routledge
605 Third Avenue, New York, NY 10158

*Routledge is an imprint of the Taylor & Francis Group, an informa business*



First edition published by Routledge 2021

*British Library Cataloguing-in-Publication Data*
A catalogue record for this book is available from the British Library

ISBN: 9781032766010 (hbk)
ISBN: 9781032766119 (pbk)
ISBN: 9781003479260 (ebk)

DOI: 10.4324/9781003479260

Typeset in Sabon
by codeMantra

*To our Loved Ones*

# CONTENTS

# TABLES

# BOXES

# FIGURES

# ACKNOWLEDGEMENTS

This second edition of *EU environmental governance: current and future challenges* is one output of the Jean Monnet Chair «The European Union and Global Green Politics: the Anthropocene Challenge» (EUGlobalGreen) under call ERASMUS-JMO-2021-HEI-TCH-RSCH, topic ERASMUS-JMO-2021-CHAIR awarded to Prof. Amandine Orsini (2022-2025).

For Chapter 11, Prof. Amandine Orsini and Eleni Kavvatha recognize support by UCLouvain Saint-Louis – Bruxelles under an FSR Project Grant (2024-2026).

For Chapter 12, Prof. Amandine Orsini recognizes support by the Fonds de la Recherche Scientifique - FNRS under Grant n° T.0020.21.

# CONTRIBUTORS

## In order of appearance

**Amandine Orsini** is Professor of International Relations at the UCLouvain Saint-Louis – Bruxelles (USL-B). She is Jean Monnet Chair on "the European Union and Global Green Politics: the Anthropocene Challenge" (EUGlobal-Green) and is the academic Director of the advanced Master on EU environmental governance (POLLEN Master) at USL-B. She has published *Essential Concepts of Global Environmental Governance* (2020) and *Global Environmental Politics. Understanding the Governance of the Earth* (2020).

**Eleni Kavvatha** is a PhD researcher in Arctic and EU Relations at UCLouvain Saint-Louis – Bruxelles funded by FSR and working under the supervision of Professor Amandine Orsini. Her research is focused on the impact of EU environmental policies on the Sámi Indigenous People in the European Arctic region. She has worked in Belgium, Norway, Iceland, Svalbard, Greenland and the Finish Lapland on themes related to non-state actors in the Arctic, Indigenous Peoples Organizations, Arctic communities, Arctic Governance, EU environmental policies, Policy complexity and Fragmentation. She is a political scientist with a Master's degree in Environmental Governance and Sustainable Development, followed by a Master's degree in Governance and Entrepreneurship in Northern and Indigenous Areas. She has co-edited the first edition of this book and has several publications related to the Arctic.

**Delphine Misonne** is Professor at UCLouvain Saint-Louis – Bruxelles, FNRS Senior Research Associate and Head of CEDRE (Environmental Law Research Center) at UCLouvain Saint-Louis – Bruxelles. She is an environmental law specialist, interested in the legal determinants of ambition in environmental law, in various fields like climate change, air pollution, nature protection, new approaches to pollution prevention. Her research portfolio also includes works on rights for nature and on global commons. She recently published in Science a paper on harnessing science, policy and law to deliver clean air, and is the author of several chronicles on climate litigation and Human Rights (Revue trimestrielle des droits de l'homme), and the co-editor of the *Grands arrêts inspirants de l'environnement*, Larcier, 2024 (landmark cases on the environment), among other publications. She is a member of the Belgian Council for Sustainable Development, as a specialist of product regulation. She is an académicienne since 2023, as a member of the Royal Academy of Belgium, Classe des Lettres et des Sciences morales et politiques.

**Camille Bertaux** graduated from the Université Libre de Bruxelles in Public and International Law (Bachelor's and Master's degrees) and pursued her studies with an LLM in International Environmental Law at the Queen Mary University of London. She is currently a doctoral student at UCLouvain Saint-Louis – Bruxelles, under the supervision of Professor Delphine Misonne and a member of the CEDRE. Her research investigates the pivotal role of World Health Organization guidelines in shaping Environmental Law, with a specific focus on water and air quality law. In addition to her doctoral studies, she works at the Université Libre de Bruxelles as a teaching assistant in Public International Law.

**Tom Delreux** is a Professor of Political Science and EU Politics at the Institut de sciences politiques Louvain-Europe (ISPOLE) of the University of Louvain (UCLouvain). He is also a Visiting Professor at the College of Europe and at Sciences Po Paris. His research interests include the EU's external relations, EU environmental policy, EU institutions and international environmental politics.

**Christopher Crellin** is a PhD student in Political Science at the Institut de Sciences Politiques Louvain-Europe (ISPOLE), UCLouvain and a Fonds de la Recherche Scientifique (FNRS) aspirant. His current research interests focus on environmental policy processes and politics, coalitions, the European Union, and political networks. During his PhD, Christopher has been a visiting scholar at the University of Bern. His work has been published in journals such as *Climatic Change* and the *JCMS: Journal of Common Market Studies*. Prior to beginning his PhD, Christopher completed an MA of Public Policy from the University of Western Australia (UWA) and an MA of European Studies from KU Leuven.

**Janani Mohan** is a PhD candidate and Gates Cambridge scholar at the University of Cambridge, Department of Politics and International Studies, researching on the security of nuclear energy facilities. She has extensive experience as a policy analyst with a background in nuclear, development, and technology policy for leading think tanks, government, and international organizations. She holds a Master's in International Policy from Stanford University and Bachelor's in Political Science from the University of California, Berkeley. Her prior experiences include collaborating with the United Nations Innovation Cell to monitor open-source information from the War in Ukraine, consulting with the United Nations Development Programme in South Africa to increase technology access, and supporting nuclear nonproliferation efforts in the United States. She is also an avid researcher and author, publishing with organizations like the Stimson Center, Bulletin of Atomic Scientists, and Smithsonian Institution.

**Loïc Cobut** holds a PhD from UCLouvain Saint-Louis – Bruxelles. His area of research covers European climate and energy policies, interest representation in the European Union and the influence of non-state actors, with a focus on renewable energy cooperatives, in Europe. He holds a Bachelor in Political Science (Université catholique de Louvain), a Master in European Studies (Université catholique de Louvain and Université Saint-Louis – Bruxelles) and a Master in European Interdisciplinary Studies (College of Europe Natolin Campus).

**Daniel Petrovics** is a postdoctoral researcher studying the role of meta-organizations in scaling community enterprises for sustainability, as part of the SCENSUS project at the Erasmus University Rotterdam. He holds MSc degrees in International Development and Urban Planning from the University of Amsterdam, and completed his PhD at the Amsterdam Institute for Social Science Research (AISSR), focusing on the development of energy communities. He has worked as a research associate at the Institute for Environmental Studies (Vrije Universiteit Amsterdam) on EU-funded projects such as NEWCOMERS (Horizon 2020) and RESILIO (Urban Innovation Action). Prior to academia, Daniel worked with the Global Reporting Initiative and consulted on sustainability policy and strategy for organizations including UNGC, UNEP, PwC, CDP, and REN21.

**Thomas Bauwens** is an Assistant Professor researching collective action for sustainability at Rotterdam School of Management, Erasmus University. Previously, he was a researcher at Utrecht University, the Swiss Federal Institute of Technology in Lausanne and Oxford University. His research contributes to understanding the role of community governance, alongside market and state institutions, in tackling sustainability transformations, as well as the scaling processes of community enterprises and their roles in fostering societal support for sustainable innovations. He is also the PI of the project "SCENSUS" ("Scaling the Societal Impact of Community Enterprises for Sustainability"), funded by the European Research Council.

**Ariel Brunner** is Senior Head of Policy at BirdLife Europe and Central Asia. He co-ordinates work on a wide range of policies ranging from nature and biodiversity conservation to climate and energy, fisheries and agriculture. He is currently a member of the EU Platform on Sustainable Finance and the board of the Renewable Grid Initiative. Over the last decade, he has been deeply involved in debates on reforming the Common Agriculture Policy and is widely recognized as a prominent expert on the environmental impacts of EU farm policies. He has chaired DG Agriculture's Advisory Group on Agriculture and Environment and has been a member of its Expert Group on Public Goods and the European Commission's High-level Steering Group on the European Innovation Partnership on agriculture. In recent years, he has played a leading role in advocacy around the European Green Deal, the EU Biodiversity Strategy and the NatureAlert campaign to defend and improve the implementation of the EU nature protection legislation.

**Marilda Dhaskali** is a Senior Agriculture & Land-Use Policy Officer at BirdLife Europe and Central Asia, where she has been shaping EU agriculture and bioenergy policy since November 2021. With extensive experience in public policy and biodiversity, Marilda has previously contributed to the Fondation pour la Recherche sur la Biodiversité (FRB) and the French Ministry of Ecological Transition. Her work bridges science and society, focusing on biodiversity, climate change and environmental policy. Marilda holds a Master's degree in Public Policy and Strategies for the Environment from AgroParisTech and a Bioengineering Master's degree from the École de Bioingénierie de Bruxelles.

**Arnaud Gane** is a PhD candidate in Political Science and Philosophy at the Research Center in Political Science (CResPo) at UCLouvain Saint-Louis – Brussels. He holds a FRESH fellowship from the Belgian National Fund for Scientific Research (FNRS). A graduate of Sciences Po Toulouse and contemporary philosophy from Jean Moulin Lyon-III University and the University of Sherbrooke (Canada), his research focuses on the European rewilding movement, its relationship with European nature conservation policies and the developments in biodiversity conservation science.

**Tiago Pires da Cruz** is a PhD candidate at the Centre Émile Durkheim – Comparative Political Science and Sociology (UMR 5116) at Sciences Po Bordeaux. His work focuses on the Ocean's environmental governance, and specifically on deep-sea extractive phenomena. He is a member of the "Deep Ocean and Societies" research group of the French National Centre for Scientific Research Sea and Ocean Research Group (GdR OMER).

**Cándido García Molyneux** is a Spanish of Counsel in the Brussels office of Covington & Burling LLP. His practice focuses on EU environmental law, product safety law and life sciences. He advises clients on legal issues concerning chemical law, environmental product regulation, waste management, climate change, renewable energies and energy efficiency. Dr García Molyneux holds a PhD in Law from the European University Institute, an LLM from the University of Georgetown and a law degree from the Autonomous University of Madrid.

**David Seke** is a Master's student in Political Science at the University of Liège and a student-researcher affiliated with the Spiral Research Centre. His research focuses on the institutionalization of the One Health agenda, interdisciplinarity, and science-policy dynamics.

**François Thoreau** is an Associate Professor at the University of Liège, Belgium and a qualified researcher at the Spiral research center. He earned his PhD in Social and Political Science from the University of Liège in 2013, focusing on responsible innovation in nanotechnologies. His research interests encompass Science and Technology Studies, Anthropology of Natures-Cultures, Genomics and the selection and reproduction of livestock. In 2020, François received a Starting Grant from the European Research Council for a project on the political anthropology of techniques related to selection and reproduction in cattle breeding (2021–2026).

**Marcus Gustafsson** is a Policy Officer at the European Commission, specializing in trade relations with Taiwan, Hong Kong and Macau, and a member of the China Team. He has extensive experience in international trade law, having previously worked as a legal advisor in the European Commission's Legal Service and as an associate at Van Bael & Bellis, where he represented clients in WTO disputes and EU trade defense investigations. He holds an LLM in International Business and Economic Law from Georgetown University, where he was awarded the Thomas Bradbury Chetwood, S.J. Prize for the highest GPA in his program. Marcus has multiple publications in leading trade law journals and has lectured at prestigious institutions worldwide. He is fluent in Swedish, English and French and is admitted to the New York Bar.

**Victor Crochet** is a lawyer, partner in Nishimura & Asahi's international trade practice. He advises clients from Europe and Asia on EU regulations, trade remedies and litigations before the EU Courts, as well as on international economic law with a particular focus on WTO and free trade agreements.

**Elyse Kneller** focuses on EU and international economic law, with a particular emphasis on trade remedy investigations. Elyse assists companies and governments in anti-dumping, anti-subsidy and safeguard investigations initiated by the European Commission and by third countries. Elyse also advises industry associations and governments on EU trade, environment and digital policies.

**Angèle Minguet** is a Belgian political scientist and advisor on sustainable development. She earned her PhD from the University of Rome La Sapienza in 2019 with a thesis titled "Environment, from Conflicts to Justice: Political Theory and Trials". Her academic journey included studies in Ancient Greek History and European Studies at institutions such as FUNDP, UCL and Saint-Louis University in Belgium. Following her studies, she worked with international organizations and NGOs focusing on Human Rights and environmental issues across Europe. Currently, she serves as a Sustainable Development advisor in the Office of the Belgian Minister of Climate, Environment, Sustainable Development and Green Deal.

**Fabienne Haubold** is a Belgian legal advisor and criminologist, specializing in criminal and environmental law. She recently served within the Office of the Belgian Minister for Climate, Environment, Sustainable Development, and Green Deal (2021-2025). She holds a degree in law from the Université libre de Bruxelles. In her role, she provides legal and strategic advice on policy, political strategy, and communication.

**Zoé Briard** holds a Bachelor's degree in Law (Université Saint-Louis Bruxelles, Belgium) as well as a double Master's degree in Law and Political Sciences – International Relations (Université Libre de Bruxelles, Belgium) and a Master of specialization in Gender Studies (multiple universities, Belgium). She is now a PhD researcher (ASP) at UCLouvain funded by a grant from the Belgian Fund for Scientific Research (FNRS). Her thesis, under the supervision of Sylvie Sarolea and Amandine Orsini, is co-hosted at the Centre Charles de Visscher pour le droit international et européen (CeDIE) within the European Rights and Migration Team (EDEM) and the UCLouvain Saint-Louis – Bruxelles Political Science Research Center (CReSPo). Her doctoral research focuses on the processes of planned relocation as an adaptation strategy to climate change.

# ABBREVIATIONS

| | |
|---|---|
| AC | Arctic Council |
| ACCESS | Arctic Climate Change, Economy and Society |
| BBNJ | Biodiversity Beyond National Jurisdiction |
| BfN | Bonn, Federal Agency for Nature |
| BSNN | Black Sea NGO Network |
| CAP | Common Agricultural Policy |
| CBAM | Carbon Border Adjustment Mechanism |
| CBD | Convention on Biological Diversity |
| CCB | Coalition Clean Baltic |
| CCS | Carbon Capture and Storage |
| CEE | Central and Eastern Europe |
| CEP | Climate and Energy Package |
| CJEU | Court of Justice of the European Union |
| COP | Conference of the Parties |
| CSD | Commission on Sustainable Development |
| DG MARE | Directorate-General for Maritime Affairs and Fisheries |
| DG GROW | Directorate-General for Internal Market, Industry, Entrepreneurship and Small and Medium Enterprises |
| DIO | Do-It-Ourselves |
| ECA | European Court of Auditors |
| ECtHR | European Court of Human Rights |
| EEA | European Environment Agency |
| EEB | European Environmental Bureau |
| EEC | European Economic Community |
| EEE | Electrical and Electronic Equipment |
| EESC | European Economic and Social Council |
| EJOLT | Environmental Justice, Liabilities and Trade |

| | |
|---|---|
| ENGOs | Environmental Non-Governmental Organisations |
| ENSREG | European Nuclear Safety Regulatory Group |
| EPA | Environmental Protection Agency |
| EPCA | Enhanced Partnership and Cooperation Agreement |
| EPR | Extended Producer Responsibility |
| ESR | Effort Sharing Regulation |
| ETS | Emissions Trading Systems |
| EU | European Union |
| FAO | Food and Agriculture Organization of the United Nations |
| FFF | Fridays for Future |
| FTAs | Free Trade Agreements |
| GAEC | Good agricultural and environmental conditions |
| GDP | Gross Domestic Product |
| GHG | Greenhouse Gas |
| HLPF | High-level political forum |
| I-ACtHR | Inter-American Court of Human Rights |
| IAEA | International Atomic Energy Agency |
| ICA | International Cooperative Alliance |
| ILO | International Labour Organization |
| IPBES | Intergovernmental Platform for Biodiversity and Ecosystem Services |
| IPCC | Intergovernmental Panel on Climate Change |
| IPP | Integrated Product Policy |
| ISA | International Seabed Authority |
| IT | Information Technology |
| JTF | Just Transition Fund |
| LULUCF | Land Use Change, and Forestry |
| LULUCF | Land Use, Land Use Change, and Forestry |
| MFF | Multiannual Financial Framework |
| MIOECSDE | Mediterranean Information Office for Environment, Culture and Sustainable Development |
| MPA | Marine Protected Area |
| NECPs | National Energy and Climate Plans |
| NGO | Non-Governmental Organization |
| NWM | Nuclear Waste Management |
| OECD | Organisation for Economic Co-operation and Development |
| OH | One Health |
| OM | One Medicine |
| OWOH | One World One Health |
| REC | Renewable Energy Community |
| RED | Renewable Energy Directive |
| RES | Renewable energy sources |
| RoHS | Restriction of the Use of Certain Hazardous Substances |

| | |
|---|---|
| SBR | Sustainable Batteries Regulation |
| SCF | Social Climate Fund |
| SDG | Sustainable Development Goal |
| SWP | Stiftung Wissenschaft und Politik |
| TEU | Treaty on the European Union |
| TFEU | Treaty on the Functioning of the European Union |
| TSD | Trade and Sustainable Development |
| TTC | Trade and Technology Councils |
| UN | United Nations |
| UNCHE | United Nations Conference on the Human Environment |
| UNFCCC | United Nations Framework Convention on Climate Change |
| UNGA | United Nations General Assembly |
| US | United States |
| WCS | Wildlife Conservation Society |
| WEEE | Waste Electrical and Electronic Equipment |
| WHO | World Health Organization |
| WOAH | World Organisation for Animal Health |
| WTO | World Trade Organization |
| YAP | Youth Action Plan |

# EU ENVIRONMENTAL GOVERNANCE REVISITED

## An introduction

*Amandine Orsini and Eleni Kavvatha*

### The European Union, still a leader and laboratory for environmental politics?

#### *From the European Union as a leader and global laboratory...*

This book is the second edition of an edited volume scrutinizing the European Union (EU) as an environmental governance actor. It has been recognized long ago that the EU, as the most advanced example of regional international organization in the world, was also particular in its capacity to pioneer environmental protection. With its 27 Member States, it evolves around three major European institutions: the Commission, the Parliament, and the Council that, together, have adopted a plethora of environmental laws (de Sadeleer 2023) including European directives, regulations, decisions, recommendations, and opinions. But not only did the EU develop a plethora of environmental legislations, it also operates as a sophisticated governance system where environmental policy is developed by different actors, institutions, and processes (Jordan and Gravey 2021) in line with the support of democracy and fundamental freedoms (Buzogány et al. 2025).

The EU's environmental governance system has been recognized to serve as a testing ground for global policy innovations, especially in the environmental domain. A key example of this is the European Green Deal, the latest global agenda of the EU for sustainable development adopted in December 2019. More importantly, from emissions trading systems to biodiversity conservation strategies, the EU's initiatives inform and inspire international efforts (Adelle et al. 2022). In addition, the EU is particularly active at the international level of policymaking. Among others, it has adopted more than 398 bilateral and multilateral agreements on environmental topics

DOI: 10.4324/9781003479260-1

(Mitchell et al. 2020) and is constantly investing efforts to create coherence across these complex institutional arrangements (Delreux and Earsom 2023). The main recent internal and external achievements of the EU are summarized in Box 0.1.

**BOX 0.1 THE EU'S LEADERSHIP IN ENVIRONMENTAL GOVERNANCE**

The EU has reinforced its leadership role through ambitious initiatives under the European Green Deal, introduced in December 2019 and expanded upon in subsequent years. Key achievements include:

- The Climate Law: Enacted in 2021, this law enshrines the EU's commitment to achieving carbon neutrality by 2050 and a 55% reduction in greenhouse gas (GHG) emissions by 2030.
- The Biodiversity Strategy for 2030: Aims to restore ecosystems, expand protected areas, and promote sustainable agriculture and forestry.
- The Circular Economy Action Plan: Focuses on resource efficiency, reducing waste, and promoting sustainable production and consumption patterns. It is one of the building blocks of the EU Green Deal.
- The Zero Pollution Action Plan: Sets targets for cleaner air, water, and soil by 2050.
- The "Fit for 55" package aims to align various policy areas with climate neutrality goals, ensuring that EU policies align with the climate goals the European Council and the European Parliament agreed upon. The legislation package presents five new initiatives across various policy areas and economic sectors: climate, energy and fuels, transport, buildings, land use, and forestry.

These policies exemplify the EU's integrated approach, linking environmental goals with economic, social, and technological advancements.

Additionally, the EU has been working to extend its influence internationally by, among others:

- Supporting the adoption of the Kunming-Montreal Global Biodiversity Framework the 15th Conference of the Parties to the Convention on Biological Diversity (COP15) at COP15 (2022) which marked a milestone in biodiversity conservation, aligning with the EU's Biodiversity Strategy.
- Through leadership at COP26 and COP27 of the climate convention it has championed the Glasgow Climate Pact and the Sharm El-Sheikh Implementation Plan, respectively. Similarly, the Loss and Damage Fund established at

COP27 (2022) reflects a growing recognition of the need for climate justice, a principle supported by the EU.

- Moreover, the EU actively engages in global partnerships, namely being involved in the High Ambition Coalition to End Plastic Pollution, which includes 65 countries committed to aiming high in the negotiations to end plastic pollution by 2040. A new agreement on plastic pollution is meant to be adopted in 2025.
- The EU has been using instruments to achieve its objectives beyond the EU borders: the EU's Carbon Border Adjustment Mechanism (CBAM), introduced in 2023, seeks to prevent carbon leakage while promoting decarbonization globally. CBAM will apply in its definitive regime from 2026, while the current transitional phase lasts between 2023 and 2025.

These efforts underscore the EU's dual role as a policy innovator and a catalyst for global environmental action.

Source: Authors' own synthesis.

While the EU remains a global standard-bearer, the first edition of this book highlighted that challenges persist. Discrepancies in national implementation, economic disparities among Member States and social disparities within states, and the repercussions of Brexit complicated policy coherence. Most importantly, at the time the first edition was released, no one could foresee the upcoming major global crises that were going to hit the EU and the world more generally. Facing such new challenges, we already knew, at the time the first edition was out, that we would need a second edition to reflect on such changes.

## ...to the disruptive changes of the post-2020 era

Five years after the first edition, it was already time to update knowledge on the EU environmental agenda. But this has been even more pressing that the release of the first edition of this edited volume happened in a context of abrupt changes: 2020 was marked by the COVID-19 pandemic, soon followed by the invasion of Ukraine by Russia, in a global environmental context already framed around a triple ecological crisis. These major disruptions raised fears and hopes, for the international agenda more generally, as well as for the environmental agenda, that this book analyzes in more depth in its different chapters.

### *The COVID-19 pandemic*

The COVID-19 expansion in early 2020 underscored the vulnerability of societies to systemic shocks, including those linked to environmental degradation. On the one hand, the pandemic froze most of the legal and policy activities of States and citizens, putting the world under brackets for a significant amount of time. On the other hand, its proven link to environmental mismanagement problems helped States recognize the interdependence of ecological and social systems. While the priority of the EU Green Deal agenda was questioned as soon as one month after its very adoption by the pandemic, the EU decided to prioritize sustainability in its pandemic recovery efforts, allocating substantial funding for green initiatives under the *NextGenerationEU* framework, among others. Simultaneously, the EU has placed greater emphasis on social inclusion, ensuring that the transition to a green economy is equitable and leaves no one behind. As a result, authors such as Bongardt and Torres (2022) see the EU Green Deal as a building block of a sustainable European economic model after the pandemic.

*The war in Ukraine*

The Russian invasion of Ukraine in February 2022 not only disrupted worldwide security but also led to a major energy crisis, especially on the European continent. According to European Central Bank experts, the first two weeks after the invasion, the prices of oil, coal, and gas went up by around 40%, 130%, and 180%, respectively (Feveile Adolfsen et al. 2022). While energy prices were skyrocketing, they again posed a risk of paralyzing Europe that lost access to affordable energy. Instead, the EU reshaped its energy strategy, highlighting the need for energy independence and accelerating the shift to renewables. The *REPowerEU* initiative, launched in May 2022 in response, aims to diversify energy supplies while reducing reliance on fossil fuels. Technological advancements, including renewable energy storage, and digital tools for environmental monitoring, have accelerated progress toward sustainability (Kuzemko et al. 2022; Von Homeyer et al. 2022).

*The sharpening of the global environmental crisis*

As reported by the United Nations Secretary-General, the world is now facing a triple environmental crisis (Guterres 2022). Since 2020, environmental issues have only intensified, intersecting with public health, security, and economic stability. Scientific reports during this period highlighted alarming trends, including unprecedented biodiversity loss, escalating GHG emissions, and the increasing likelihood of crossing critical planetary boundaries (Richardson et al. 2023). The global temperature rise continues to approach 1.5°C, with severe consequences for climate stability, biodiversity, and human livelihoods. Moreover, tipping points—such as the melting of polar ice sheets and the Amazon rainforest's transition to savannah—have become more immediate concerns. These international developments could have discouraged States facing what appears like a never-ending environmental degradation curse. And yet, they did not stop the rule-creating power of the EU for greater environmental protection on the continent (Peeters and Misonne 2022) and worldwide (Fermeglia 2022).

Without embracing a naïve view according to which EU environmental objectives have always been able to bypass geopolitical tensions nor a pessimistic view of geopolitical tensions wiping up the environmental agenda of the EU, this book aims at questioning the different ways the EU has continued to adapt, leveraging its multilevel governance structure to address complex, cross-sectoral challenges and pursuing its role of a global standard setter (Eckert 2021).

## Objectives and structure of this book

### *What the second edition offers*

As the EU faces an unprecedented convergence of environmental, social, and economic challenges, its role as a leader in environmental governance becomes increasingly important to scrutinize. This second edition provides a comprehensive overview of the EU's evolving strategies, successes, and obstacles, offering valuable insights for policymakers, academics, and practitioners alike. This updated edition considers key developments in EU environmental governance from 2020 to 2025, including the post-pandemic green recovery, heightened global commitments to combat environmental challenges, and advancements under the European Green Deal.

More precisely, this book pursues a triple objective. First, it is meant to provide empirical insights on real-world challenges with regards to environmental problems and on concrete policy solutions developed to tackle them. Each chapter therefore details the realities of the environmental issue at hand, citing scientific reports, explaining direct and indirect environmental impacts, and presenting the current state of knowledge on the issues. Each chapter also details the legal and political instruments adopted to tackle them, often at different scales: each chapter refers to existing treaties, agreements, and legislations as well as to policy measures developed to implement them such as incentives, bans, or new decision-making procedures.

Second, each chapter provides an interdisciplinary perspective on the environmental problem at hand. Synthetizing existing research, authors draw from natural sciences and science & technology studies to explain the corresponding environmental issue; and from politics and law to develop the adopted solutions. This book is actually a collection of contributions from scholars and practitioners across very different disciplines, sometimes who combine training, education, and experience from several disciplinary backgrounds. The term "governance", chosen for the title of this book, is therefore favored to reflect this diversity of perspectives. The chapters in this book show how policymakers and lawyers are themselves struggling to know whether policies will be legally acceptable and laws politically feasible, all this in a context that will actually solve environmental issues.

Third, each chapter analyzes its given topic from a diversity of scales. Some global problems, such as climate change, require localized solutions, such as energy communities. Some apparently local problems, such as batteries recycling, are actually developed by the EU to counter worldwide dynamics of competition for new materials that can damage the environment. All chapters deal with different vertical scales (local, national, European, global) while many of them also deal with interactions on horizontal scales across topics (environmental, economic, health, technological, etc.). This reflects the current state of environmental problems that lie at the intersection of constantly evolving dynamics.

### *Chapter abstracts*

Through various chapters, the contributors offer insights into diverse sectors, such as climate policy, waste management, agriculture, biodiversity, and international environmental governance, emphasizing the need for integrated solutions, continuous adaptation, and collaboration at both the EU and global levels.

More precisely, in Part 1 (Air Quality and EU Climate Policy), Chapter 1, by Delphine Misonne and Camille Bertaux, discusses how the environment relates to Human Rights (and even possibly other rights) in Europe today and why such discussion matters, including in relation to the European Green Deal and the challenges of climate neutrality. It also assesses the stand of play and discussions around the explicit recognition of a right to a healthy environment at the scale of the EU or even Council of Europe. Chapter 2, by Tom Delreux and Christopher Crellin, offers a comprehensive overview of the EU's climate policies, focusing primarily on mitigation efforts aimed at reducing GHG emissions. It examines the key developments and policy innovations that have shaped the EU's climate policy, particularly in the wake of the European Green Deal adopted in 2019, which set a bold target for climate neutrality by 2050. This chapter traces the evolution of EU climate policy from the 1990s to 2024, analyzing the interplay between EU and international climate frameworks, the setting of targets, and the adoption of diverse policy instruments, including regulatory, market-based, and informational tools. Special attention is given to the EU Emissions Trading System (ETS) and its expansion, as well as the integration of national climate policies through the Effort Sharing Regulation. This chapter also explores new mechanisms such as the Social Climate Fund, the forthcoming ETS-2 for buildings and transport, and the European Climate Law and discusses the EU's funding instruments, including the Innovation Fund and the Just Transition Fund.

Part 2 (Energy Policy) starts with Chapter 3, where Janani Mohan discusses how the EU governance of nuclear waste offers an interesting comparative environment for how domestic attitudes toward nuclear energy affect waste policies. Since 2015, EU countries have been required to submit triennial national reports on their implementation of the EU's Radioactive Waste and Spent Fuel Management Directive. This chapter focuses on reports from EU countries with nuclear energy and explores how domestic attitudes cause their nuclear waste policies to increasingly diverge, as well as the impact of this divergence on coordination at EU level. This chapter considers the development of underground spent-fuel repositories as a case sample for understanding the impact of attitudinal differences on governance. This study then offers key findings for EU governance to more effectively mitigate challenges and encourage solutions to nuclear waste. In Chapter 4, Loïc Cobut, Daniel Petrovics, and Thomas Bauwens discuss renewable energy communities (RECs) which were officially recognized at the EU level in 2018. This recognition introduced several provisions now embedded in the revised Renewable Energy Directive. In this chapter, they analyze the implementation of these provisions in three Central and Eastern European Member States—Czechia, Lithuania, and Poland—a region that remains underexplored in both practice and academic literature. The provisions address various aspects of RECs, including their autonomy, internal governance principles, citizen participation, purpose, respect for local proximity, adherence to International Cooperative Alliance principles, and the establishment of controlling authorities. To assess their implementation, they use the concept of polycentric governance, which highlights how climate and energy governance increasingly incorporates non-state actors that foster experimentation through innovative activities while operating under overarching rules, such as those established by the EU. Their analysis reveals that the transposition of REC provisions reflects diverse models of polycentric governance across Central and Eastern Europe, while their findings indicate that poor implementation risks exposing RECs to business capture, undermining their intended role of supporting citizen participation and advancing energy democracy.

In Part 3 (Agriculture and Biodiversity), Chapter 5, written by Ariel Brunner and Marilda Dhaskali, revolves around the collapse of biodiversity, which is, along with climate breakdown, a clear and present danger to our civilization's survival. The first cause of biodiversity loss, globally and in Europe, is agricultural expansion and intensification. In the EU, farming practices have been heavily influenced by the Common Agricultural Policy (CAP), one of the world's biggest systems of public subsidies. Thirty years of attempted reforms have shifted the CAP from paying for production to untargeted income support and, in theory, to rewarding farmers for delivering public goods such as biodiversity and watershed management. Unfortunately, the effective environmental payments and schemes are still severely underfunded, while the bulk of the CAP is still distributed with virtually no-strings attached. The recently concluded reform cycle has failed to change this situation. The European Green Deal, with its Farm to Fork and Biodiversity Strategy, promised to address some of the direst impacts of the farming system and attempt a more holistic approach to food and farming. It has however triggered a massive backlash in 2024, by the dominant farmers' unions and right of center parties. According to the authors, tackling the climate and biodiversity crisis requires deep societal transformations, and agriculture will not be exempt. EU policies will have to address more than subsidy reform. From enforcement of environmental laws to new food policies to reduce meat and dairy overconsumption, a new set of policy debates is set to steer controversy, bring change, and transform the sector. In the box written by Arnaud Gane, under the title "Rewilding European Landscape", rewilding is presented as a bold conservation strategy aimed at restoring self-sustaining ecosystems in Europe, especially on abandoned farmland. While growing in popularity, the approach faces legal, social, and conceptual challenges. Despite limited EU legislative recognition, financial tools like the LIFE programme and the new Nature Restoration Law could support its expansion. In Chapter 6, Tiago Pires da Cruz examines how the European Commission shifted from a "blue" growth approach—focused on resource exploitation—to a more environmentally cautious "green" stance. European environmental governance in international ocean policy is analyzed as not just a rational calculation of pros and cons but a process of EU interest formation. When the EU Commission sought to position itself as an ocean governance leader, a precautionary approach was eventually announced, in the context of a new high-seas sustainability agreement. The study uses process-tracing, document analysis, stakeholder interviews, and direct observation. This chapter applies critical geography and John Hannigan's four ocean governance narratives and concludes with a nuanced shift. It leads to the proposal of a "turquoise" classification for the EU's mixed strategy.

In Part 4 (Waste and Health), Chapter 7 by Candido Garcia Molyneux discusses the New Batteries regulation and how it serves as an example of the EU's move on waste policy issues from dealing merely from producer responsibility issues to a more circular approach where strategic autonomy considerations are also paramount. Moreover, this chapter gives examples of the challenges and opportunities that the new requirements will pose for businesses. Chapter 8 by David Seke and François Thoreau attempts a review of "One Health" (OH) policies and their framings by international institutions. Drawing on a comprehensive review of scientific literature, including social sciences and the history of science, the authors trace the historical development of the OH discourse, highlighting its evolution. This chapter demonstrates how the concept has expanded to integrate previously separate dimensions. Utilizing grey literature, such as reports and policies from international institutions, the authors examine the emergence of OH as a central concern in public policy. In the final section, they explore the implications of adopting an integrated OH agenda for environmental governance, particularly within the EU context. This chapter concludes with a discussion of future directions for this evolving framework.

In Part 5 (Reconciling environmental concerns with market imperatives), Chapter 9 focuses on EU Trade and Environmental Policies. Marcus Gustafsson, Elise Kneller, and Victor Crochet look at the intersection between EU trade and environmental policy. In the context of a weakened rules-based global order and world trade regime, several WTO Members, including the EU, are relying ever more on unilateral rather than multilateral solutions to achieve their policy aims. In the EU's case, this includes multiple sustainability initiatives, such as the Carbon Border Adjustment Mechanism and regulations on deforestation and corporate due diligence. The trend toward enhanced sustainability considerations is also visible in the EU's partnerships with other trading partners. Under pressure from the European Parliament, sustainability chapters in bilateral free trade agreements have consistently grown in importance and have even threatened to torpedo certain negotiations, such as with MERCOSUR. The EU has also launched new and innovative partnership models, such as the Coalition of Trade Ministers on Climate, Raw Materials Club, and the Global Arrangement on Sustainable Steel and Aluminium. These international efforts have been complemented by extensive new industrial policies aimed at building a new green industrial base. These policies create important spill-over effects in the trade sphere by leading to additional trade protection that hinders imports of goods necessary for the green transition. All of these examples highlight the tension between sustainability objectives and the trade regime's traditional objectives of efficiency and competitiveness. Moreover, given the strategic importance of the green transition, green goods and technologies have been incorporated into the EU's plans to boost domestic supply security and supply chain resilience through its Economic Security Strategy. Going forward, the EU will have to carefully balance continued economic growth and openness through trade, and adherence to global rules, with these multiple new objectives and imperatives linked to the green transition. In Chapter 10, Angèle Minguet and Fabienne Haubold revisit the concept of "environmental conflict" and update previous discussions on the topic, exploring its evolution and the EU's legislative responses. The first section examines a contemporary example of environmental conflict, providing a refined definition and updated data on the phenomenon. The second section reviews the key EU legislative texts introduced since 2020, aimed at preventing environmental conflicts, including the precautionary principle, the CS3D, the CBAM, and the regulation on deforestation. The third section reflects on recent case law examples related to environmental conflict. This chapter concludes by highlighting the increasing global momentum for environmental justice, exemplified by cases such as PFAS pollution in Belgium and climate litigation across Europe. It emphasizes the need for greater accountability, transparency, and international cooperation to address environmental degradation and mitigate future conflicts, with a focus on the role of citizens, civil society organizations, and the private sector in promoting more sustainable practices. The box entitled "Planned Relocations as an Adaptation Strategy to Climate Change" written by Zoé Briard, discusses planned relocations as a climate change adaptation tool, highlighting their structured nature and growing relevance in Europe. While examples exist, such as in Germany and Austria, EU policy mostly overlooks internal relocations. The author emphasizes the need for clearer governance, legal safeguards, and better integration into EU climate strategies.

Finally, Part 6 (Upcoming challenges for the European Union) includes Chapter 11, where Eleni Kavvatha provides a comprehensive analysis of the EU's role in Arctic governance, highlighting the region's strategic importance and the evolution of EU policy over the past two decades. Initially focused on environmental issues such as pollution and climate change, the EU's Arctic policy has expanded to address economic, social, and geopolitical dimensions. This chapter explores the key pillars of EU Arctic policy, including climate change mitigation, sustainable resource management, and the protection of indigenous rights, particularly those of the Sámi people. It also discusses the challenges the EU faces in the Arctic, such as geopolitical tensions, multilateral cooperation complexities, and policy coherence, as well as its limited influence as an observer in the Arctic Council. This chapter further examines the EU's response to rapidly changing environmental conditions in the Arctic and its efforts to promote a sustainable and peaceful Arctic through initiatives like the European Green Deal. The conclusion emphasizes the EU's role in balancing environmental, economic, and geopolitical objectives while navigating significant challenges in the region's governance. Chapter 12 by Amandine Orsini explores the complex relationship between European youth and EU environmental policy, focusing on youth's participation in the policymaking process. Despite the EU's ambitious environmental agenda, youth often express dissatisfaction with how their concerns are addressed, particularly regarding climate change and sustainability. This chapter examines the mechanisms through which youth engage with EU environmental policy, such as the National Youth Councils and the UN Youth Delegates program, while identifying their limitations. Based on interviews with youth delegates at key international environmental meetings, this chapter reveals that while these mechanisms are viewed positively, youth still feel underrepresented, often resorting to protests as a means to make their voices heard. This chapter concludes that more inclusive and effective participation mechanisms are needed and that youth should be given greater attention in policymaking, considering their future role in implementing environmental policies. It also suggests further research on youth involvement in other policy areas and international contexts to better understand and improve youth engagement in environmental governance.

A short conclusion reflects on the successes and challenges of the EU environmental governance, putting all chapters in perspective.

## References


Adelle, C., Biedenkopf, K., & Torney, D. (dir.). (2022). *European Union External Environmental Policy. Rules, Regulation and Governance Beyond Borders*. Cheltenham: Edward Elgar.

Bongardt, A., & Torres, F. (2022). The European Green Deal: More than an Exit Strategy to the Pandemic Crisis, a Building Block of a Sustainable European Economic Model. *Journal of Common Market Studies*, *60*(1), 170–185. https://doi.org/10.1111/jcms.13264

Buzogány, A., Parks, L., & Torney, D. (2025). Democracy and the European Green Deal. *Journal of European Integration*, *47*(2), 135–154. https://doi.org/10.1080/07036337.2025.2459859

Delreux, T., & Earsom, J. (2023). The European Union as an Actor Navigating International Regime Complexes. *Politics and Governance*, *11*(2), 1–5. https://doi.org/10.17645/pag.v11i2.6901

de Sadeleer, N. (2023). Environmental Law in the EU: A Pathway toward the Green Transition. In da Glória Garcia, M., & Cortês, A. (Eds.), *Blue Planet Law. Sustainable Development Goals Series* (pp. 21–34). Cham: Springer. https://doi.org/10.1007/978-3-031-24888-7_2

Eckert, S. (2021). The European Green Deal and the EU's Regulatory Power in Times of Crisis. *Journal of Common Market Studies*, *59*(S1), 81–91. https://doi.org/10.1111/jcms.13241

Fermeglia, M. (2022). Recent Developments in EU Environmental Policy and Legislation. *Journal of European Environmental & Planning Law* *19*(4): 369–378. https://brill.com/view/journals/jeep/19/4/article-p369_007.xml

Feveile Adolfsen, J., Kuik, F., Lis, E. M., & Schuler, T. (2022). The Impact of the War in Ukraine on Euro Area Energy Markets. *ECB Economic Bulletin,* 4/2022. https://www.ecb.europa.eu/press/economic-bulletin/focus/2022/html/ecb.ebbox202204_01~68ef3c3dc6.en.html

Guterres, A. (2022). Ambitious Action Key to Resolving Triple Planetary Crisis of Climate Disruption, Nature Loss, Pollution, Secretary-General Says in Message for International Mother Earth Day. UN Press Release. G/SM/21243, 21 April 2022. https://press.un.org/en/2022/sgsm21243.doc.htm

Jordan, A., & Gravey, V. (Eds.). (2021). *Environmental Policy in the EU*. London: Routledge. https://doi.org/10.4324/9780429402333

Kuzemko, C., Blondeel, M., Dupont, C., & Brisbois, M. C. (2022). Russia's War on Ukraine, European Energy Policy Responses & Implications for Sustainable Transformations. *Energy Research & Social Science*, *93*, 102842. https://doi.org/10.1016/j.erss.2022.102842

Mitchell, R. B., Andonova, L. B., Axelrod, M., Balsiger, J., Bernauer, T., Green, J. F., … Morin, J.-F. (2020). What We Know (and Could Know) about International Environmental Agreements. *Global Environmental Politics*, *20*(1), 103–121. https://doi.org/10.1162/glep_a_00544

Peeters, M., & Misonne, D. (2022). The European Union and Its Rule-Creating Force on the European Continent for Moving to Climate Neutrality by 2050 at the Latest. In *Research Handbook on Climate Change Mitigation Law* (pp. 59–102) Cheltenham, UK and Northampton, Massachusetts, USA: Edward Elgar Publishing. https://doi.org/10.4337/9781839101595.00010

Richardson, K., Steffen, W., Lucht, W., Bendtsen, J., Cornell, S. E., Donges, J. F., … Rockström, J. (2023). Earth beyond Six of Nine Planetary Boundaries. *Science Advances*, *9*(37), eadh2458. https://doi.org/10.1126/sciadv.adh2458

von Homeyer, I., Oberthür, S., & Dupont, C. (2022). Implementing the European Green Deal during the Evolving Energy Crisis. *Journal of Common Market Studies*, *60*(S1), 125–136. https://doi.org/10.1111/jcms.13397

# PART 1

# Air quality and EU climate policy

# 1

# THE ENVIRONMENT AS A HUMAN RIGHT

## New challenges, also in Europe?

*Delphine Misonne and Camille Bertaux*

In July 2022, the General Assembly of the United Nations (UNGA) recognized at last the right to a clean, healthy and sustainable environment as a Human Right (Box 1.1).

**BOX 1.1 GENERAL ASSEMBLY OF THE UNITED NATIONS, RESOLUTION NO 76/300. THE HUMAN RIGHT TO A CLEAN, HEALTHY AND SUSTAINABLE ENVIRONMENT.**

The General Assembly:

1 (...) Recognizes the right to a clean, healthy and sustainable environment as a Human Right;
2 Notes that the right to a clean, healthy and sustainable environment is related to other rights and existing international law;
3 Affirms that the promotion of the Human Right to a clean, healthy and sustainable environment requires the full implementation of the multilateral environmental agreements under the principles of international environmental law;
4 Calls upon States, international organizations, business enterprises and other relevant stakeholders to adopt policies to enhance international cooperation, strengthen capacity-building and continue to share good practices in order to scale up efforts to ensure a clean, healthy and sustainable environment for all (UNGA, 2022).

DOI: 10.4324/9781003479260-3

This was a long-awaited development for environmental and Human Rights defenders (Knox, 2020). Indeed, such a Human Right, even if already recognized in one way or another in many countries at the national level, had never been truly proclaimed worldwide until that moment.

This chapter highlights several of the challenges that are brought forward by such articulation between "Human Rights" and "the environment" in the current European context. It explores several introductory and clarifying questions that may arise for those new to the field of environmental law and policy[1]: does attributing individuals a right to a healthy environment mean that a choice had to be made between "rights" and "duties" and would it be more fruitful to keep it to duties? Is a Human Right appropriate or is it too individual, or too anthropocentric? Do rights for nature create risks for humans?

This chapter examines the evolving relationship between Human Rights and the environment in Europe, focusing on the challenges and opportunities arising from the recent recognition of a Human Right to a healthy environment. It begins by exploring the historical context of environmental law in the European Union (EU), tracing its development from the early 1970s following the ecological crisis. This chapter then addresses key questions surrounding the attribution of environmental rights, including the tension between rights and duties, the appropriateness of Human Rights frameworks and the emerging concept of "rights for nature". It also considers the role of the European Court of Human Rights (ECtHR) and recent legal developments that highlight the intersection of climate change and Human Rights protections. The discussion extends to the global shift toward recognizing environmental rights and examines the potential for these shifts to inform EU policies and legislation. Finally, this chapter concludes by reflecting on the future of environmental rights in Europe, emphasizing the need for comprehensive and dynamic approaches to address the intertwined crises of climate change and biodiversity loss, while ensuring the protection of both human and ecological well-being.

## From the early 1970s to the first steps of EU action on the environment

European Union (EU) law and policy on the environment started developing in the early 1970s in the aftermath of the ecological crisis of the 1960s, also due to the impetus and legitimacy of the 1972 United Nations Conference on the Human Environment (UNCHE), which was initiated by Sweden as early as 1968 with a memorandum mentioning "that the changes in the natural surroundings, brought about by man, had become an urgent problem for developed as well as developing countries, and that these problems could only be solved through international co-operation" (*Digital Archives of the United Nations: Declaration of the United Nations Conference on the Human Environment*, 2012). The call launched a process that culminated internationally in the city of Stockholm with the June 1972 UNCHE.

Its output, the Stockholm Declaration, paved the way for a coordinated global policy at the intersection of development, equity and environmental issues. It also rapidly gained influence for activating the promotion of the first rules on the environment at a country or regional level, at a time when capacities, policies and even institutions in so many countries were still non-existent or in their prime infancy.

Quite importantly, the Declaration sowed the first seeds linking in its own way the words "man", "fundamental right", "environment" and "quality" under the necessity of the compromises of that moment, while also specially opting for not going as far as recognizing a new Human Right: "various proposals for a direct and thus unambiguous reference to an environmental human right were rejected" (Handl, 2013, p. 3). The first principle of the Declaration reads as follows, showing not only the interesting intersection between issues of environment and development, but also concerns of equity and justice for present and future generations:

> Man has the fundamental right to freedom, equality and adequate conditions of life, in an environment of a quality that permits a life of dignity and well-being, and he bears a solemn responsibility to protect and improve the environment for present and future generations. In this respect, policies promoting or perpetuating apartheid, racial segregation, discrimination, colonial and other forms of oppression and foreign domination stand condemned and must be eliminated.
>
> *(United Nations Conference on the Human Environment in Stockholm, 1972)*

It is only months later, in a Council meeting in Paris on 19 and 20 October 1972, that heads of States of the European Economic Community (EEC) (former EU), composed at the time of six countries, declared in the same vein that

> economic expansion is not an end in itself: its first aim should be to enable disparities in living conditions to be reduced. It must take place with the participation of all the social partners. It should result in an improvement in the quality of life as well as in standards of living. As befits the genius of Europe, particular attention will be given to intangible values and to

protecting the environment so that progress may really be put at the service of mankind.

*(Council Declaration of November 1973, 1973, p. 2)*

This was a key moment for the future European environmental policy. Community institutions were invited to establish, before 31 July 1973 (the EEC would enlarge by that time to nine Members), a first pragmatic "program of action", which initiated the adoption of the first dedicated legislations and policies at the EEC level. There was so much work pending on the plate: legislation and policies only existed very partially in a few countries (Germany, Denmark), and this was also already feared as a regrettable source of potential competition distortion.

The word environment will appear in the Treaty of Rome only years later, in 1987, with the Single Act, not in relation to any devoted right but as the title and content of the first environmental chapter on the environmental action of the Union. These historical elements explain how EU environmental policy started on a path that was not originally driven by Human Rights concerns.

## EU law and its ambition for the environment

### *How is EU law structured on the environment?*

EU law on the environment must be approached in two steps. First, via primary law, consisting of the founding Treaties (Treaty on European Union (TEU) and Treaty on the Functioning of the European Union (TFEU)) and containing the essential rules on which the constituting powers (Member States) agreed to allocate competences to the Union, establish European institutions and organize the interface between the Union, themselves and even their citizens and enterprises. Second, via secondary law that assembles the fruit of what the EU as such is producing through the functioning of its own institutions (among which are the European Commission, the European Parliament and the European Council). That EU legislation is currently applicable at the scale of 27 Member States, either directly (via the Regulations) or upon the condition of a transposition process at the Member State level (the case of Directives), except if provisions are so clear and unconditional that they are deemed directly applicable to the benefit of citizens. A reader unfamiliar with EU law and policy must be aware of the following exceptional feature, not found in other international or intergovernmental cenacles pursuing a goal of market integration: the EU is a lawmaker as such, with the capacity to organize and govern large fields of competences at the scale of the Union, while also leaving tasks and competences to its Member States, depending on the chosen degree of Europeanization and the way attributions are framed under primary law.

**BOX 1.2 ARTICLE 191 TREATY ON THE FUNCTIONING OF THE EUROPEAN UNION**

1 Union policy on the environment shall contribute to pursuit of the following objectives:

- preserving, protecting and improving the quality of the environment,
- protecting human health,
- prudent and rational utilization of natural resources,
- promoting measures at international level to deal with regional or worldwide environmental problems, and in particular, combating climate change.

2 Union policy on the environment shall aim at a high level of protection taking into account the diversity of situations in the various regions of the Union. It shall be based on the precautionary principle and on the principles that preventive action should be taken, that environmental damage should as a priority be rectified at source and that the polluter should pay (...) (Treaty on the Functioning of the European Union).

For instance, searching for specific rules on the protection of water bodies and water quality in the European Treaties will not be fruitful; they are to be found in secondary law, in specific policies and legislations. However, those must abide by all rules set in primary law, among which are the rules on their necessary level of ambition (Misonne, 2015, 2023; Sikora, 2020). The Court of Justice of the EU (CJEU) could be called upon to pronounce itself on the legality of their content. Member States must also interpret them in due regard to the requisites of primary law.

Ambition, distribution of powers and attributions are among the main aspects that can be found today in the Lisbon Treaties, which do contain many provisions on the environment (that also include climate change issues): general provisions (Art. 3 TEU), integration provisions (Art. 11 TFEU), a full environmental chapter Art. 191–193 TFEU), other chapters mentioning the environment (such as Art. 194 TFEU on the energy competence). However, none of them mention any wording referring expressly to the notion of the "right to a healthy environment". The wording does not even appear in the environmental section of the EU Charter of Fundamental Rights, which has the same legal force as the Treaties. In Chapter 5 dedicated to "Solidarity", Art. 37 on environmental protection keeps it to:

> A high level of environmental protection and the improvement of the quality of the environment must be integrated into the policies of the Union and ensured in accordance with the principle of sustainable development.

### *The absence of explicit language on a right to a healthy environment in the treaties*

The EU policy on the environment has not been built originally around that specific prism of Human Rights (one of them is related to the existence of a concurring organization, the Council of Europe, with its European Convention on Human Rights and its own Court of Human Rights located in Strasbourg[2]), but on a series of principles (prevention, precaution, polluter-pays, correction by priority at source) and on the assertion of a main goal, approached in an objective rather than a subjective way, through the concept of a "high level of protection".

The need to pursue and even achieve a high level of protection is an original feature of EU law. It finds its origin in a defensive attitude to the integration process: concerns were raised that the approximation of laws meant to accelerate the internal market (maximal harmonization), based on a qualified majority quorum (depriving Member States of their former veto power), could lead to lower environmental protection by comparison to preexisting national rules or to what would be promoted at the national level.

That ambition is also reflected in Art. 3(3) of the TEU, which is highlighted by the CJEU as providing that the EU works in particular for a high level of protection and improvement of the quality of the environment (Case C-444/15, *Nostra Onlus*, 2016, para. 42; Case C-24/19, *Windmills Aalter*, 2020, para. 47). The notion is likewise explicitly anchored in Art. 114 and Art. 191 of the TFEU and in Art. 37 of the Charter of Fundamental Rights of the European Union (Charter). The aim of a "high level of protection" is not the exclusive attribute of the environmental competences of the EU. It is also an explicit feature of the public health chapter (Art. 168 TFEU, Art. 35 Charter) and of consumer protection attributions (Art. 169 TFEU, Art. 38 Charter) under EU primary law.

Interestingly, the Court now interprets Art. 37 of the Charter in light of Art. 52(2) of the Charter, which "provides that rights recognized by the Charter for which provision is made in the Treaties are to be exercised under the conditions and within the limits defined by those Treaties" (Case C-444/15, *Nostra Onlus*, paras. 61–62), while that article was so far interpreted as belonging to another category of requirements, less prescriptive. In *Nostra Onlus*, the CJEU, by referring to Art. 52(2) and not to Art. 52(5) of the Charter (devoted to principles that *may* be implemented), thus interprets Art. 37 in a way that not only blurs the differences between rights and principles under the Charter but also refrains from diluting the related obligations under the TEU and TFEU.

### The presence of explicit language on rights in secondary EU law

That absence of explicit mention of a "right" in relation to the environment in primary law does not bar secondary law from investing and mobilizing a right-based language. There are indeed several sets of landmark rules allowing rights to humans on environmental issues in the EU, explicitly or implicitly, on substance or from a more procedural angle. Interpretations alongside that kind of line can also evolve through case law and the authority of the CJEU.

On procedural aspects and under the influence of the Aarhus Convention on Access to Information, Public Participation in Decision-making and Access to Justice in Environmental Matters,[3] EU legislation developed dramatically in order to ease access to environmental information, impose public participation in decision-making or even, more recently, ease access to justice for NGOs at national level on environmental issues. In those texts and because of the language of the Aarhus Convention, references to rights are explicit even if kept abstract (it does not say, for instance: the right of every person of present and future generations). Concretely, one of the objectives of Directive 2003/4/EC of the European Parliament and of the Council of 28 January 2003 on public access to environmental information is "to guarantee *the right* of access to environmental information held by or for public authorities and to set out the basic terms and conditions of, and practical arrangements for, its exercise". The objective of Regulation 1367/2006 of the European Parliament and of the Council of 6 September 2006 (as modified in 2021; the Aarhus Regulation) on the application of the provisions of the Aarhus Convention is also to contribute to the obligations arising from that Convention

> by guaranteeing *the right* of public access to environmental information received or produced by Union institutions or bodies and held by them, and by setting out the basic terms and conditions of, and practical arrangements for, the exercise of *that right*.

Other sets of rules do not explicitly use the language of rights, even if their content is necessarily related. This can be observed in legislation on substantial (by contrast to procedural) issues, like those dedicated to the quality of the environment as such or to products or services that shall be made available to the public. The important Directive 2020/2184 of the European Parliament and of the Council of 16 December 2020 on the quality of water intended for human consumption contains 16 references to the notion of "right" in its preamble, as it was initiated by the European citizens' initiative on the right to water ('the Right2Water initiative'); the content of the main provisions does not allude to it and prefers a less sensitive vocabulary, which is the language of duties: the Directive

> concerns the quality of water intended for human consumption for all in the Union. The objectives of this Directive are to protect human health from the adverse effects of any contamination of water intended for human consumption by ensuring that it is wholesome and clean, and to improve access to water intended for human consumption.

In another register, the preamble of the EU Climate Law (Regulation 2020/1119 of the European Parliament and the Council of 30 June 2021 establishing the framework for achieving climate neutrality) indicates that

> the Regulation respects the fundamental rights and observes the principles recognized by the Charter of Fundamental Rights of the European Union, in particular, Article 37 thereof which seeks to promote the integration into the policies of the Union of a high level of environmental protection and the improvement of the quality of the environment in accordance with the principle of sustainable development,

but it does not expand on the notion of rights in its binding content and subtly avoids any allusion to the recipient of its own rules that are anyway governed by primary law depending upon the chosen legal base (the environment, human health, natural resources, solving global problems, as mentioned above concerning the environmental chapter):

> This Regulation establishes a framework for the irreversible and gradual reduction of anthropogenic greenhouse gas emissions by sources and enhancement of removals by sinks regulated in Union law. This Regulation sets out a binding objective of climate neutrality in the Union by 2050 in pursuit of the long-term temperature goal set out in point (a) of Article 2(1) of the Paris Agreement, and provides a framework for achieving progress in pursuit of the global adaptation goal established in Article 7 of the Paris Agreement. This Regulation also sets out a binding Union target of a net domestic reduction in greenhouse gas emissions for 2030.

### *The sensitivity of a right-based approach*

What difference would it make to refer more explicitly to a concrete right to a healthy environment under EU law or to a clean and sustainable environment, as mentioned by the UN General Assembly?

An illustration of the added value but also sensitivity of the right-based dimension can be found in the case law on air quality. In the early 1990s, the CJEU decided that a Member State could not implement a directive on air or even water quality through mere (non-binding) technical rules, as it was important for the citizens and potential project developers to know not only their duties but also their subjective rights. Those air and water aspects actually concerned the health of citizens: whenever the exceeding of limit values was capable of endangering human health, the individuals concerned were in a position to rely on those rules in order to assert their rights (*Commission v. Germany*, 1991a, para. 16; *Commission v. Germany*, 1991b, paras. 14 and 19).

This opened the door much later to the *Janecek* jurisprudence, according to which natural or legal persons directly concerned by a risk that the limit values or alert thresholds may be exceeded must be in a position to require the competent authorities to draw up an action plan on air quality where such a risk exists, if necessary by bringing an action before the courts having jurisdiction (*Janecek*, 2008, para. 39). It was a logical consequence arising from the *TA Luft* case law here above mentioned: whenever the failure to observe the measures required by the directives which relate to air quality and drinking water, and which are designed to protect public health, could endanger human health, the person concerned must be in a position to rely on the mandatory rules included in those directives. That air quality and the health of individuals are related was the main message, transforming EU legislation into a topic concerning the people, not only the realm of administrative discretion.

In another central case, *Deutsche Umwelthilfe v. Freistaat Bayern*, the Grand Chamber decided that failure to adopt the measures required by the ambient air quality directive would endanger human health, making the right to an effective remedy all the more important (*Janecek*, 2008, para. 39). But that right is not absolute and has to be balanced against other rights, such as the right to liberty of an office holder in the exercise of its authority. Rights were either linked to individuals without any further precisions (a consequence of health-related legislation) or are approached in relation to other items than an explicit mention of the words "right to a healthy environment" *per se* (the right to an effective remedy). In another case, *JP. v. Ministre de la transition écologique*, the Grand Chamber asserted – quite surprisingly in the face of the ascending line pursued so far – that the right thus recognized by the Court in its case law, stemming in particular from the principle of effectiveness of EU law, effectiveness to which affected individuals are entitled to contribute by bringing administrative or judicial proceedings based on their own particular situation, "does not mean that the obligations resulting from Article 13(1) and Article 23(1) of Directive 2008/50 and the analogous provisions of the earlier directives were intended to confer individual rights on interested persons", for the purpose of the principle of State liability for loss and damage caused to individuals as a result of breaches of EU law for which the State can be held responsible.

### Why don't we keep it to duties?

The answer relates to the degree of reactivity of society, groups or even individuals to possible failures or infringements made by public authorities and to the need to enlarge the spectrum of stakeholders and claimants. The notion of "rights" needs to be approached in all its richness and variety, also with regard to its dedicated beneficiaries and collective or individual dimensions (Misonne, 2020, p. 409). Answers in that regard can be found in other cenacles, like the Court of Human Rights in Strasbourg or at the domestic level (see for instance, across the Atlantic, the case *Held v. Montana* (*Rikki Held et al. v. State of Montana et al.*, 2023)), where a court ruled that a provision of the Montana Environmental Policy Act violated the plaintiffs' right to a clean and healthful environment under the Montana Constitution. The provision prohibited consideration of greenhouse gas (GHG) emissions and corresponding climate change impacts in environmental reviews. The judge found that such limitation was unconstitutional because the state must actively protect, in a preventative manner, the constitutional right to a clean and healthful environment. Therefore, the State must not deprive citizens or decision-makers of the possibility to rely on remedies and adequate procedures, like impact assessment.

Rights mean duties and they cannot be disconnected, as inferred by the abundant case law of the European Court on Human Rights in Strasbourg regarding climate and the environment: positive duties (one must act) and negative duties (one must abstain). This is true not only for States but also for enterprises, as recently confirmed by the Court of Appeal of The Hague, *Shell v. Milieudefensie* of 12 November 2024,[4] asserting that Shell has an obligation toward citizens to limit its $CO_2$ emissions, based on the Human Right to protection from dangerous climate change. According to the Court of Appeal, it is first and foremost up to governments to ensure the protection of Human Rights, but indirectly those rights also have a bearing on the social standard of care which companies like Shell must observe.

### Human Right language on the environment in Strasbourg

#### *New challenges in Strasbourg*

The recent developments in the case law of the ECtHR on climate change provide valuable insights on the current challenges in Europe regarding the future of the right to a clean, healthy and sustainable environment as a Human Right.

Such a right, while recognized in one form or another in many national constitutions in Europe (often expressed as a duty bore by public authorities), has not yet formally been inserted in the text of the Convention of Human Rights at the level of the Council of Europe. The right to a healthy environment has nonetheless repeatedly been at the center of discussions at the Council of Europe: the Parliament Assembly of the Council called multiple times for the adoption of an additional protocol aiming to anchor the right to a healthy environment and render such a right enforceable at the European Court of Human Rights (see i.a. PACE, Resolution No 2396, 2021). Such calls have been hitherto left without a positive response.

Despite the absence of an enforceable right to a healthy environment *per se*, there is an abundant case law (now some 300 environment-related cases) at the Strasbourg Court dedicated to controlling the respect of the Convention by its Member States on environmental-related duties. The pivotal bases of the Court's jurisprudence related to the environment are "mainstream" Human Rights such as the right to private and family life and home (Art. 8) (*Lopez Ostra v. Spain*, 1994), the right to life (Art. 2) (*Oneryildiz v. Turkey*, 2004) or the right to a fair trial (Art. 6) (*L'Erablière v. Belgium*, 2009). In the pollution case *Tătar v. Romania* in 2009, and probably because that right was established under the Constitution of the defending State, Romania, the Court concluded that the Romanian authorities failed in their obligations to carry out a prior and satisfactory assessment of the possible risks of the polluting activity in question and take adequate measures capable of protecting the rights of the individuals concerned in due regard of their private life and their home "and, more generally, their enjoyment of a healthy and protected environment" (*Tătar v. Romania*, 2009, para. 112). In its case law, the Court makes a distinction between, on the one hand, the rights protected under the Convention and, on the other hand, the weight of environmental concerns in the assessment of legitimate aims and the weighing-up of rights and interests in the context of the application of the Convention.

While abundantly dealt before national jurisdictions since 2015 and the landmark case *Urgenda* in The Netherlands (and its confirmation by the Dutch Supreme Court in 2019), the issue of climate change has only very recently been brought to the fore of the Strasbourg Court at the occasion of other lawsuits, in several judgments and decisions of non-admissibility.

### *New insights on the right to a healthy environment*

The case *Verein KlimaSeniorinnen Schweiz and Others v. Switzerland* of April 2024 brings interesting new perspectives, while raising new questions, as the Court decides to open a new track, different from its environmental acquis, and explains why. One of the consequences of that new "way" is of utmost importance as the Court opens access to justice to NGOs on climate issues, taking as a main lever the Aarhus Convention. In that case, the recent recognition of the right to a healthy environment at a world level is discussed, even if it does not become central to the argument. Confirming the approaches adopted by Dutch and Belgian jurisdictions, the Court finds in Art. 2 and 8 of the Convention sufficient ground for condemning the State and partly welcoming the NGO's claims. It also welcomes a breach of Art. 6.

Quite unexpectedly thus, instead of pursuing and consolidating its own way as already drawn on environmental issues, the Grand Chamber explains that it felt the "need to develop a more appropriate and tailored approach as regards the various Convention issues" arising in the climate change context not addressed by Court's existing environmental case law (*Verein KlimaSeniorinnen Schweiz and Others v. Switzerland*, 2024, para. 422). Because of several fundamental differences (it is not a local but a complex global problem, the Court says),

> it would be neither adequate nor appropriate to follow an approach consisting in directly transposing the existing environmental case-law to the context of climate change. The Court considers it appropriate to adopt an approach that both acknowledges and takes into account the particularities of climate change and is tailored to address its specific characteristics. In the present case, therefore, while drawing some inspiration from the principles set out in the Court's existing case-law, the Court will seek to develop a more appropriate and tailored approach as regards the various Convention issues which may arise in the context of climate change.
>
> *(Verein KlimaSeniorinnen Schweiz and Others v. Switzerland, 2024, para. 422)*

The main concern of the Grand Chamber in provoking that divide was probably to break new ground while at the same time offering the guarantee that it would still bar the possibility of an *actio popularis* on environmental issues. But altogether, it cannot cut the *cordon ombilical* with the broader environmental dimension, keeping, for instance, the reference to the Aarhus Convention as a main building block.

On the right to a healthy environment, the Grand Chamber recalls and explains that upon the invitation of the Human Rights Council formulated in its Resolution 48/13 of 8 October 2021, the UNGA adopted its Resolution 76/300 on the Human Right to a clean, healthy and sustainable environment on 28 July 2022, that it was adopted with 161 votes in favor (of the 169 Member States present), 8 abstentions and no votes against, and mentions that 45 of the 46 Member States of the Council of Europe voted in favor. It also highlights that in a Recommendation adopted on

29 September 2021, entitled "Anchoring the right to a healthy environment: Need for enhanced action by the Council of Europe", the Parliamentary Assembly of the Council of Europe recommended that the Committee of Ministers draw up an additional protocol to the European Convention on Human Rights and an additional protocol to the European Social Charter on the right to a safe, clean, healthy and sustainable environment, based on the terminology used by the UN. A step to which the Committee of Ministers replied on 4 October 2022 with a recommendation to adopt a new protocol to the European Convention on Human Rights, indicating that it had instructed the Steering Committee for Human Rights to undertake other possible work, including the preparation of a study on the need for and feasibility of a further instrument or instruments on Human Rights and the environment. The first pages of the ruling are indeed an excellent syllabus listing all decisions and documents that were recently adopted with regard to the right to a healthy environment as a Human Right, even on other continents (Africa and America) and in the case law of other supreme jurisdictions.

Most importantly, the Court explains that recent international initiatives for the recognition of the Human Right to a clean, healthy and sustainable environment (see, in particular, UN General Assembly Resolution 76/300, and Committee of Ministers Recommendation CM/Rec(2022)20, both cited above) should be understood "from the perspective of the Convention" (*Verein KlimaSeniorinnen Schweiz and Others v. Switzerland*, 2024, para. 448), which is actually dual, as it recalls: while the Court has on occasion referred to "the right of the people concerned [...] to live in a safe and healthy environment" (*Di Sarno and others v. Italy*, 2012, para. 110; *Tătar v. Romania*, 2009, para. 112), this language cannot be understood without regard to the distinction that must be made between, on the one hand, the rights protected under the Convention and, on the other hand, the weight of environmental concerns in the assessment of legitimate aims and the weighing-up of rights and interests in the context of the application of the Convention (*Verein KlimaSeniorinnen Schweiz and Others v. Switzerland*, 2024, para. 447). It decides that such a development

> forms part of the international-law context in which the Court assesses Convention issues before it (*Demir and Baykara v. Turkey*, 2008, para. 76), notably as regards the recognition by the Contracting Parties of a close link between the protection of the environment and human rights,

meaning by that it will bear on its interpretations. But the Court does not want to answer Norway's question who, in its intervention to the case, wanted to hear the Court saying that the recent UN recognition would not have any legal effect and would only be political: it is not up to her "to determine whether the general trends regarding the recognition of such a right give rise to a specific legal obligation (see paragraph 372 above concerning the arguments raised by the intervening Norwegian Government)" (*Verein KlimaSeniorinnen Schweiz and Others v. Switzerland*, 2024, para. 448).

While explaining that its role is only to interpret the Convention, it also reasserts that the interpretation and application of the rights provided for under the Convention can and must be influenced both by factual issues and developments affecting the enjoyment of the rights in question and also by relevant legal instruments designed to address such issues by the international community:

> the Court has consistently held that the Convention should be interpreted, as far as possible, in harmony with other rules of international law. Moreover, a failure by the Court to maintain a dynamic and evolutive approach would risk rendering it a bar to reform or improvement.
>
> *(Verein KlimaSeniorinnen Schweiz and Others v. Switzerland, 2024, para. 455)*

### *New insights on the right to climate protection*

In *KlimaSeniorinnen*, the Court acknowledges once again that individuals or groups of individuals might see their rights (as already recognized under the Convention) affected by the absence or insufficiency of actions taken by national authorities in the fight against climate change. One of the major developments induced by that landmark case lies in the interpretation of Art. 8 on the right to private life and home: "Article 8 must be seen as encompassing a right for individuals to effective protection by the State authorities from serious adverse effects of climate change on their life, health, well-being and quality of life" (*Verein KlimaSeniorinnen Schweiz and Others v. Switzerland*, 2024, para. 519). Quite interestingly, that protection against climate change did not need a specific right to a healthy environment for blooming up; as in the existing case law of the same Court on the environment, it could be built on the basis of existing provisions of the Convention for the protection of Human Rights.

Another crucial development concerns access to justice for climate NGOs before the Strasbourg Court, which is truly revolutionary (the Court used to provide access to associations solely when Individual Rights could not be protected by affected individuals). The universal nature of climate change and the related repercussions on future generations led the ECtHR to open up the way for environmental associations' legal actions. The Court underlines that associations are sometimes the best option to benefit from efficient protection of Individual Rights, especially concerning climate that requires efforts through generations, as addressed in the preamble of the UNFCCC. The Aarhus Convention played a decisive role in the reasoning (*Verein KlimaSeniorinnen Schweiz and Others v. Switzerland*, 2024, paras. 489–491). Despite the general jurisprudence that requires associations to have a distinct affection from the one an individual might suffer, the Court establishes that in the particular context of climate change, associations have direct access to its proceedings:

> there has been an evolution in contemporary society as regards recognition of the importance of associations to litigate issues of climate change on behalf of affected persons. Indeed, climate-change litigation often involves complex issues of law and fact, requiring significant financial and logistical resources and coordination, and the outcome of a dispute will inevitably affect the position of many individuals.
>
> *(Verein KlimaSeniorinnen Schweiz and Others v. Switzerland, 2024, para. 497)*

The Court also organizes the criteria that must be met before its own jurisdiction, would individuals want to call on the Strasbourg judges to investigate the adequacy of the protection Member States are offering – or not – on climate change and associated risks. To be admitted before the Court, an individual victim must demonstrate (i) severe exposure to climate change adverse effects resulting from governmental inaction or insufficient actions and (ii) an urging need for individual protection arising from insufficient national measures to reduce such harm (ibid, para. 487). These conditions require a high degree of exposure to climatic risks to the extent that the victims' health should be critically endangered by the related adverse consequences and demands an urging need for protection (ibid, paras. 448 and 533). The threshold for individual victims will thus become difficult to reach, the idea being to avoid lawsuits dedicated to the general interest. But a victim of a flood or very high temperature (due to inappropriate urban planning for instance), with associated health consequences, could fit in these conditions.

The Court underlines the urgency of combating climate change not only in due regard to its impact on citizens presently but also in the future: States should take "adequate action notably through suitable general measures" to mitigate the grave risks of irreversible damage to Human Rights of "individuals who are currently affected by climate change, but also those individuals within their jurisdiction whose enjoyment of Convention rights may be severely and irreversibly affected in the future in the absence of timely action" (ibid, para. 499).

The causal link between observed damage and climate change has always been challenging in the prospect of climate litigation, especially due to the cumulative impact of GHGs across time and their possible attribution to a myriad of sources, falling or not under the province of national authorities.

The ECtHR stresses that

> there has been an evolution of scientific knowledge, social and political attitudes and legal standards concerning the necessity of protecting the environment, including in the context of climate change. There has also been a recognition that environmental degradation has created, and is capable of creating, serious and potentially irreversible adverse effects on the enjoyment of human rights. This is reflected in the scientific findings, international instruments and domestic legislation and standards, and is being recognized in domestic and international caselaw.
>
> *(Verein KlimaSeniorinnen Schweiz and Others v. Switzerland, 2024, para. 431)*

It decides that it "cannot ignore the above-noted developments and considerations" and

> needs to take into account the existing and constantly developing scientific evidence on the necessity of combating climate change and the urgency of addressing its adverse effects, including the grave risk of their inevitability and their irreversibility, as well as the scientific, political and judicial recognition of a link between the adverse effects of climate change and the enjoyment of (various aspects of) human rights.
>
> *(Verein KlimaSeniorinnen Schweiz and Others v. Switzerland, 2024, para. 434)*

As a matter of fact, the Court affirms,

> there are sufficiently reliable indications that anthropogenic climate change exists, that it poses a serious current and future threat to the enjoyment of human rights guaranteed under the Convention, that States are aware of it and capable of taking measures to effectively address it, that the relevant risks are projected to be lower if the rise in temperature is limited to 1.5°C above pre-industrial levels and if action is taken urgently, and that current global mitigation efforts are not sufficient to meet the latter target.
>
> *(Verein KlimaSeniorinnen Schweiz and Others v. Switzerland, 2024, para. 436)*

The Court therefore acknowledges the link between climate change and its threats to the enjoyment of Human Rights.

Extraterritoriality of environmental and climate issues is another well-known challenge in climate litigation and this is exacerbated under Human

Rights treaties which tend to offer protection to individuals under the jurisdictions or control of the concerned States, which might be at odds with global climate harms. As stated by Alan Boyle,

> the obvious problems are the multiplicity of states contributing to the problem and the difficulty of showing any direct connection to the victims. The inhabitants of sinking islands in the South Seas may justifiably complain of human rights violations, but who is responsible?
>
> *(Boyle, 2012)*

The Court plays in various ways with that dimension. On the one hand, it asserts that even if "climate change is undoubtedly a global phenomenon which should be addressed at the global level by the community of States, [...] each State has its own share of responsibilities to take measures to tackle climate change" (*Verein KlimaSeniorinnen Schweiz and Others v. Switzerland*, 2024, para. 442), therefore confirming individual responsibility of each state in tackling climate change challenges. But when confronted with individual claims against a state from citizens located outside the given country (like Portuguese children acting on climate change against Belgium and 32 other states in *Duarte Agostinho and others v. Portugal and 32 Other States*, 2024), the Court is very prudent and does not cross the Rubicon.

The protection of Human Rights against climate change has very concrete consequences as ordered by the Grand Chamber, which provides an interesting recipe for optimal climate governance and legislation: States must "act in good time and in an appropriate and consistent manner when devising and implementing the relevant legislation and measures" on climate change (*Verein KlimaSeniorinnen Schweiz et autres c. Suisse*, 2024, para. 550). They must adopt a timeline to achieve carbon neutrality, quantify their remaining carbon budget based on the time frame and set intermediate GHG emissions reduction targets, for instance (*Verein KlimaSeniorinnen Schweiz and Others v. Switzerland*, 2024, para. 550). The Court also imposes an obligation of reporting in the process of compliance with the overarching GHG emissions reduction target. Those are clear and concrete "positive obligations" – as revisited in a climate change context – under Art. 8 of the Convention.

The Human Right to a healthy environment developed as now promoted at the UN level also reaffirms the importance of a stable climate for a healthy environment and the enjoyment of all Human Rights (UNGA, 2022).

## Human Rights versus rights of nature?

At a time where even the right for humans to a healthy environment and protection against climate change is only at this very moment receiving a boost due to the recent developments, rights are also claimed for nature *as such*, and even already attributed to natural entities across the world and in Europe. Those trends must be seen as complementary and not in opposition (Schoukens & Bouquelle, 2024). Additionally, both are, despite appearances, quite Western-centered (the very idea of allocating a "right to" in order to enhance protection or assert a status). What rights for nature can challenge, though, is the approach to the vast acquis, especially in Europe, on duties toward the natural world and elements. Even if already very strong on the protection of water and natural habitats, current legislations could be challenged in their "spirit" if they concern elements that are not only seen as resources or objects but that could also become the *titulaires* of rights, with various consequences too as to their representativity and mode of governance.

In jurisdictions and contexts where citizens can claim protection against their States through Human Rights actions, the protection is often still quite limited. As reasserted by the Grand Chamber of the ECtHR in *KlimaSeniorinnen* (and in opposition to EU law), the Court does not offer "a general protection of the environment" (*Kyrtatos v. Greece*, 2003, para. 51; *Verein KlimaSeniorinnen Schweiz and Others v. Switzerland*, 2024, para. 445). Yet, the legal practice on another continent, America, offers interesting insights and opens the door to new interpretations, where the (human) right to a healthy environment could encompass more ecocentric aspects. In 2020, the Inter-American Court of Human Rights (I-ACtHR) ruled over the right to a healthy environment that

> the right to a healthy environment 'constitutes a universal value'; it 'is a fundamental right for the existence of humankind,' and that 'as an autonomous right [...] it protects the components of the environment, such as forests, rivers and seas, as legal interests in themselves, even in the absence of the certainty or evidence of a risk to individuals. This means that nature must be protected, not only because of its benefits or effects for humanity, 'but because of its importance for the other living organisms with which we share the planet'.
>
> *(Lhaka Honhat v. Argentina, 2020, para. 203)*

The I-ACtHR therefore articulates humans and the environment as part of the right to a healthy environment in a complementary manner: on the one hand, the environment is essential to humankind, as a foundation for fulfilling basic Human Rights, such as the right to life and to health – for instance, without healthy water, there is no bearable life (WHO, 2017) – and on the other, the environment exists for itself as foundation to other living organisms – for example, freshwater fish can only thrive in healthy waters (Mamun & An, 2022). When implementing the right to a healthy environment, nature rights and Human Rights should thus not be seen as contradictory but more as complementary, especially since humans are not separated from the environment they are living in (Bertaux, 2024). As stated by the International Court of Justice, years ago, the environment is "the living space, the quality of life and the very health of human beings" (*Nuclear Weapons*, 1996, paras. 29–35).

## Conclusion

The Member States and the EU are under evolving dynamics regarding the relationship between Human Rights and the environment. Recent developments at the global level, with the UN resolution on the right to a clean, healthy and sustainable environment, are creating new opportunities for revisiting such a relationship. In the meantime, law does not stay still and advances on various fronts are being observed, taking existing duties (the important and essential acquis), long-recognized Human Rights, the newest transnational inspiration and even the broad scientific consensus as stepping stones to stretch boundaries. Human Rights are meant to protect citizens against the acts and omissions of States and, to some extent, private actors. In a tense and increasingly adverse political context, addressing environment and climate issues under the prism of rights (and related duties) is of the utmost importance, also on the European continent.

## Notes

1 Questions that were raised by students during the 2024 POLLEN program at UCLouvain Saint-Louis – Bruxelles.
2 The scope, functioning, members and main project of the Council of Europe are different from the one of the EU, even if both are international organizations active in the European area.
3 UNECE, 1998. United Nations, Treaty Series, vol. 2161.
4 *Milieudefensie et al. v. Royal Dutch Shell*, Court of Appeal, The Hague, ECLI:NL:GHDHA:2024:2099.

## References

Bertaux, C. (2024). Une conception étendue du droit de propriété collective et des droits inhérents au mode de vie des populations autochtones. In D. Misonne & M.-S. de Clippele (Eds.), *Les grands arrêts inspirants du droit de l'environnement*. Larcier, 452–470.

Boyle, A. (2012). Human Rights and the Environment: Where Next? *European Journal of International Law*, 23(3), 613–642, https://doi.org/10.1093/ejil/chs054.

*Commission v. Germany*, Case C-361/88 (European Court of Justice 30 May 1991, a).

*Commission v. Germany*, Case C-59/89 (European Court of Justice 30 May 1991, b).

*Council declaration of November 1973 on the Programme of Action of the European Communities on the Environment.* (1973). Official Journal C 112, pp. 0001–0002.

*Demir et Baykara v. Turkey*, 34503/97 (European Court of Human Rights (Grand Chamber) 12 November 2008).

*Deutsche Umwelthilfe eV v. Freistaat Bayern*, C-752/18 (European Court of Justice (Grand Chamber) 19 December 2019).

*Digital Archives of the United Nations: Declaration of the United Nations Conference on the Human Environment, 1972.* (2012). United Nations Audiovisual Library of International Law.

*Di Sarno and Others v. Italy*, 30765/08 (European Court of Human Rights 10 January 2012).

*Duarte Agostinho and Others v. Portugal and 32 Other States*, 39371/20 (European Court of Human Rights (Grand Chamber) 9 April 2024).

UNGA. (2022). Resolution No 76/300. *The Human Right to a Clean, Healthy and Sustainable Environment.* https://digitallibrary.un.org/record/3983329?ln=en&v=pdf.

Handl, G. (2013). On the Stockholm and Rio Declarations. *United Nations Audiovisual Library of International Law*, https://legal.un.org/avl/pdf/ha/dunche/dunche_f.pdf.

*Janecek v. Freistaat Bayern*, C-237/07 (European Court of Justice 25 July 2008).

*JP. v. Ministre de la transition écologique*, C-61/21 (European Court of Justice 22 December 2022).

Knox, J. H. (2020). Constructing the Human Right to a Healthy Environment. *Annual Review of Law and Social Science*, 16, 79–95, https://doi.org/10.1146/annurev-lawsocsci-031720-074856.

*Kyrtatos v. Greece*, 41666/98 (European Court of Human Rights 22 May 2003).

*L'Erablière v. Belgium*, 49230/07 (European Court of Human Rights 24 February 2009).

*Lhaka Honhat v. Argentina* (Inter-American Court of Human Rights 6 February 2020).

*Lopez Ostra v. Spain*, 16798/90 (European Court of Human Rights 9 December 1994).

Mamun, M., & An, K.-G. (2022). Key Factors Determining Water Quality, Fish Community Dynamics, and the Ecological Health in an Asian Temperate Lotic System. *Ecological Informatics*, 72, 101890, https://doi.org/10.1016/j.ecoinf.2022.101890.

Misonne, D. (2015). The Importance of Setting a Target: The EU Ambition of a High Level of Protection. *Transnational Environmental Law*, 4, 1–26, https://doi.org/10.1017/S2047102514000284.

Misonne, D. (2020). Arm Wrestling around Air Quality and Effective Judicial Protection: Can Arrogant Resistance to EU Law-related Orders Put You in Jail? *Journal for European Environmental & Planning Law*, 17, 409–425, https://doi.org/10.1163/18760104-01704004.

Misonne, D. (2023). High Level of Environmental Protection. In J. Van Zeben, S. Garben, & L. Gormley (Eds.), *Oxford Encyclopedia of EU Law*. Oxford University Press.

*Nostra Onlus v. Comune di Venezia e.a.*, Case C-444/15 (European Court of Justice 21 December 2016).

*Nuclear Weapons*, Report 1996/226 (International Court of Justice (Advisory Opinion) 1996).

*Öneryildiz v. Turkey*, 48939/99 (European Court of Human Rights 30 November 2024).
Parliamentary Assembly of the Council of Europe. (2021). Resolution No 2396. *Anchoring the Right to a Healthy Environment: Need for Enhanced Action by the Council of Europe.*
*Rikki Held et al. v. State of Montana et al.,* cdv-2020-307 (First Judicial District Court Lewis and Clark County 14 August 2023).
Schoukens, H., & Bouquelle, F. (Eds.). (2024). *The Right to a Healthy Environment in and beyond the Anthropocene*. Edward Elgar.
Sikora, A. (2020). *Constitutionalisation of Environmental Protection in EU Law*. Europa Law Publishing.
*Tătar v. Romania*, 67021/01 (European Court of Human Rights 29 January 2009).
*United Nations Conference on the Human Environment in Stockholm*. (1972). Pub. L. No. A/CONF.48/14/Rev.1.
*Verein KlimaSeniorinnen Schweiz and Others v. Switzerland*, 53600/20 (European Court of Human Rights (Grand Chamber) 9 April 2024).
WHO. (2017). *Guidelines for Drinking Water Quality*. WHO.
*Windmills Aalter*, C-24/19 (European Court of Justice (Grand Chamber) 25 June 2020).

# 2

# EU CLIMATE POLICY

*Tom Delreux and Christopher Crellin*

## Introduction

Since the late 1990s, climate change has become the most important and politicized issue on the environmental agenda of the European Union (EU). This chapter presents an overview of the EU's climate policies, with particular attention being paid to the main developments and policy innovations since the publication of the European Green Deal in 2019. The Green Deal, with the overarching target to achieve 'climate neutrality' in the EU by 2050, has provided a new impetus to the climate acquis, leading to a series of reformed and novel policies. This chapter takes stock of the climate policy architecture as it existed at the end of the ninth term of the European Parliament in April 2024. The scope of this chapter is limited to the EU's climate-focused mitigation policies (i.e. reducing greenhouse gas emissions [GHGs]). Although the Green Deal led to an enhanced attention for adaptation (i.e. adjusting to current and future effects of climate change), mitigation policies have been the prominent focus in EU climate policies. Furthermore, despite the emphasis in the Green Deal to mainstream climate change in all EU policy fields, this chapter does not discuss the climate-related policies and the integration of climate change in other policy fields such as agriculture, energy, cohesion, or finance policy (Dupont 2016; Campins Eritja and Fernández-Pons 2024).

This chapter is structured as follows. It provides a brief historical overview of EU climate policy from the 1990s until today, focusing on the main political dynamics and drivers behind the evolution of this policy field, as well as the major milestones that have shaped it. Particular attention is paid to the European Green Deal. It then presents the main architecture of EU

DOI: 10.4324/9781003479260-4

climate policy, first discussing the targets that the EU has adopted, followed by an overview of the main policies that have been put in place to achieve these targets and successively discussing the EU-level Emissions Trading System (ETS); the effort sharing between Member States which serves as the umbrella under which they develop climate policies at the national level; a range of harmonized EU standards; absorption policies (which are intended to absorb the remaining emissions); and a number of funding instruments. Overall, this chapter shows how the EU has progressively built a complex and dynamic climate policy framework, with the European Green Deal serving as a crucial milestone in reinforcing and expanding mitigation efforts while fostering greater policy coherence across multiple sectors.

## Brief historical overview of EU climate policy

After a few non-binding calls for action by the EU institutions, a failed attempt at an EU carbon tax, and a first range of modest climate policies in the 1990s, the actual breakthrough in the development of EU climate policies occurred in the early 2000s. In the two and a half decades that followed, the scope of climate policies expanded and their level of ambition increased, leading to today's dense web of policy frameworks both at the European level and at the level of the Member States (Delreux and Happaerts 2016; Zito 2024). This is the result of both incremental policy reforms and pioneering policy innovation. Four main political dynamics have characterized the evolution of EU climate policy.

First, EU climate policies have not developed in an international vacuum but rather in close interaction with global climate governance (Oberthür and Pallemaerts 2010; Delreux and Happaerts 2016). The interdependence between European and international climate policy is not only a question of historical development, but is also an important characteristic of contemporary EU climate policy-making. On the one hand, the EU presented itself as a 'leader' in multilateral climate negotiations (with varying levels of success) and has contributed to the global climate regime (Tobin et al. 2023). In the run-up to both the United Nations Framework Convention on Climate Change's 15th Conference of the Parties (COP) in Copenhagen (2009) and COP21 in Paris (2015), the EU adopted climate targets (see further), which were not only applicable in the EU itself but which also served as a 'leading by example' strategy of the EU for the upcoming international negotiations. On the other hand, global climate governance has also incentivized, facilitated, and legitimated the development of EU climate policies. At the end of the 1990s, the Kyoto Protocol played an important role in triggering the breakthrough in the EU's climate policies. The policies that the EU adopted at that time were put in place to achieve its international Kyoto targets. Similarly, the Paris Agreement – and again the EU's aspiration to achieve its international commitments – served as a catalyst for policy reforms and new policy initiatives at the EU level.

Second, EU climate policy is conducted via a variety of policy instruments (Oberthür and von Homeyer 2023). While regulatory instruments, which are 'command-and-control' instruments that set standards and norms, are the most frequently used in EU environmental policy more broadly, they have been supplemented in the field of climate policy by other types of policy instruments, which create more flexibility for societal and economic regulatory addressees (Moore et al. 2021). The main example is the ETS, which is a market-based instrument that employs market mechanisms and economic incentives to achieve emissions reductions in certain economic sectors. Likewise, the EU has established a number of funds to achieve climate targets. The mix of policy instruments also consists of softer informational instruments and procedural instruments.

Third, climate policy has become the most politicized and salient issue in the environmental domain in the EU (Biedenkopf and Delreux 2023). The consequences of climate policies for broader societal and economic systems, combined with the increased call for mainstreaming climate concerns in other policy fields (see below), have put climate change at the centre of the political debate in Europe, both at the EU and at the national level. By the mid-2000s, climate change was firmly on the EU's agenda and the wider political context enabled climate policies to be advanced. While the development of climate policies was more constrained during the economic and financial crises of the late 2000s and early 2010s, it received a new impetus in 2018–2019, enabling the adoption of the European Green Deal and the succeeding climate policy development. This increased politicization was triggered by events such as the adoption of the 2015 Paris Agreement, the publication of various Intergovernmental Panel on Climate Change (IPCC) reports demonstrating the urgency of the issue, and the success of Fridays for Future protests (Dupont et al. 2024).

Fourth, the climate acquis consists of several targets that the EU has set for itself (in close interaction with global climate governance), which have driven the adoption of actual legislation with the objective to achieve said targets (Dupont et al. 2024). The EU has set targets to be achieved in 2012, 2020, 2030, and 2050, with new targets for 2040 forthcoming (see further). The 2007 'climate and energy package', with targets to be achieved in 2020, has led to a series of legislative acts being adopted in 2008. Similarly, the 2030 targets adopted in 2014 resulted in another legislative step forward. The 2019 European Green Deal, with its reinforced targets for 2030 and 2050, also triggered legislative reform and innovation through a target package, usually referred to as 'Fit for 55'.

The European Green Deal has indeed been a major milestone for EU climate policy. While it is best known for its long-term target to establish climate neutrality – and thus 'net-zero emissions' – (see further), it is much broader and more comprehensive. It calls for a systemic approach, whereby all societal systems (such as the food system, the energy system or the transport system) need to be approached comprehensively. In other words, one of the main messages of the Green Deal is that addressing climate change requires a systemic approach and no longer an approach from the perspective of environmental policy only. Therefore, the underlying idea behind the Green Deal is that climate concerns – and the climate-neutrality objective – need to be mainstreamed and integrated into all EU policies and actions. With its systemic ambition and transversal approach, the Green Deal indeed goes beyond mere climate policies. It has led to the adoption of several climate-related strategies in other policy domains, such as the 'Farm to Fork Strategy' (on agriculture), a strategy on 'Sustainable and Smart Mobility' (transport) or the 'Sustainable Europe Investment Plan' (green finance).

The Green Deal is technically speaking a Communication from the European Commission and is therefore not legally binding. It rather sketches the broad rationale behind the 2050 target and announces policy initiatives that the Commission intended to table in the following years. During the 2019–2024 von der Leyen Commission, this has resulted in a broad range of policy initiatives which intended both to reform and strengthen existing policies and to introduce new ones. The implementation of the European Green Deal has not been an easy ride, as the EU was confronted with a range of crises and unforeseen events in that period, which had the potential to undermine the Green Deal's objectives. However, remarkably, it has initially proven to be rather crisis-resilient (Bäckstrand 2022; von Homeyer et al. 2022; Rayner et al. 2023). These crises have even served as enablers rather than obstructions for ambitious climate action, particularly on two occasions. First, the COVID-19 pandemic and the economic recovery policies have not forced back the Green Deal's ambitions or its implementation. Even in contrast, the 2021–2027 Multiannual Financial Framework (MFF) and the related economic recovery package NextGenEU, which were adopted during the pandemic, got a stronger climate earmarking than what was initially foreseen. Second, although the Russian invasion of Ukraine and the related energy crisis and increased gas prices in Europe were seen as a potential threat to the Green Deal, this external shock did not break its dynamic. For instance, the Commission's 'REPowerEU' programme, intending to phase out the import of fossil fuel from Russia, maintained a key focus on the green transition. Yet, towards the end of the first von der Leyen term, the implementation of the Green Deal seems to have less resilience towards the farmer protests. Policy proposals on the greening of agricultural policy and on pesticides were weakened or even withdrawn as a result of the increased contestation of such measures.

## The architecture of EU climate policy

The overarching climate policy architecture consists of a set of overall targets that the EU has set for itself and that are to be achieved by a specific year (Box 2.1) and a range of different policies that are intended to enable the EU to reach these targets. Five categories of policies can be distinguished: (i) the EU-level market-based ETS; (ii) EU-level effort sharing policies under which Member States develop national climate policies; (iii) regulatory policies that harmonize standards at the EU level; (iv) policies that regulate the absorption of GHGs (as opposed to the previous categories which aim to reduce emissions); and (v) funding instruments.

### BOX 2.1 EU TARGETS

All the EU targets have in common that they have been proposed by the European Commission and politically endorsed by the heads of state and government in the European Council; that they have triggered legislative processes resulting in legislative climate policies; and that they are – with the major exception of the 2030 and 2050 targets since 2021 – not legally binding but rather political intentions. While the earliest targets did not correspond to what climate science called for, the later targets – and particularly those established in the framework of the European Green Deal – are more and more aligned with climate science.

The 2012 target was the EU's emission reduction target as stipulated in the Kyoto Protocol. According to this international treaty, the EU (at that time with 15 Member States) had to reduce its emissions by 8% compared to 1990 levels by 2012. It was this target – or international commitment – that triggered the first real climate EU policies in the late 1990s and early 2000s. The EU easily achieved this target, with emissions in the EU having decreased by around 15% in 2012.

In 2007, the European Council endorsed the so-called 'climate and energy package', consisting of a threefold target to be achieved by 2020: to reduce GHGs by 20% (and even 30% if other developed countries would make similar commitments in the context of global climate negotiations); to reach 20% of renewable energy in the EU's energy consumption; and to increase energy efficiency in the EU by 20%. These were indeed the 20-20-20 targets to be achieved by 2020. Despite the fact that not every Member State separately achieved the three targets, the EU overall did, as overachieving Member States compensated for underachieving Member States.

In 2011, the European Commission presented its 'Roadmap for moving to a competitive low carbon economy in 2050'. Here the Commission proposed to reduce GHGs by at least 80% by 2050, with differentiated reduction scenarios for various economic sectors. Although that target was not legally binding, it was significant. The EU was indeed the first actor in the world with such a long-term target, which demonstrated that such targets could not be achieved with current policy paradigms, nor by environmental actors and policies alone.

Three years later, in 2014, the Commission presented new targets to be achieved by 2030. At that time the EU was on track to achieve its 2020 targets but considered it necessary to have a mid-term milestone for heading to the 2050 target, and to demonstrate leadership in the run-up to the crucial COP15 where the Paris Agreement was to be adopted. The Commission opted for an update of the threefold target of the 2007 'climate and energy package', but now to be achieved in 2030: at least 40% reduction of emissions; 27% renewables; and a 27% increase in energy efficiency. In the following years, these targets were updated to 55%, 32%, and 32.5%, respectively (Kulovesi and Oberthür 2020).

The 2019 European Green Deal reinforced the targets again. Moreover, the updated targets became legally binding via the so-called 'European Climate Law' in 2021 (Rayner et al. 2023; Kulovesi et al. 2024). With regard to targets, the Climate Law has a threefold importance. First, it establishes a 'climate-neutrality' target for 2050. Climate neutrality refers to 'net-zero' emissions, which means that as much GHGs that are still emitted should be absorbed, for instance, via technology that either captures and stores GHGs or removes them from the atmosphere, or via developing natural 'carbon sinks' (such as forests) which absorb carbon. It is the start towards 'net-zero' emissions in 2050 that was called by Commission President von der Leyen 'Europe's man on the moon moment'. Second, the Climate Law codifies the updated 2030 emission reduction target: the target that was set in 2014 (40% emission reduction) was upgraded to 55%. It is that target that subsequently triggered a strengthening of a broad set of climate policies: the legislative package known as 'Fit for 55' – referring to making the EU's climate policies ready for the upgraded 2030 target. Third, the Climate Law included a process for setting a 2040 target. In 2024, the Commission presented its assessment and suggested a 90% target for 2040, but the final target would only be decided upon by the von der Leyen II Commission in 2025.

Sources: Kulovesi and Oberthür (2020, pp. 151–166); Rayner et al. (2023).

### *Emissions Trading System*

The ETS is a cornerstone of the EU's climate policy. Based upon an EU-wide carbon market, it is the policy instrument that addresses GHG emission reductions from large European industries, intra-European aviation and maritime transport. Combined, they represent currently approximately 40% of all emissions in the EU. Aiming to put into practice the 'polluter pays' principle, the underlying idea of the ETS is that large industries pay for the 'environmental externality' of their GHGs. It indeed puts a price on emissions and uses market forces to incentivize emission reductions (Wettestad and Jevnaker 2018; Meadows et al. 2019).

The ETS covers the following sectors: electricity and heat generation, energy-intensive industry sectors (such as oil refineries, steel, iron, aluminium, metals, cement, lime, glass, ceramics, pulp, paper, cardboard, acids, and bulk organic chemicals), intra-European aviation and maritime transport (the latter since 2024). The starting point of the ETS is that each emitter included in the system – i.e. a large company from these sectors – needs to hold an allowance for every ton of GHG (mostly carbon dioxide) it emits. The reduction of these emissions is then achieved via a 'cap and trade' system. 'Cap' refers to the fact that there is a quantitative limit on emissions and, consequently, on the available allowances (because every emission needs to be covered by an allowance). Importantly, the cap is decreasing every year. Indeed, the number of available allowances annually shrinks by 4.3% (and from 2028 onwards by 4.4%). The consequence of the decreasing cap is that there are fewer allowances available every year. As emissions need to be covered by allowance, emissions should, as a result, decrease as well.

The incentive structure behind the decrease of emissions is the 'trade' part of 'cap and trade'. The ETS functions with a market where emission allowances can be bought and sold. The rationale is here that industries that emit less will be rewarded because they can sell their surpluses on the market. By contrast, the system economically sanctions those industries that do not invest in green technologies, because they will need to buy additional allowances on the market. As allowances become more and more scarce (as a result of the annual reduction factor), they will also become more expensive. This market-based mechanism thus intends to incentivize large industries to reduce their emissions.

More than half of all allowances are currently auctioned. The money generated by the auctioning is for the Member States, which have to spend at least half of these revenues for climate-related purposes. The remaining allowances are given for free. The reasoning is here that free allowances are a response to the risk of so-called 'carbon leakage', which is the relocation of European industries to third countries with less stringent climate policies (i.e. where they would not have to pay for emitting). In other words, free allowances aim to counter the competitive disadvantage of European industries. Power-intensive manufacturing industries, such as cement or steel producers, are the most vulnerable to carbon leakage. Free allowances are thus given primarily to these companies. In the following decade, free allowances will gradually fade out, as carbon leakage will be addressed by another climate policy instrument, namely the Carbon Border Adjustment Mechanism (CBAM) (see further).

The ETS is operational since 2005 and it has been reformed three times. The current, fourth, phase of the ETS runs from 2021 to 2030. For a long time, the ETS's effectiveness has been challenged by the fact that the price of allowances was too low in order to incentivize industries to reduce their emissions (Wettestad and Jevnaker 2016). The oversupply of allowances on the market was the basis of this problem. Several ETS reforms have been adopted in an attempt to increase the price of an allowance, but it took until the fourth phase of the ETS before the price increased to a level that the ETS is finally performing. Nonetheless, although the price is currently significantly higher than in the past, it is still rather volatile.

In the context of the implementation of the European Green Deal, two new policy instruments have been adopted that are related to the initial ETS. First, CBAM is a climate-related unilateral trade instrument that has been in force since 2023 but will be fully operational from 2034 onwards (Wettestad 2023; Fernández-Pons 2024; see also Chapter 9). It requires that exporters from non-EU countries pay a carbon price at the EU border that corresponds to what European producers have to pay for an ETS allowance. If that non-EU producer has already paid a 'carbon price' domestically, that is taken into account in the calculation of the price of the CBAM certificate.

Just like free allowances under the ETS, CBAM also aims to counter the carbon leakage risk. It intends to neutralize the competitive disadvantage of EU industries, which have to pay for their emissions under the ETS, vis-à-vis non-EU industries, which do not face similar regulations in their country. CBAM will gradually replace free allowances under the ETS. While both address carbon leakage and aim to counterbalance the competitive disadvantage of EU industries, CBAM also incentivizes third countries to adopt climate-friendly measures, such as domestic emissions trading.

CBAM currently only applies to imports from specific energy-intensive sectors (cement, iron, steel, aluminium, chemicals, fertilizers, refineries, hydrogen, and electricity), which are the sectors that can benefit from free allowances under the ETS (see above). As the current scope of CBAM is limited, CBAM only covers a small proportion of the goods imported into the EU. Its effect for the EU's main trading partners is currently 'very small' or 'almost negligible' (Wettestad 2025, p. 10). Yet, CBAM's scope might be extended in the future.

Second, a new ETS – the so-called 'ETS-2' – will be operational from 2027 or 2028 onwards. This is a separate carbon market that will expand the scope of emissions trading to the sectors of road transport and the heating of buildings. A difference with the initial ETS is that it is not the individual car users or households that will have to buy an allowance, but rather the fuel suppliers. The latter will be obliged to buy allowances to cover the emissions of the use of the fuel they will sell to consumers. It is, however, to be expected that the fuel suppliers will pass on that cost to car users and households, but this is meant to incentivize the consumers to reduce their fossil energy consumption. Yet, this immediately raises questions on the social dimension of increasing energy prices, particularly for vulnerable groups in society. Following the creed that the green transition should 'leave no one behind', the revenues from the auctioning of allowances under the ETS-2 will go to the Social Climate Fund (SCF), which will support these socially vulnerable groups (see further).

### *Effort sharing and national policies*

GHG emissions from sectors that are not covered by the ETS – currently comprising approximately 60% of GHGs in the EU – are regulated under the so-called 'Effort Sharing Regulation' (ESR) (Peeters and Athanasiadou 2020). This currently includes emissions from agriculture, road transport, buildings, waste, and small industry. As mentioned before, road transport and buildings will switch from the ESR regime to the ETS regime in 2027 or 2028 with the establishment of ETS-2. The ESR, as amended in 2023 under the Fit for 55 package, imposes a collective target of 40% emissions reduction in the ESR sectors by 2030, compared to 2005 levels, as well as binding individual Member State emissions reduction targets, which are planned to meet the overarching collective target. The revised 40% ESR target is an increase of 10% points compared to the previous 2018 collective target.

The national targets, which decrease year on year, are the maximum level of emissions each Member State can generate per year in the ESR sectors. The individual targets primarily reflect the Member States' gross domestic product (GDP) per capita so that Member States with lower levels of GDP per capita have less stringent targets compared to Member States with higher levels of GDP per capita. Yet, GDP per capita is not the only determinant of the target, as the latter is adjusted to take into account cost-effectiveness for richer Member States. The current national emissions reduction targets range from 10% (Bulgaria) to 50% (Denmark, Germany, Luxembourg, Finland, and Sweden). Besides a national target to be achieved in 2030, Member

States also have a 'linear reduction trajectory' with annual targets that serve as stepping stones to their 2030 target.

The ESR incorporates five flexibility instruments to assist Member States in reaching their individual targets and to ensure the cost-effectiveness of emissions reductions. First, a Member State can bank a surplus of its emissions in a certain year for use in a future year when its emissions are lower than its annual target. Second, a Member State can borrow a part of its annual emissions target from a future year to use it in a current year if its emissions exceed its annual target. Third, a Member State can trade a surplus of its emissions in a certain year to another Member State. Fourth, select Member States, which have emission reduction targets significantly above both the EU average and their cost-effective reduction potential, have the option to access allowances from the EU ETS to offset a certain shortage in emissions reductions in certain ESR sectors. Currently, Belgium, Denmark, Ireland, Luxembourg, Malta, The Netherlands, Austria, Finland, and Sweden have the option to access ETS allowances. Finally, Member States have the possibility to offset a certain shortage in emission reductions by accessing emission credits related to the land use sector. Alternatively, emissions reductions from the effort sharing sectors can be used to compensate for emissions from the land use sector (Peeters and Athanasiadou 2020).

Importantly, to reach their individual targets, Member States are responsible for implementing policies at the national level in the sectors covered by the ESR. Hence, while the ESR targets – both collectively and individually – are determined at the EU level, the policies to implement them are adopted at the national level. Member States have national room for manoeuvre to choose the appropriate measures across the sectors covered by the ESR, such as policies to promote public transport, support energy efficiency in buildings or promote more climate-friendly agriculture (Böhringer 2014).

Member States are required to adopt and submit integrated 'national energy and climate plans' (NECPs), in which they report on the progress they make to achieve their ESR target (Kulovesi and Oberthür 2020). The NECPs play a key role in the Member States' planning of their compliance with the ESR (Peeters and Athanasiadou 2020). On the basis of the NECPs, the Commission monitors the progress and publishes annual assessments of progress as well as more comprehensive and formal compliance checks. If the Commission finds that a Member State has not made significant progress, the Member State must submit a revised action plan outlining additional actions to be taken and a strict timetable. Briefly, the governance approach behind the ESR follows a bottom-up logic, with national leeway for Member States and a monitoring Commission, as opposed to a hierarchical top-down approach.

### *Harmonized EU-level standards*

For the EU to reach its 2030 and 2050 emissions reduction targets, additional and significant sectoral emissions reduction efforts are required. Such traditional policies, which set regulatory standards at the EU level, have been part of the EU's climate policy portfolio since the 2000s. Yet under the Fit for 55 package, several sectoral pieces of climate legislation have been revised and strengthened, including setting emissions reduction standards for cars and vans, the aviation sector, and the maritime sector; regulating fluorinated GHGs in certain products on the EU market; and setting emissions reduction standards in the building sector. These harmonized standards at the EU level have a twofold objective. First, they aim to have a significant influence on reducing the EU's overall GHGs (Gheuens 2023). Second, setting EU-wide standards creates a level playing field for a well-functioning EU internal market.

A first example of such a regulatory measure is legislation on fluorinated greenhouse gases (F-gases). These are gases used in everyday products such as refrigeration, air conditioning, and aerosols or in the pharmaceutical and cosmetic industries. F-gases, such as hydrofluorocarbons and perfluorocarbons, currently account for approximately 2.5% of total GHGs in the EU, but their greenhouse gas effect is much stronger than that of carbon dioxide. Yet, they can remain in the atmosphere for thousands of years (Simöes 2023). As part of the Fit for 55 package, the revised F-gas Regulation aims to prevent F-gas emissions equivalent to 300 million tonnes of carbon dioxide by 2050 through the updating of measures regarding the containment, use, recovery, and destruction of F-gases. It is foreseen that hydrofluorocarbons will be reduced by 95% by 2030 compared to 2015 and completely phased out in the EU by 2050. In addition, the revised regulation also introduces new restrictions, applied from 2025 to 2035, ensuring that F-gases are only used in new equipment where no suitable alternative is available or that only the most climate-friendly F-gases are used.

A second example are regulatory measures on $CO_2$ emissions from the road transport sector, which is responsible for a quarter of the EU's GHGs and the only sector where emissions have been rising between 1990 and 2019 (Birel et al. 2024; Gheuens 2023). Since 2009 and 2011, respectively, the EU has regularly revised and incrementally strengthened the regulation on the $CO_2$ emission standards of new passenger cars and vans sold on the EU market. The regulations remain technologically neutral and put the responsibility on the automobile producers to develop appropriate technological solutions to meet the EU's targets (Dyrhauge 2021). As part of Fit for 55, the Regulation was revised and it sets a more ambitious target for 2030 (93.6 $CO_2$/km for cars; 153.9 for vans) as well as a new target for 2035 (49.5 $CO_2$/km for cars; 90.6 for vans) (Erbach 2023). These targets do not apply to individual vehicle models, but to the fleet of the automobile producers. The most salient element of the Regulation is a ban on the sale of new vehicles with internal combustion engines from 2035 onwards (Birel et al. 2024). However, the Regulation gives reference to e-fuels or $CO_2$-neutral fuels, which enables the sale of new internal combustion engine vehicles on the EU market if they run on e-fuels (Transport and Environment 2023).

### *Absorption policies*

While the previously discussed ETS, ESR, and harmonized standards aim at the reduction of GHGs, the EU has also developed two core approaches that deal with the absorption of GHGs that are already in the atmosphere. This is related to the 2050 climate-neutrality target, which prescribes 'net-zero' emissions by 2050, implying that the remaining emissions are to be 'absorbed' (see above). The first absorption approach is through natural processes, regulated in the Land Use, Land Use Change, and Forestry (LULUCF) Regulation. The second aims to deal with carbon absorption through technological processes to primarily deal with hard-to-abate emissions.

First, the LULUCF sector covers over three-quarters of the EU's territory and includes, for example, the management of cropland, grassland, wetlands, and forests, as well as land use change activities such as afforestation, deforestation, or draining of peatlands. The LULUCF sectors function as a carbon sink, as more carbon is absorbed and stored in biomass, soils, and harvested wood products than it emits (Savaresi et al. 2020). However, the sink function has decreased in recent years, due to fires and soil disturbances for example, which emit carbon into the atmosphere (Jensen 2023).

Designed to complement the EU ETS and the EU ESR, LULUCF policies were introduced onto the EU agenda in preparations for the UNFCCC COP15 in Copenhagen, resulting in the 2018 LULUCF Regulation (Savaresi et al. 2020; Sotirov et al. 2021). Under the Fit for 55 package, the revised LULUCF Regulation aims to enforce the sustainable management of forests and soils, and for the first time, systematically use carbon sinks to reach the 2030 and 2050 targets (Jensen 2023). Without a solid LULUCF base to absorb carbon, the EU's 2050 climate-neutrality target and the internal and external credibility of the EU may be questioned (Chikh M'hamed and Sprinz 2024). The revised Regulation moves away from the 'no debit rule', which requires Member States to ensure that GHGs do not exceed GHG removals in the LULUCF sector (Savaresi et al. 2020). It introduces an EU-wide target of 310 million tonnes of carbon dioxide equivalent of net removals by 2030, which is implemented through binding net removal targets for each Member State's LULUCF sector.

Second, while land-based carbon sinks – and thus absorption via natural processes – are proven at scale and are the primary absorption approach for the EU, the EU also foresees the use of technological solutions to capture and store carbon, primarily for hard-to-abate emissions from certain sectors like agriculture, transport, and some industrial processes, such as cement and steel production. The EU has a Carbon Capture and Storage (CCS) Directive since 2009. Since 2021, in the context of the Green Deal, three Communications and a Regulation related to CCS have been tabled to further develop the EU's CCS policy framework. As a lack of $CO_2$ storage space has been identified, which creates a bottleneck for the uptake of CCS technologies, the EU's technological absorption framework aims to improve the EU's carbon storage capacity, incentivize the uptake of carbon management technologies, and avoid greenwashing (Jensen 2022, 2024).

### *Funding*

The last category of climate policies is funding instruments to support the EU's transition to climate neutrality by 2050 as well as global efforts to combat climate change. The EU aims to mobilize 1 trillion EUR in sustainable investments through the European Green Deal Investment Plan, primarily financed through the MFF and NextGenEU, as well as revenues of the ETS auctioning. Under the 2021–2027 MMF and the NextGenEU temporary recovery support package combined, the total amount of funds dedicated to combat climate change is 1.8 trillion EUR (de la Porte and Jensen 2021; Bocquillon et al. 2024). Within the MFF and NextGenEU, the InvestEU programme and the Just Transition Fund (JTF) are specific mechanisms for funding climate change projects (D'Alfonso 2021). Additionally, the MFF provides funds for mainstreaming climate in other policy fields, such as cohesion policy. The most notable updates to climate financing in the context of the European Green Deal and 'Fit for 55' was the introduction of the JTF and the SCF. These are two eco-social financial policies, integrating environmental and social policy goals by accounting for the negative social consequences of the transition (Crespy and Munta 2023).

The JTF aims to support carbon-intensive regions, which are economically most affected by the transition to climate neutrality, such as regions where coal mining is still an important economic activity. Its budget is 17.5 billion EUR (10 billion EUR from NextGenEU and 7.5 billion EUR from the MFF) (Kyriazi and Miro 2022). The JTF funds two types of transitional activities: green investments that foster the modernization of local economies; and social investments that address the negative impacts of the transition on workers and employment (Crespy and Munta 2023). The allocation of JTF funds to Member States is determined by the Member States' levels of industrial emissions in carbon-intensive regions, employment figures in fossil fuel industries, and degrees of economic development. Based on these criteria, Poland and Germany are the main recipients of JTF funds. To access the funds, Member States need to devise Territorial Just Transition Plans outlining how they intend to use the funds.

With the introduction of the ETS-2, the SCF was established. Due to the foreseen increases in prices for households generated by the ETS-2 (see above), the SCF aims to provide direct income support and investment funding to low-income households and small companies (Crespy and Munta 2023; Kyriazi and Miro 2022). The SCF is financed by ETS-2 revenues. The allocations of SCF funds to Member States are determined based on a number of criteria including total population, gross national income per capita, and several indicators of the population at risk of poverty and a country's economic dependence on carbon. Based on the criteria, Poland, France, Italy, Spain, and Romania are expected to be the main recipients of SCF funding. Like the JTF, to access the SCF funds, Member States must draw up national Social Climate Plans.

The funding instruments not only target internal addressees, but the EU also contributes to international climate financing, which plays an important role in helping developing countries mitigate and adapt to climate change. The Paris Agreement set a goal for developed countries to contribute $100 billion per year to international climate finance until 2025. Jointly, the EU and its Member States are the largest providers of public climate finance (Oberthür and Dupont 2021). Since 2013, the EU has more than doubled the funds raised to help developing countries mitigate and adapt to the impact of climate change. In 2022, the EU contributed 28.5 billion EUR in funding to developing countries to support mitigation and adaptation actions.

## Conclusion

The EU has developed, through a combination of incremental updates and new policy developments, one of the world's most advanced climate mitigation policy portfolios over the past three decades. The most recent update and surge in climate policy activity came in the form of the European Green Deal and subsequent Fit for 55 package. The Green Deal takes a systemic and just approach in striving for the overarching objective of climate neutrality by 2050 (von Homeyer et al. 2022). Through the subsequent Fit for 55 package, concrete policies have been adopted to implement the Green Deal's ambitions. Taken together, Fit for 55 has expanded the scope, breadth, and depth of the EU's climate policies as well as introduced new policies, including the CBAM, ETS-2, the SCF, and JTF. Since their introduction, the EGD and 'Fit for 55' package have also proceeded, and in some instances been propelled forward, during a period where the EU has had to face multiple crises, including the COVID-19 pandemic, the Russian invasion of Ukraine and the energy crisis that unfolded thereafter.

Against this backdrop, this chapter, first, briefly outlined the historical developments of EU climate policy from the early 1990s until 2024. It then proceeded to discuss the key aspects of the EU's climate policy framework with added attention given to the most recent updates. Here, the EU emissions reduction targets and the subsequent policies implemented with the overarching aim of reaching the set targets were outlined. The key policies focused on included the European Climate Law and the emissions reduction targets; the EU ETS and the more recent developments of ETS-2 and CBAM; the ESR; sectoral and product EU standard setting (such as on F-gases and on cars and vans); the EU's absorption policies (LULUCF and CCS); and the climate financing policies.

Despite being the most developed and complex globally, the current EU climate policy framework is not ambitious enough to lead to climate neutrality in 2050. The EU needs to considerably ramp up its efforts to combat climate change to achieve the European target of climate neutrality and the international target of limiting global warming to at least 2°C, as set by the Paris Agreement.

## References

Bäckstrand, K. (2022). 'Towards a Climate-Neutral Union by 2050? The European Green Deal, Climate Law, and Green Recovery' in Engelbrekt, A., Ekman, P., Michalski, A., & Oxelheim, L. (eds.), R*outes to a Resilient European Union: Interdisciplinary European Studies*, Cham, Palgrave Macmillan, pp. 39-62.

Biedenkopf, K., & Delreux, T. (2023). 'European Integration, the Environment and Climate Change' in Segers, M., & Van Hecke, S. (eds.), *The Cambridge History of the European Union*, Cambridge, Cambridge University Press, pp. 418–441.

Birel, T., Breeman, G., van Buitenen, A., & Vijver, M. (2024). 'Defueling the Impasse: EU Political Discourse on e-fuels' in *Energy Policy*, 187, p. 114022.

Bocquillon, P., Brooks, E., & Maltby, T. (2024). 'Talkin' Bout a Revolution? Institutional Change in the EU Recovery and Resilience Facility: The Case of Climate Policy' in Journal of Common Market Studies, 62(6), pp. 1633–1653. DOI: 10.1111/jcms.13536.

Böhringer, C. (2014). 'Two Decades of European Climate Policy: A Critical Appraisal' in *Review of Environmental Economics and Policy*, 8(1), pp. 1–17.

Campins Eritja, M., & Fernández-Pons, X. (eds.) (2024). *Deploying the European Green Deal. Protecting the Environment beyond the EU Borders*, Abingdon, Routledge.

Chikh M'hamed, S., & Sprinz, D. (2024). 'The Keys to the EU's Climate Neutrality Goal' in Dyrhauge, H., & Kurze, K. (eds.), *Making the European Green Deal Work: EU Sustainability Policies at Home and Abroad*, Abingdon, Routledge, pp. 60–75.

Crespy, A., &Munta, M. (2023). 'Lost in Transition? Social Justice and the Politics of the EU Green Transition' in Transfer: *European Review of Labour and Research*, 29(2), pp. 235–251.

D'Alfonso, A. (2021). *InvestEU Programme: The EU's New Investment Support Scheme*, EPRS Briefing. https://www.europarl.europa.eu/thinktank/en/document/EPRS_BRI(2020)659364.

de la Porte, C., & Jensen, M. (2021). 'The Next Generation EU: An Analysis of the Dimensions of Conflict behind the Deal' in *Social Policy & Administration*, 55(2), pp. 388–402.

Delreux, T., & Happaerts, S. (2016). *Environmental Policy and Politics in the European Union*, London, Palgrave Macmillan.

Dupont, C. (2016). *Climate Policy Integration into EU Energy Policy. Progress and Prospects*, Abingdon, Routledge.

Dupont, C., Moore, B., Boasson, E., Gravey, V., Jordan, A., Kivimaa, P., Kulovesi, K., Kuzemko, C., Oberthür, S., Panchuk, D., Rosamond, J., Torney, D., Tosun, J., & von Homeyer, I. (2024). 'Three Decades of EU Climate Policy: Racing toward Climate Neutrality?' in *WIREs Climate Change*, 15(1), pp. 1–12.

Dyrhauge, H. (2021). 'Discourses about EU Transport Decarbonisation: Towards a Paradigm Shift?' in *The International Spectator*, 56(3), pp. 71–86.

Erbach, G. (2023). *$CO_2$ Emission Standards for New Cars and Vans: 'Fit for 55' Package*, EPRS Briefing. https://www.europarl.europa.eu/thinktank/en/document/EPRS_BRI(2022)698920.

Fernández-Pons, X. (2024). 'Conditioning Access to the European Union Market on Carbon Footprint. The Carbon Border Adjustment Mechanism' in Campins Eritja, M., & Fernández-Pons, X. (eds.), *Deploying the European Green Deal. Protecting the Environment beyond the EU Borders*, Abingdon, Routledge, pp. 16–33.

Gheuens, J. (2023). 'Putting on the Brakes: The Shortsightedness of EU Car Decarbonization Policies' in *Npj Climate Action*, 2(1), pp. 1–9.

Jensen, L. (2022). *Sustainable Carbon Cycles: Promoting Removal, Storage and Recycling*, EPRS Briefing. https://www.europarl.europa.eu/thinktank/en/document/EPRS_BRI(2022)733679.

Jensen, L. (2023). *Revision of the LULUCF Regulation: Strengthening the Role of the Land Use, Land-Use Change and Forestry Sector in Climate Action*, EPRS Briefing. https://www.europarl.europa.eu/thinktank/en/document/EPRS_BRI(2021)698843.

Jensen, L. (2024). *A Union Certification Framework for Carbon Removals*, EPRS Briefing. https://www.europarl.europa.eu/thinktank/en/document/EPRS_BRI(2023)739312.

Kulovesi, K., & Oberthür, S. (2020). 'Assessing the EU's 2030 Climate and Energy Policy Framework: Incremental Change toward Radical Transformation?' in *Review of European, Comparative & International Environmental Law*, 29(2), pp. 151–166.

Kulovesi, K., Oberthür, S., van Asselt, H., & Savaresi, A. (2024). 'The European Climate Law: Strengthening EU Procedural Climate Governance?' in *Journal of Environmental Law*, 36(1), pp. 23–42.

Kyriazi, A., & Miro, J. (2022). 'Towards a Socially Fair Green Transition in the EU? An Analysis of the Just Transition Fund Using the Multiple Streams Framework' in *Comparative European Politics*, 21(1), pp. 112–132.

Meadows, D., Vis, P., & Zapfel, P. (2019). 'The EU Emissions Trading System' in Delbeke, J., & Vis, P. (eds.), *Towards a Climate-Neural Europe. Curbing the Trend*, Abingdon, Routlegde, pp. 66–94.

Moore, B., Benson, D., Jordan, A., Wurzel, R., & Zito, A. (2021). 'Governing with Multiple Policy Instruments?' in Jordan, A., & Gravey, V. (eds.), *Environmental Policy in the EU. Actors, Institutions and Processes*. Fourth edition, Abingdon, Routledge, pp. 299–316.

Oberthür, S., & Dupont, C. (2021). 'The European Union's International Climate Leadership: towards a Grand Climate Strategy?' in *Journal of European Public Policy*, 28(7), pp. 1095–1114.

Oberthür, S., & Pallemaerts, M. (2010). 'The EU's Internal and External Climate Policies: an Historical Overview' in Oberthür, S., & Pallemaerts, M. (eds.), *The New Climate Politics of the European Union*, Brussels, VUB Press, pp. 27–63.

Oberthür, S., & von Homeyer, I. (2023). 'From Emissions Trading to the European Green Deal: the Evolution of the Climate Policy Mix and Climate Policy Integration in the EU' in *Journal of European Public Policy*, 30(3), pp. 445–468.

Peeters, M., & Athanasiadou, N. (2020). 'The Continued Effort Sharing Approach in EU Climate Law: Binding Targets, Challenging Enforcement?' in *Review of European, Comparative & International Environmental Law*, 29(2), pp. 201–211.

Rayner, T., Szulecki, K., Jordan, A., & Oberthür, S. (2023). 'The Global Importance of EU Climate Policy: An Introduction' in Rayner, T., Szulecki, K., Jordan, A., & Oberthür, S. (eds.), *Handbook on European Union Climate Change Policy and Politics*, Cheltenham, Edward Elgar Publising, pp. 1–21.

Savaresi, A., Perugini, L., & Chiriacò, M. (2020). 'Making Sense of the LULUCF Regulation: Much ado about Nothing?' in *Review of European, Comparative & International Environmental Law*, 29(2), pp. 212–220.

Simöes, H.M. (2023). *Review of the Regulation on Fluorinated Greenhouse Gases: 'Fit for 55' Package*, EPRS Briefing. https://www.europarl.europa.eu/thinktank/en/document/EPRS_BRI(2022)733673.

Sotirov, M., Winkel, G., & Eckerberg, K. (2021). 'The Coalitional Politics of the European Union's Environmental Forest Policy: Biodiversity Conservation, Timber Legality, and Climate Protection' in *Ambio*, 50(12), pp. 2153–2167.

Tobin, P., Torney, D., & Biedenkopf, K. (2023). 'EU Climate Leadership: Domestic and Global Dimensions' in Rayner, T., Szulecki, K., Jordan, A., & Oberthür, S. (eds.), *Handbook on European Union Climate Change Policy and Politics*, Cheltenham, Edward Elgar Publishing, pp. 187–200.

Transport and Environment (2023). *How to Prevent an e-fuels Loophole Undermining the EU Car $CO_2$ Law*, Position Paper, September 2023, https://www.transportenvironment.org/wp-content/uploads/2023/09/2023_09_Efuel_imp_del_act_position_paper_final-min.pdf.

Von Homeyer, I., Oberthür, S., & Dupont, C. (2022). 'Implementing the European Green Deal during the Evolving Energy Crisis' in *Journal of Common Market Studies*, 60(S1), pp. 125–136.

Wettestad, J. (2023). 'Proactive Prevention of Carbon Leakage? The EU Carbon Border Adjustment Mechanism' in Rayner, T., Szulecki, K., Jordan, A., & Oberthür, S. (eds.), *Handbook on European Union Climate Change Policy and Politics*, Cheltenham, Edward Elgar Publising, pp. 231–245.

Wettestad, J. (2025). 'The EU's Carbon Border Adjustment Mechanism: Shaped and Saved by Shifting Multi-Level Enforcement?' in *Review of Policy Research*, 42, pp. 508–529. DOI: 10.1111/ropr.12597.

Wettestad, J., & Jevnaker, T. (2016). *Rescuing EU Emissions Trading. The Climate Policy Flagship*, London, Palgrave Macmillan.

Wettestad, J., & Jevnaker, T. (2018). 'EU Emissions Trading. Frontrunner – and «black sheep»?' in Wettestad, J., & Gulbrandsen, L. (eds.), *The Evolution of Carbon Markets: Design and Diffusion*, Abingdon, Routledge, pp. 30–52.

Zito, A. (2024). 'European Union' in Fiorino, D., Eisenstadt, T., & Ahluwalia, M. (eds.), *Elgar Encyclopedia of Climate Policy*, Cheltenham, Edward Elgar Publishing, pp. 429–433.

# PART 2
# Energy policy

# 3

# SAFETY, SUSTAINABILITY, AND SYNCHRONIZATION

## The Gap in EU Policies on Nuclear Waste Management

*Janani Mohan*

> It should be an ethical obligation of each Member State to avoid any undue burden on future generations in respect of spent fuel and radioactive waste.
>
> —European Atomic Energy Community (Euratom, 2011)

Finland's Onkalo nuclear waste repository, the world's first final nuclear waste storage site, is expected to start receiving spent fuel within the next few years (Vattenfall, 2023). Onkalo is intended to store spent fuel from Finland's nuclear energy reactors for 100,000 years, a period of time that is sufficient for the associated radiation hazards to neutralize (Benke, 2023). As the International Atomic Energy Agency (IAEA) Director General Rafael Grossi commented: "Everybody knew of the idea of a geological repository for high-level radioactive nuclear waste, but Finland did it" (IAEA, 2020). Indeed, most of the European Union (EU) countries are far behind in securing permanent waste repositories. Out of the 13 EU countries with nuclear energy programs, only three countries—Finland, Sweden, and France—are expected to open permanent underground spent-fuel repositories by the 2030s (Simicevic, 2023). The other EU countries are still entirely reliant on interim storage facilities that will require replacement within the next century.

Such interim storage facilities for nuclear waste pose a significant burden for present and future generations within the EU. "Nuclear waste" includes both radioactive waste and spent fuel generated during nuclear energy production. When nuclear fuel undergoes fission in reactors, the process splits the atoms in the fuel to generate energy, triggering a chain reaction that produces radioactive byproducts. Only a small percentage of these byproducts is considered "high-level waste," which requires between 1,000 and 10,000 years to return

DOI: 10.4324/9781003479260-6

to its original radiation levels (World Nuclear Association, 2024a). In fact, due to fuel reprocessing, France categorizes only 0.2% of its waste as high-level. However, all nuclear waste poses a serious risk to human health, environmental conditions, and the security of dual-use materials. If mishandled, such waste can remain dangerous to humans for thousands of years, leading to chronic health issues, birth defects, and even fatalities (US Energy Information Administration (EIA), n.d.). Developing solutions beyond interim nuclear waste facilities is therefore a pressing issue to secure the EU from the possible consequences of nuclear waste in the present and for the future.

However, a permanent EU solution still faces numerous challenges. The EU's commitment to nuclear waste management (NWM) is a fundamental element of its role as a regional organization. Indeed, one of the foundational treaties for the EU created and tasked Euratom "to establish uniform safety standards to protect the health of workers and of the general public" from nuclear energy by regulating radioactive materials like nuclear waste (European Union, 1957). Yet, the EU lacks clear standards for NWM, as the energy mix of Member States is set at the national level. EU policies thus revert to an open method of coordination that relies on reporting and soft law, rather than binding legislation or technical requirements for national programs. This leads to divergent practices between Member States. For example, some EU countries are expected to inaugurate permanent waste repositories between the 2040s and 2100s, decades after the Onkalo site in Finland (Simicevic, 2023).

For the EU to address this critical gap in its NWM, policies to encourage, if not require, timeframes and standards for Member States to develop permanent spent-fuel storage are needed. This chapter proposes a novel "three-S" model, where the EU should adopt policies that set standards for three S's: safety, sustainability, and synchronization. As discussed in further detail below, *safety* policies address preventing radiological incidents from waste, *sustainability* maintaining national NWM programs, and *synchronization* coordinating transnationally for mutual benefit. These three S's should expedite the development of long-term storage sites within the EU by improving NWM in Member States.

To support the adoption of this three-S model, this chapter provides a brief history of EU policies on nuclear waste management (NWM), drawing conclusions for the importance of consistency to address historical EU challenges in NWM. This chapter then considers the current inconsistencies in national frameworks, suggesting that the differences are driven by varying government support for nuclear energy in EU Member States. To identify trends in the national frameworks, this chapter discusses programs within the 13 EU states with recent nuclear energy programs: Belgium, Bulgaria, Czechia, Finland, France, Germany, Hungary, The Netherlands, Romania, Slovakia, Slovenia, Spain, and Sweden. These trends underscore the necessity of an EU response to address the differences in nuclear waste programs, while enabling EU states to retain their sovereign and diverse attitudes toward nuclear energy itself. The EU's adoption of safety, sustainability, and synchronization policies could therefore prove vital to improving NWM throughout the region and beyond.

## A brief history of EU policies on nuclear waste management

On March 25, 1957, the initial members of the EU signed the Treaty establishing the European Atomic Energy Community (Euratom), one of three foundational treaties forming instead of founding the Union itself (European Parliament, n.d.). Although the Euratom Treaty's stated purpose was to create "conditions necessary for the speedy establishment and growth of nuclear industries," the Treaty also opened other areas for discussion, notably including language on NWM (European Union, 1957). EU nuclear waste governance today stems from Article 2(b) of the Euratom Treaty, which states that the EU shall "establish uniform safety standards [in nuclear industries] to protect the health of workers and of the general public and ensure that they are applied." Article 30 in the Euratom Treaty further defines these standards to include the specific context of ionizing radiation, as found in nuclear waste. In addition, Article 37 provides an initial framework for the development of such standards, noting that:

> Each Member State shall provide the [European] Commission with such general data relating to any plan for the disposal of radioactive waste in whatever form as will make it possible to determine whether the implementation of such plan is liable to result in the radioactive contamination of the water, soil, or airspace of another Member State. The Commission shall deliver its opinion within six months.

The Euratom Treaty thus brought NWM and safety into the core fabric of EU governance by including elements of these in its initial definition of the organization's role as a regional governing body. However, although the treaty discussed the development of waste standards, it left uncertain how standards should be drafted and, more importantly, what standards should include. In fact, the treaty primarily mentions "radioactive waste" in its annexes when outlining necessary research, industrial activities, and materials subject to regulations on the "nuclear common market." These gaps were unsurprising because at the time of the Euratom Treaty's ratification, nuclear energy—and therefore nuclear waste—was still limited within the region (World Nuclear Association, 2024b). Although nuclear waste was therefore a concern for the young EU, other areas of nuclear governance like nonproliferation were more immediately pressing given Cold War tensions.

**TABLE 3.1** EU Regulations on Nuclear Waste

| *Year* | *EU Directive* | *Key Objective* |
|---|---|---|
| 1987 | Council Decision 87/600/Euratom | Established a framework for exchanging information during radiological emergencies. |
| 1989 | Council Directive 89/618/Euratom | Defined objectives for informing the public of emergency health protection procedures. |
| 1996 | Council Directive 96/29/Euratom | Proposed safety standards for workers and the public from ionizing radiation, which is found in nuclear waste. |
| 2003 | Directive 2003/35/EC | Invited public participation in environmental planning efforts, which could include those on nuclear safety. |
| 2003 | Council Directive 2003/122/Euratom | Regulated high-activity sealed radioactive sources and orphan sources. |
| 2006 | Directive 2006/21/EC | Discussed waste management in the context of all extractive industries, which includes radioactive materials. |
| 2006 | Council Directive 2006/117/Euratom | Created a community system for supervising and controlling radioactive waste movement across borders. |
| 2006 | Commission Recommendation 2006/851/Euratom | Set standards for financial resource management when decommissioning nuclear installations and radioactive waste. |

Author's Own Elaboration.

In fact, it was not until toward the end of the Cold War that further EU regulations on nuclear waste would occur. Such efforts include a handful of EU Council directives, Court of Justice decisions, and Commission recommendations—noteworthy efforts are presented in Table 3.1.

Beyond the above directives, the EU also engaged in nuclear waste more broadly, via other internal and external efforts. The EU Court of Justice recognized the initial Euratom Treaty on health and safety as providing the Commission with "considerable scope" in preventing nuclear contamination (Euratom, 2011). This scope is reflected in engagements between the EU and the IAEA, including the Joint Convention on the Safety of Spent Fuel Management and on the Safety of Radioactive Waste Management. Unlike other directives that focused on specific elements of NWM, the Joint Convention provided a comprehensive legal instrument to set global standards for spent fuel and radioactive waste (IAEA, 1997). The Convention also offered legal definitions for key terms. *Radioactive waste* includes "radioactive material…for which no further use is foreseen…and which is controlled as radioactive waste by a regulatory body." *Radioactive waste management* is "all activities, including decommissioning activities, that relate to the handling, pretreatment, treatment, conditioning, storage, or disposal of radioactive waste, excluding off-site transportation." This Joint Convention was then followed by updates to the IAEA's Fundamental Safety Principles in 2006, a process which was supported by the EU (IAEA, 2006). These updated principles provided further rationale behind safety standards for NWM.

While Euratom therefore had robust engagement internationally in setting specific nuclear waste standards, the EU itself did not have a single, cohesive regulatory framework internally to do the same. Toward the end of the 2000s, several EU entities therefore supported initiatives to review the EU's efforts in nuclear energy policy to determine a consolidated plan of action for the future of radioactive waste management and, more broadly, nuclear safety. In 2007, the European Parliament adopted a Resolution on Assessing Euratom, where Member States agreed that "supervision of nuclear energy use in Europe, in view of the very specific characteristics of this energy source, requires the maintenance of a dedicated legal framework such as the Euratom Treaty" (European Parliament, 2007). This resolution was closely followed by Commission Decision 2007/530/Euratom to establish the European Nuclear Safety Regulators Group (ENSREG), as a high-level group to "reach a common understanding of the areas of nuclear safety and radioactive waste management" (Euratom, 2007; ENSREG, n.d.).

While ENSREG exists to this date, it had two immediate contributions to discussions on nuclear safety. Initial recommendations from ENSREG were provided publicly in the Council Resolution on Spent Fuel and Radioactive Waste Management (2008) and Council Conclusions on the report by ENSREG (EU Council, 2009b). The first Council Resolution noted that (i) Member States are responsible for their own spent fuel and national waste management, (ii) states must establish and publish plans on their waste management practices, (iii) states must implement policies on waste management transparently, (iv) Euratom should utilize the review mechanism noted in the Joint Convention, and (v) such a peer-review mechanism should enable an exchange of views and identification of best practices (EU Council, 2009a). The Conclusions asked ENSREG to provide further good practice guidance on transparency, improve qualified staff and knowledge in nuclear safety, and identify areas of Council Directive 2009/71/Euratom (2009) where common action by ENSREG could benefit Member States (EU Council, 2009b). This referenced directive was vital to the development of EU nuclear waste governance, because it established clear standards for safety at nuclear installations (Euratom, 2009). Directive 2009/71/Euratom defined *nuclear safety* as the "achievement of proper operating conditions, prevention of accidents and mitigation of accident consequences, resulting in protection of workers and the general public from dangers arising from ionizing radiations from nuclear installations." It also required Member States to establish national frameworks, regulatory authorities, licensee structures, educational opportunities, transparency efforts, and reporting mechanisms by July 2011.

The initial deadline in 2011 coincided with the passage of Council Directive 2011/70/Euratom, the foremost regulation on NWM in use by Euratom today. This Directive established "a community framework for the responsible and safe management of spent fuel and radioactive waste" (Euratom, 2011). Only a handful of months after the Fukushima nuclear disaster in Japan, Euratom utilized this momentum to require Member States to establish national legislative, regulatory, and organizational frameworks ("national frameworks") with specific functions, including: (i) national programs for the implementation of nuclear waste policy (on spent fuel and radioactive waste management); (ii) national arrangements for safety in NWM; (iii) licensing systems for waste management activities and facilities; (iv) systems for appropriate control, including management, inspections, documentation, and reporting; (v) enforcement actions to ensure compliance with regulations; (vi) allocation of responsibility to bodies involved in different aspects of the waste management process; (vii) national requirements for public information and participation; and (viii) financing schemes to support these national initiatives. This Directive created reporting obligations for Member States to submit a report to the Commission on their implementation in August 2015, and every three years thereafter.

Council Directive 2011/70/Euratom's reporting mechanism has since provided critical information on the status of NWM in the EU via regular Member State reports and program reviews provided by states with nuclear energy or fuel reporting obligations. These reports have, as of December 2024 occurred three times in 2015, 2018, and 2021. In addition, the EU Commission must draft a report on the overall implementation of this directive utilizing information from these reporting obligations. While these reporting mechanisms still form the backbone of EU NWM policy, the EU has also continued relevant discussions. For example, Council Directive 2013/59/Euratom (2013) repealed several prior directives and provided new standards for protection against ionizing radiation exposure (Euratom, 2013). In addition, the EU has recently funded various programs to support radioactive waste management in the context of Ukraine. The War in Ukraine generated renewed EU interest in nuclear safety and waste and has led to discussions on potential safety protocols and requirements with the IAEA (IAEA, 2022).

## The EU's governance challenge

Even with these historic directives, the EU today lacks the policy structures requisite for ensuring effective NWM across all Member States. In the previous edition of this book, Grégoire Lits (2020) discussed the social acceptability of technical solutions, focusing on national competencies and lengthiness of planning as key issues for nuclear waste governance. These issues reflect the primary challenge that the EU faces today: achieving long-term NWM across Member States. Because of its current policy to leave the details of implementation and energy mixes to the states, the EU has shied away from setting technical requirements for NWM, instead establishing requirements for reporting and administration. However, these requirements are unfortunately insufficient for ensuring long-term NWM, due to the fact that interim storage facilities and other temporary solutions achieve the minimal standards set by the EU.

Out of the 13 EU Member States with current or recent nuclear energy programs, only three—Finland, France, and Sweden—are currently constructing long-term nuclear waste facilities capable of storing high-level waste. As depicted in Table 3.2, the other EU countries that have ongoing

**TABLE 3.2** EU States' Nuclear Waste Management Program Status

| *Countries* | *Interim Facilities* | *Long-term Facilities* |
|---|---|---|
| Belgium | On-site at nuclear power plants and at government-run processing facilities. | Initiating construction |
| Bulgaria | Reactor pools and two spent-fuel storage facilities. | Under construction |
| Czechia | Three radioactive waste disposal facilities. | Planned |
| Finland | Pool-type storages at the power plant sites. | Scheduled to open in the mid-2020s* |
| France | Stainless steel containers at nuclear plants. | Scheduled to open in 2035* |
| Germany | Interim storage at plants as well as central storage facilities. | Under construction |
| Hungary | Spent rods in reactor pools and in a dry storage facility at the nuclear power site. | Under construction |
| Romania | Dry condition at the Dry Intermediate Storage Facility. | Planned |
| Slovakia | Reactor pools at the nuclear power plant, before transportation to an interim spent-fuel storage facility. | Under construction |
| Slovenia | On-site at its nuclear plant. | Under construction |
| Spain | In pools at the power plants in on-site individual storage facilities. | Under construction |
| Sweden | Central interim storage at a nuclear power plant | Initiating construction* |
| The Netherlands | Specially designed buildings while awaiting a final disposable facility. | Planned |

*Able to Hold High-Level Waste

construction of long-term facilities are only capable of storing low- and intermediate-level waste, not to mention that many of these facilities' completion dates are still decades away. In the meantime, EU Member States have been utilizing a variety of interim facilities that range from reactor pools or spent-fuel storage sites at nuclear power plants to centralized storage locations. This diversity in interim and long-term solutions across the region points to the lack of technical coordination at the EU level. While certain EU countries are utilizing NWM strategies considered at the forefront of technology today, other EU countries will take a lengthy period to achieve comparable results, which could leave those countries vulnerable. Given that gold standards for long-term and high-level waste storage can be agreed upon scientifically, Europeans could benefit from the EU setting requirements based on technical and operational best practices.

To consider how the EU can effectively facilitate long-term NWM, understanding why countries have different NWM strategies is important. A study of the national reports published by Member States under Council Directive 2011/70/Euratom provides further insights into how variations in the national frameworks of Member States impact their NWM. For example, while certain governments are involved solely in oversight, in other instances they handle waste disposal, develop expertise in personnel, finance nuclear waste funds, and advocate transparency. Interestingly, the reports suggest that a driving force behind these variations are governmental attitudes toward nuclear energy within Member States. In particular, a trend exists where countries that are skeptical of nuclear energy tend to have more government engagement. By contrast, although the relation between government engagement and NWM effectiveness is complex, countries with fewer concerns about nuclear energy tend to engage the private sector more and this can improve their ability to implement cutting-edge, long-term NWM solutions.

To understand this relation, this chapter divides the EU Member States with nuclear energy programs into four categories. Category I includes countries that are expanding their nuclear energy programs with strong government support. Category II includes countries that are expanding their nuclear programs currently but are also experiencing significant issues. Category III includes countries that are expanding their nuclear programs currently but have a history of phasing out nuclear energy. Category IV includes countries that are entirely phasing out their programs. These four categories have been selected to broadly reflect a scale from countries with the most government support toward nuclear energy (Category I) to those with the least (Category IV). As depicted in the reports, these categories in turn align with government engagement on nuclear waste, with higher-category countries performing better.

Based on these definitions, Category I countries in the EU include Bulgaria, Czechia, Finland, and France, as these countries are all enthusiastically expanding their nuclear energy programs. Bulgaria has two nuclear reactors that generate approximately one-third of its electricity needs (World Nuclear Association, 2024c). In addition, the Parliament formalized an agreement on the installation of two new nuclear reactors in 2024, which would be provided by the United States to replace its current dependency on Russia (Yordanov, 2024; Nikolav, 2024). Czechia has six nuclear reactors that generate approximately one-third of its electricity (World Nuclear Association, 2024d). The country plans to increase its nuclear energy consumption in the future to generate at least 50% of its power from nuclear sources and has agreed to build two new reactors (Ministry of Industry and Trade, 2018; Euronews, 2024). Finland has five nuclear reactors, which generate approximately one-third of its electricity (World Nuclear Association, 2024e). Finnish reactors are "among the world's most efficient," due to numerous upgrades and good maintenance practices since installation. Although Finland had plans for developing two more reactors, these plans were delayed following the War in Ukraine, due to the role of Russia in their installation (Vanttinen, 2022). France has 58 operable reactors at 18 commercial nuclear power plants, the most of any EU Member State (World Nuclear Association, 2024b). The French government strongly supports nuclear energy development, with a 2005 law establishing nuclear power as central to French clean energy (France, 2005). In 2023, the Parliament agreed to the construction of six new reactors (Reuters, 2023). Support for nuclear power is also part of France's foreign policy vision to serve as a global leader in nuclear energy (Messad, 2024).

Category II countries include Hungary, Romania, and Slovakia, as these countries are interested in expanding their programs but are also facing difficulties due to issues with dependency on Russia, financing, and technical expertise. Hungary has four nuclear reactors that generate approximately 50% of its electricity (World Nuclear Association, 2024f). While the Hungarian government supports nuclear energy, its attempts to rely on Russian support for developing new facilities have led to backlash, including legal action by the European Commission over concerns about its fuel dependency. Construction on new reactors is expected to start in 2025 given this slow legal process (Gizinska and Sadecki, 2023; Power Technology, 2023). Romania has two nuclear reactors that contribute approximately 20% of its electricity (World Nuclear Association, 2024g). While government support for nuclear energy is strong, nuclear reactor maintenance and development has faced financial issues, requiring external financing from the International Monetary Fund, the EU, and the United States. Romania's 2021 Integrated National Plan for Energy and Climate Change agreed to construct two new nuclear reactors by 2031, as well as refurbish its existing reactors (Romania, 2020). Slovakia has five nuclear reactors that contribute approximately 50% of its electricity (World Nuclear Association, 2024h). During its 2004 accession to the EU, shutting down two of its nuclear power plants was a pre-condition due to concerns over their safety. Although Slovakia's government entered agreements to complete new nuclear reactors, those installations were delayed due to difficulties with passing EU safety standards—and a sixth nuclear reactor in the country is still under construction.

Category III countries include The Netherlands, Slovenia, and Sweden, as these countries were previously attempting to phase out nuclear energy before recently reversing their course. The Netherlands only has one nuclear reactor, which generates a small proportion of its electricity (World Nuclear Association, 2024i). The Dutch government was cautious toward nuclear energy with efforts to phase out its program occurring at sporadic intervals since 1994. However, in 2021, The Netherlands changed its stance when its government decided to build two nuclear power stations (World Nuclear News, 2021). This switch appears to stem from its objective to achieve climate neutrality by 2050. Slovenia has one nuclear reactor, which is owned jointly with Croatia (World Nuclear Association, 2024j). Its government has a history of vacillating on nuclear energy; a proposal for a new nuclear reactor was submitted in 2010, but the government is yet to make a decision beyond the necessity of nuclear energy to its "path to a carbon-free future" (World Nuclear News, 2024). Sweden has six nuclear reactors producing approximately 40% of the country's electricity (World Nuclear News, 2022). A 1980 Swedish referendum decided to phase out nuclear energy by 2010 (Bergenas, 2009). However, in 2010, the government abandoned this policy and decided to replace existing reactors with safer models (World Nuclear News, 2010). More recently, in 2022, the government agreed to new investments in nuclear energy, including guaranteeing power plants "the right to operate and produce electricity as long as the facilities are in good condition and operated safely."

Category IV countries include Belgium, Germany, and Spain, as these countries are phasing out their nuclear energy programs due to limited government support for nuclear energy: Belgium currently has five nuclear reactors. In 2003, the Belgian government passed a federal act to prohibit new nuclear power plants and limit the lifetime of existing plants up to 2025 (Belgium, 2003). However, Belgium delayed its nuclear energy phaseout plans, allowing two reactors to continue operating until 2035, due to increased energy prices from the War in Ukraine (Shaw, 2023; Euronews, 2022). Germany shut down its last three nuclear power plants in 2023 (Germany, 2011). Germany decided to phase out its nuclear energy program following the nuclear disaster in Fukushima in 2011. Due to the energy crisis following the War in Ukraine, this phase out was extended until 2023. Spain has seven nuclear reactors, contributing approximately one-fifth of its electricity (World Nuclear Association, 2024k). In 2023, the Spanish government decided to phase out all nuclear reactors by 2035 (Lombardi, 2023). This phaseout plan was in fact a central component of the prior Spanish election (Edie, 1999; Euractiv, 2023).

As classified above, these four categories for EU Member States offer a framework to understand the impact of government support for nuclear energy on NWM. This is evidenced by the duties of the Member State regulatory agencies, as described in reporting under Council Directive 2011/70/ Euratom. In particular, these regulatory agencies can view their primary role as both overseeing NWM and directly handling waste disposal (A) or only overseeing NWM while relying on private companies for disposal (B). Training initiatives to upskill NWM personnel can likewise be conducted by the government (C), conducted by private entities based on training provided by the government (D), or provided and conducted by private entities with government oversight (E). In addition, funding sources for NWM also provide an example of the varying extent of government participation, given that funds and organization budgets are sometimes primarily allocated by the government (F), by taxes on produced electricity (G), or by fees owed by licensees (H). Even transparency reflects this question of the government's role: whether the government itself is transparent (J), or whether the government also requires licensees and other private entities to propagate transparency (K).

Table 3.3 identifies the roles of regulatory government agencies within each EU Member State with nuclear energy. While nuances exist between each country in the four categories, a trend occurs of lesser government engagement in countries that are more in favor of nuclear energy and consequently more government engagement in countries with greater concerns over nuclear energy. As a result, Category I countries are the most likely to have government regulatory bodies only involved in oversight, not waste disposal; to leave private entities responsible for training programs; to rely on private entities to fund waste management; and to engage private entities as actors in transparency initiatives. This empowerment of private entities is reversed in higher category countries, with Category IV countries more likely to rely on an expansive government role. Such a trend is expected given that EU Member State governments with greater concerns about nuclear energy are more likely to desire a larger role to prevent nuclear waste accidents. In addition, this could reflect greater trust in Category I countries in the ability of the private sector to conduct NWM than in Category IV countries.

The reason that the size of government engagement and role for the private sector are important is because consultations by the government with external entities such as the private sector and the public are critical for creating cutting-edge technical solutions that prioritize the long-term needs of the population. For example, in Finland's case, its Onkalo nuclear waste repository's design stemmed from technical ideation by the private Swedish Nuclear Fuel Management Company (Vattenfall, 2023). Another example is

**TABLE 3.3** Differences in Government Roles in Waste Management

| *Countries* | *Regulatory Role* | *Training Programs*[a] | *Funding Sources*[a] | *Transparency*[a] |
|---|---|---|---|---|
| *Category (I)* | | | | |
| Bulgaria | (B) oversight | (D) mix, (E) private | (H) private | (J) govt[b] |
| Czechia | (B) oversight | (E) private | (H) private | (J) govt, (K) private |
| Finland | (B) oversight | (C) govt, (D) mix, (E) private | (F) govt, (H) private | (J) govt |
| France | (B) oversight | (E) private | (H) private | (J) govt, (K) private |
| *Category (II)* | | | | |
| Hungary | (A) both | (C) govt, (D) mix | (H) private | (J) govt, (K) private |
| Romania | (A) both | (E) private | (H) private | (J) govt |
| Slovakia | (B) oversight | (E) private | (F) govt, (H) private | (J) govt, (K) private |
| *Category (III)* | | | | |
| Netherlands | (A) both | (C) govt, (E) private | (F) govt, (H) private | (J) govt |
| Slovenia | (B) oversight | (C) govt, (E) private | (F) govt, (G) tax, (H) private | (J) govt |
| Sweden | (A) both | (C) govt, (E) private | (F) govt, (H) private | (J) govt |
| *Category (IV)* | | | | |
| Belgium | (A) both | (C) govt | (F) govt, (H) private | (J) govt |
| Germany | (A) both | (C) govt, (E) private | (F) govt, (H) private | (J) govt |
| Spain | (A) both | (C) govt, (E) private | (G) tax, (H) private | (J) govt |

Data gathered in national reports under Council Directive 2011/70/Euratom.

[a] Note that these are based on the conceptualization of states on their own roles in reports—in other words, what states view as their primary expertise development, financial resources, and transparency initiatives.

[b] Public participation during licensee hearings falls under government transparency, not private entity transparency, as this transparency is driven by government actors.

France's La Hague facility, which leads efforts in Europe to reprocess spent nuclear fuel, a practice that is also useful for long-term NWM as it reduces the amount of high-level nuclear waste requiring storage. The facility is operated by the Orano group, a private company that has access to external expertise via its work to build nuclear energy facilities in other countries (Orano, n.d.). Although it is possible to create long-term solutions without private consultations, these cases suggest that having the correct technical understanding is beneficial, as well as often beyond the skill set of a national government.

However, the fact that government attitudes toward nuclear energy impact NWM initiatives underscores the difficulties that the EU will face when improving its current requirements. Specifically, the EU in itself does not have a singular attitude toward nuclear energy, rather allowing Member States to have very strong and differing perspectives. Given this lack of agreement on nuclear energy itself, generating agreement at the EU level on technical standards for long-term NWM could also prove difficult. To improve NWM, the EU will have to not only expand its own technical competencies to better understand the scientific landscape but also adjust the perspective that Member States are primarily responsible for all their energy decisions. The lack of technical competencies can be addressed via consultations with scientists and private nuclear energy firms. Meanwhile, overcoming this latter concern of national responsibility can be achieved by drawing from the EU's precedence in areas outside of nuclear waste. For example, the EU has set science-based standards for genetically modified foods, requiring rigorous safety assessments (European Union, n.d.a). Similarly, the EU must set scientific standards for nuclear waste to ensure that it likewise does not negatively impact human health and the environment.

## Safety, sustainability, and synchronization

The primary conundrum for EU nuclear waste requirements is therefore how to enable Member States to establish their own levels of government engagement in NWM, while still ensuring the same effectiveness across the region. This chapter proposes a novel three-S model, where the EU must consider policy solutions across three S's: safety, sustainability, and synchronization (Table 3.4). *Safety* policies should address whether national safety standards are adequate for preventing accidental radiation leakage from nuclear waste and securing waste for long-term disposal. *Sustainability* policies should address the resiliency of national programs to fund, resource, and maintain nuclear waste disposal processes in the future. And, *synchronization* policies should establish further practices and venues for transnational cooperation on nuclear waste. In addition, instead of simply setting reporting requirements across these three S's, the EU should set baseline standards and timelines for what countries must achieve under each of these solution spaces. For

**TABLE 3.4** Areas for the EU to Expand its NWM Requirements

| | |
|---|---|
| Safety | What are the baseline requirements to ensure that nuclear waste (low-, intermediate-, and high-level) is appropriately stored? |
| Sustainability | What are the baseline requirements for ensuring that NWM is effective in the long term? |
| Synchronization | What are the baseline requirements to enable EU Member States to share technical expertise on NWM? |

example, the EU could suggest that countries need to start construction on a long-term storage facility for high-level nuclear waste no later than a specific year, while also setting basic technical requirements to ensure safety, funding and maintenance requirements to ensure sustainability, and mechanisms for sharing best practices between Member States to ensure synchronization.

These three solution areas for EU policy expansion each address different vulnerabilities identified in current NWM practices. Safety requirements should address issues related to the technical specifications of nuclear waste facilities. For example, these requirements could include not only the use of underground storage but also suggestions for the depth, type of rock, and capacity of such storage. In addition, safety requirements could also establish a list of the primary concerns and risks associated with nuclear waste, gaining consensus on this list among EU Member States. This list could then inform decision-making both at the EU and Member State levels on safety practices in NWM. One requirement that the EU could consider adopting first is the need for countries to start construction on long-term nuclear waste repositories that adequately store low-, intermediate-, and high-level waste. Alongside this requirement, the EU could also ask countries to assess and address the potential vulnerabilities of their current interim nuclear storage sites, with special consideration for ongoing geopolitical escalation in the region.

Secondly, sustainability requirements should address issues related to the long-term operation of nuclear waste storage sites. For example, these requirements could include how to ensure adequate funding, maintenance, and operation of sites over time. The requirements should also address resiliency—in particular, how storage sites will be maintained even during times of economic, security, and other shocks. Another element of sustainability is the security of sites themselves from terrorist attacks and other threats. While certain sustainability efforts will no doubt overlap with safety, these requirements should focus on operations, rather than technology. One requirement the EU could consider first is that countries must increase their reprocessing of spent nuclear fuel as this improves the sustainability of nuclear energy industries in the long term. The EU could also suggest that countries interested in scaling nuclear power should identify other practices for increased NWM to meet future energy needs.

Finally, synchronization requirements should address issues regarding the gap in the sharing of best practices between EU Member States. A key reason for these requirements is that certain EU countries currently have more technical expertise and enabling pathways to share this expertise can only benefit the EU as a whole in the long term. Although these requirements in some ways might be the hardest to enforce, one possibility is creating a platform where entities in various Member States can share reports on best practices with each other, expanding on current reporting obligations under Council

Directive 2011/70/Euratom. The EU could also incentivize private firms like the one in Sweden that developed the underground storage methodology to consult with or offer services to other areas within the region. Such incentives would likely have to be financial in nature—however, the EU has precedence in encouraging transnational economic activity that could prove beneficial in this space.

These three solutions could address some of the critical issues in the EU's current regulation of nuclear waste. Although these solutions move beyond existing EU practices in nuclear waste, they reflect established standards-setting practices in other EU policy areas. This will be important because as with any expansion of EU policy, further requirements for nuclear waste will likely face backlash, not only from countries in the EU with nuclear energy programs, but also from those that oppose nuclear energy itself. To overcome some of these potential counterarguments, the EU could utilize the critical nature of NWM for protecting human health, which is a fundamental right of people within the Union. Such a human-centric approach could garner consensus for a more directed pathway forward.

## Achieving effective nuclear waste management

Although the three S's are only a preliminary framework for addressing the EU's policy gap in managing nuclear waste, they could ameliorate some of the inconsistencies between Member State policies today, benefitting efforts throughout the EU and beyond.

EU efforts in NWM have important implications for EU climate and clean energy policies. Effective NWM impacts the ability of EU countries to achieve climate neutrality by 2050, which is required under the European Climate Law (European Union, n.d.b). This is especially true because some EU countries are considering increased nuclear energy production as their primary means of achieving neutrality, due to the difficulties with constructing renewable energy facilities. Effective NWM could also have a positive impact on consumer prices, because nuclear energy causes minimal deforestation and releases less carbon—and the EU currently taxes products with these high environmental impacts (European Union, n.d.c). Improving NWM could therefore enable member countries to achieve the EU's climate change objectives within the required timeframes.

Beyond these internal objectives, the EU's policies on NWM can also have a meaningful impact on global standards for nuclear waste. The EU funds numerous initiatives at the IAEA on waste, including programs that focus on overcoming regional issues and on developing national frameworks (IAEA, n.d.) These initiatives also involve efforts where EU scientists train professionals working on nuclear waste disposal in other regions of the world, such as Latin America and Africa. Outside of the IAEA, the EU bilaterally cooperates on civilian nuclear safety with Argentina, Australia, Canada, Japan, Kazakhstan, Ukraine, Uzbekistan, and the United States (European Union, n.d.d). Such agreements include joint training, research collaborations, and sharing best practices, which could improve NWM efforts across the globe. Indeed, as the EU improves its own NWM in terms of long-term solutions, this could strengthen global nuclear safety standards as well as international norms on nuclear energy as a viable clean energy source.

As noted by Euratom, the "ethical obligation" of EU Member States to avoid "any undue burden on future generations" in nuclear waste sets a clear vision of how NWM can address the long-term risks of radiation utilizing permanent technical solutions. The Onkalo final spent-fuel repository is only the start of a new era in NWM. EU adoption of safety, sustainability, and synchronization in future NWM policies could enable better management to store—and in store for—its nuclear waste.

## References


Belgium. (2003, January 31). *Law of 31 January 2003 on the phase-out of nuclear energy for the purposes of the industrial production of electricity*. https://www.oecd-nea.org/law/legislation/belgium.pdf

Benke, E. (2023, June 14). *Finland's plan to bury spent nuclear fuel for 100,000 years*. BBC. https://www.bbc.com/future/article/20230613-onkalo-has-finland-found-the-answer-to-spent-nuclear-fuel-waste-by-burying-it

Bergenas, J. (2009, November 10). *Sweden reverses nuclear phase-out policy*. Nuclear Threat Initiative. https://www.nti.org/analysis/articles/sweden-reverses-nuclear-phase-out/

Edie. (1999, November 5). *Spain: Socialists promise nuclear shutdown if elected*. https://www.edie.net/spain-socialists-promise-nuclear-shutdown-if-elected

EIA. (n.d.). *Nuclear explained: Nuclear power and the environment*. https://www.eia.gov/energyexplained/nuclear/nuclear-power-and-the-environment.php

EU Council. (2009a, January 7). *Council resolution on spent fuel and radioactive waste management*. https://data.consilium.europa.eu/doc/document/ST-17438-2008-REV-1/en/pdf

EU Council. (2009b, November 10). *Council conclusions on the report by the European Nuclear Safety Regulators Group*. https://www.consilium.europa.eu/uedocs/cms_data/docs/pressdata/en/ecofin/111009.pdf

Euractiv. (2023, June 27). *Spain's election frontrunners plan U-turn in nuclear power phase-out*. https://www.euractiv.com/section/energy-environment/news/spains-election-frontrunners-plan-u-turn-in-nuclear-power-phase-out

Euratom. (2007, July 17). *2007/530/Euratom: Commission decision of 17 July 2007 on establishing the European High Level Group on Nuclear Safety and Waste Management*. https://data.europa.eu/eli/dec/2007/530/oj

Euratom. (2009, June 25). *Council directive 2009/71/Euratom of 25 June 2009 establishing a community framework for the nuclear safety of nuclear installations*. https://data.europa.eu/eli/dir/2009/71/oj

Euratom. (2011, July 19). *Council directive 2011/70/Euratom of 19 July 2011 establishing a community framework for the responsible and safe management of spent fuel and radioactive waste*. https://data.europa.eu/eli/dir/2011/70/oj

Euratom. (2013, December 5). *Council directive 2013/59/Euratom of 5 December 2013 laying down basic safety standards for protection against the dangers arising from exposure to ionising radiation, and repealing Directives 89/618/Euratom, 90/641/Euratom, 96/29/Euratom, 97/43/Euratom and 2003/122/Euratom*. https://data.europa.eu/eli/dir/2013/59/oj

Euronews. (2022, March 19). *Nuclear energy: Belgium postpones phase-out by 10 years due to Ukraine war*. https://www.euronews.com/2022/03/19/nuclear-energy-belgium-postpones-phase-out-by-10-years-due-to-ukraine-war

Euronews. (2024, August 4). *Czech government invests six billion euros in new nuclear reactors*. https://www.euronews.com/2024/04/08/czech-government-invests-6-billion-euros-in-new-nuclear-reactors

European Nuclear Safety Regulators Group – ENSREG. (n.d.). https://www.ensreg.eu

European Parliament. (2007, May 10). *Resolution on assessing Euratom – 50 years of European nuclear energy policy*. https://www.europarl.europa.eu/doceo/document/TA-6-2007-0181_EN.html

European Parliament. (n.d.) *Euratom treaty*. https://www.europarl.europa.eu/about-parliament/en/in-the-past/the-parliament-and-the-treaties/euratom-treaty

European Union. (1957, March 25). *Treaty establishing the European Atomic Energy Community (Euratom)*. https://data.europa.eu/eli/treaty/euratom/sign

European Union. (n.d.a). *GMO legislation*. https://food.ec.europa.eu/plants/genetically-modified-organisms/gmo-legislation_en

European Union. (n.d.b). *European climate law*. https://climate.ec.europa.eu/eu-action/european-climate-law_en

European Union. (n.d.c) *Regulation on deforestation-free products*. https://environment.ec.europa.eu/topics/forests/deforestation/regulation-deforestation-free-products_en

European Union. (n.d.d). *Nuclear safety*. https://energy.ec.europa.eu/topics/nuclear-energy/nuclear-safety_en

France. (2005, July 13). *Loi n° 2005–781 du 13 juillet 2005 de programme fixant les orientations de la politique énergétique*. https://www.legifrance.gouv.fr/eli/loi/2005/7/13/ECOX0400059L/jo/texte

Germany - Federal Office for the Safety of Nuclear Waste Management. (2011, June 30). *The nuclear phase-out in Germany*. https://www.base.bund.de/en/nuclear-safety/nuclear-phase-out/nuclear-phase-out_content.html

Gizinska, I., & Sadecki, A. (2023, July 4). *Russia's nuclear project in Hungary: France's growing role*. Centre for Eastern Studies. https://www.osw.waw.pl/en/publikacje/osw-commentary/2023-07-04/russias-nuclear-project-hungary-frances-growing-role

IAEA. (1997, December 24). *Joint convention on the safety of spent fuel management and on the safety of radioactive waste management*. https://www.iaea.org/sites/default/files/infcirc546.pdf

IAEA. (2006). *Fundamental safety principles*. https://www-pub.iaea.org/MTCD/publications/PDF/Pub1273_web.pdf

IAEA. (2022, April 4). *IAEA and EU senior officials discuss nuclear safety and security in Ukraine and further areas of cooperation*. https://nucleus.iaea.org/sites/zodiac/SitePages/IAEA-and-EU-Senior-Officials-Discuss-Nuclear-Safety-and-Security-in-Ukraine-and-Further-Areas-of-Cooperation.aspx

IAEA. (n.d.). *European union co-funded projects.* https://www.iaea.org/about/partnerships/european-union/european-commission-funded-projects

International Atomic Energy Agency. (2020, November 26). *Finland's spent fuel repository a 'game changer' for the nuclear industry, Director General Grossi says.* International Atomic Energy Agency. https://www.iaea.org/newscenter/news/finlands-spent-fuel-repository-a-game-changer-for-the-nuclear-industry-director-general-grossi-says

Lits, G. (2020). *The European governance of radioactive waste: a story of stakeholders' involvement* in Orsini, Amandine, and Elena Kavvatha, eds. EU Environmental Governance: Current and Future Challenges. Routledge, 2020.

Lombardi, P. (2023, December 27). *Spain confirms nuclear power phase-out, extends renewable projects deadlines.* Reuters. https://www.reuters.com/business/energy/spain-confirms-nuclear-power-phase-out-extends-renewable-projects-deadlines-2023-12-27

Messad, P. (2024, April 25). *Macron at Sorbonne: Energy key to French president's 'new European paradigm'.* https://www.euractiv.com/section/energy-environment/news/macron-at-sorbonne-energy-key-to-french-presidents-new-european-paradigm

Ministry of Industry and Trade. (2018). *Possibilities of involvement of the Czech industry in upcoming NNS projects in the CR and other countries in individual projects, maintenance of professional background of suppliers from the CR for the construction of NNS and future service.* https://www.mpo.gov.cz/assets/en/energy/new-nuclear-build/participation-of-the-czech-industry-in-nuclear-projects/2018/10/Participation-of-the-Czech-Industry-in-Nuclear-Projects.pdf

Nikolav, K. (2024, February 14). *Bulgaria's* two *US nuclear reactors to cost under $14* billion. Euroactiv. https://www.euractiv.com/section/politics/news/bulgarias-two-us-nuclear-reactors-to-cost-under-14-billion

Orano. (n.d.). *Le Hague.* https://www.orano.group/en/nuclear-expertise/orano-s-sites-around-the-world/recycling-spent-fuel/la-hague/unique-expertise

Power Technology. (2023, September 25). *Rosatom to pour first concrete at Hungary nuclear plant by 2025.* https://www.power-technology.com/news/rosatom-hungary-nuclear-plant

Reuters. (2023, March 21). *French parliament votes nuclear plan with large majority.* https://www.reuters.com/world/europe/french-parliament-votes-nuclear-plan-with-large-majority-2023-03-21/

Romania. (2020, April). *The 2021–2030 integrated national energy climate plan.* https://energy.ec.europa.eu/system/files/2020-06/ro_final_necp_main_en_0.pdf

Shaw, A. (2023, December 14). *Belgium to extend life of two nuclear power plants.* https://www.power-technology.com/news/belgium-agree-to-extend-life-of-two-nuclear-power-plants

Simicevic, V. (2023, June 15). *European Union nations grapple with nuclear waste storage.* Physics Today. https://doi.org/10.1063/PT.6.2.20230615a

Vanttinen, P. (2022, August 22). *Finnish energy company to take Russia's nuclear giant to court.* Euractiv. https://www.euractiv.com/section/energy-environment/news/finnish-energy-company-to-take-russias-nuclear-giant-to-court

Vattenfall. (2023, August 29). *Finland to open the world's first final repository for spent nuclear fuel.* https://group.vattenfall.com/press-and-media/newsroom/2023/finland-to-open-the-worlds-first-final-repository-for-spent-nuclear-fuel

World Nuclear Association. (2024a, August 12). *Radioactive waste – Myths and realities*. https://world-nuclear.org/information-library/nuclear-fuel-cycle/nuclear-waste/radioactive-wastes-myths-and-realities

World Nuclear Association. (2024b, May 21). *Nuclear power in France*. https://world-nuclear.org/information-library/country-profiles/countries-a-f/france

World Nuclear Association. (2024c, April 3). *Nuclear power in Bulgaria*. https://world-nuclear.org/information-library/country-profiles/countries-a-f/bulgaria

World Nuclear Association. (2024d, July 23). *Nuclear power in Czech Republic*. https://world-nuclear.org/information-library/country-profiles/countries-a-f/czech-republic

World Nuclear Association. (2024e, June 3). *Nuclear power in Finland*. https://world-nuclear.org/information-library/country-profiles/countries-a-f/finland

World Nuclear Association. (2024f, April 2). *Nuclear power in Hungary*. https://world-nuclear.org/information-library/country-profiles/countries-g-n/hungary

World Nuclear Association. (2024g, March 5). *Nuclear power in Romania*. https://world-nuclear.org/information-library/country-profiles/countries-o-s/romania

World Nuclear Association. (2024h, March 28). *Nuclear power in Slovakia*. https://world-nuclear.org/information-library/country-profiles/countries-o-s/slovakia

World Nuclear Association. (2024i, March 25). *Nuclear power in the Netherlands*. https://world-nuclear.org/information-library/country-profiles/countries-g-n/netherlands

World Nuclear Association. (2024j, March 28). *Nuclear power in Slovenia*. https://world-nuclear.org/information-library/country-profiles/countries-o-s/slovenia

World Nuclear Association. (2024k, January 12). *Nuclear power in Spain*. https://world-nuclear.org/information-library/country-profiles/countries-o-s/spain

World Nuclear News. (2010, June 18). *New phase for Swedish nuclear*. https://www.world-nuclear-news.org/NP_New_phase_for_Swedish_nuclear_1806101.html

World Nuclear News. (2021, December 15). *Nuclear makes a comeback in the Netherlands*. https://world-nuclear-news.org/Articles/Nuclear-makes-a-comeback-in-the-Netherlands

World Nuclear News. (2022, October 17). *New Swedish government seeks expansion of nuclear energy*. https://www.world-nuclear-news.org/Articles/New-Swedish-government-seeks-expansion-of-nuclear

World Nuclear News. (2024, January 31). *Slovenia aiming for referendum on new nuclear this* year. https://www.world-nuclear-news.org/Articles/Slovenia-aiming-for-referendum-on-new-nuclear-this

Yordanov, M. (2024, August 30). *Parliament ratifies on first reading agreement with US for strategic cooperation in development of nuclear energy in Bulgaria*. Bulgarian News Agency. https://www.bta.bg/en/news/economy/635253-parliament-ratifies-on-first-reading-agreement-with-us-for-strategic-cooperation

# 4

# THE TRANSPOSITION OF RENEWABLE ENERGY COMMUNITY PROVISIONS IN CENTRAL AND EASTERN EUROPE

## An evidence of varied polycentric governance systems

*Loïc Cobut, Daniel Petrovics, and Thomas Bauwens*

### Introduction

Since the Schuman Declaration of 1950, European integration has been closely linked with the energy sector. The EU's liberalisation and unbundling of the energy market in the 1990s opened up space for new types of actors. Consequently, European Union (EU) energy policy has increasingly recognised the role of non-state actors in energy governance. In parallel, the EU has been a driving force behind institutional decentralisation in the energy sector. Since 2019, with the Clean Energy Package (CEP), the EU has taken concrete steps to recognise and integrate renewable energy communities (RECs) into the European energy system. RECs are initiatives aimed at including citizens as active participants in the energy transition. They are based on principles of open and voluntary participation, with member-controlled governance focused on environmental and social objectives (Bauwens et al., 2022). The recognition of RECs has arguably further concretised the polycentric governance of EU energy policy (Petrovics et al., 2022; Bauwens, 2017). Indeed, what emerges is a governance model that favours local action, experimentation, mutual adjustment, and trust between actors. In such a framing, the EU may be seen as the actor who designs the overarching rules and orchestrates transition processes from the top (Jordan et al., 2018a; Abbott et al., 2015).

The EU's support for alternative forms of participation signals a positive democratic shift, especially in an energy sector long dominated by incumbents (Bauwens et al., 2016). Historically, EU energy policies have reflected the influence of major incumbents and state interests, with liberalisation efforts only partially opening the markets to new entrants. Stirling (2019) defines incumbency as the concentration of power and privilege, while

DOI: 10.4324/9781003479260-7

Johnstone et al. (2017) highlight "deep incumbency", describing a tightly woven alliance of state and corporate interests (Brisbois, 2019). Despite efforts at liberalisation, dominant players like EDF (France), ENEL (Italy), and E-ON (Germany) still present barriers for new actors like RECs, which struggle to establish themselves in markets shaped by these entrenched interests (Smith et al., 2016).

RECs operate democratically and often bring higher standards of internal governance than traditional energy players. RECs claim that they not only embody energy decentralisation, but also champion energy democracy. Burke and Stephens (2017, p. 35) define energy democracy as a movement that "offers a set of visionary organising principles that provide guidance for democratically restructuring the energy and electricity sectors through the processes of shifting from fossil-fuel-based systems to renewable energy systems". Moreover, RECs contribute to decarbonisation by generating renewable energy and helping close the estimated $1 trillion investment gap needed to limit global warming to below 2°C (Buchner et al., 2017).

Rescoop.eu, the European Federation of renewable energy communities, is the primary representative body of RECs at the EU level. It counts a growing network of 2,250 energy communities (Rescoop.eu, 2024a). RECs are characterised by pluralism and heterogeneity. Bauwens et al. (2022) analysed this diversity and identified that 183 definitions have been used to refer to "community energy, energy communities, and related notions" (p. 13). Rescoop.eu serves as the umbrella federation for these diverse models of RECs. These RECs are mostly concentrated in Northern and Western Europe, with fewer found in Southern Europe and still fewer in Central and Eastern Europe (CEE). As of late 2024, CEE countries collectively hosted 12 RECs, up from three in 2020 (Capellán-Perez et al., 2020). The war in Ukraine has likely influenced this growth by driving citizens in these Member States to consider RECs as a means to address sustainability and energy independence simultaneously (Zindulková, 2023). Still, CEE Member States remain underexplored in both practice and academic literature (Capellán-Perez et al., 2020). CEE Member States are interesting because the slow emergence of RECs may be linked to their past as former communist countries, where people remain suspicious of the idea of the cooperative structure (Capellán-Perez et al., 2020).

This chapter addresses two key gaps in the scientific literature. First, it aims to fill the geographical gap by focusing on CEE Member States that are underrepresented in existing research. Second, it explores the EU's ambition to promote local action through the energy sector, an objective whose transposition into policy varies significantly across Member States. Specifically, we examine the transposition of REC provisions from the Renewable Energy Directive in CEE countries, aiming to analyse how these provisions support polycentric governance in EU energy policy.

## The integration of non-state actors in EU energy policy

In this first section, we analyse EU energy policy through the lens of the inclusion of non-state actors. This can help better understand the legal context in which the inclusion of RECs in EU secondary law took place.

Energy policy became part of the European integration process as early as the Schuman Declaration in 1950. However, EU energy policy initially focused on the traditional state and industry actors of the sector. It was not until the late 1990s that the EU began to consider non-state actors. The 2000s subsequently created the legal and institutional conditions necessary for new energy actors such as RECs to emerge. Climate goals, a 20% increase in renewable energy sources (RES), along with the liberalisation of energy markets, helped establish a formal, institutionalised role for RECs within the energy sector.

A key initial step in this context was the 1993 Council Decision 1993/500/EC (Council of the European Union, 1993), which launched the Altener programme. This programme underscored the commitment of Member States to stabilise $CO_2$ emissions by 2000 to 1990s levels. Altener promoted "training and information activities with regard to RES at a level as close as possible to operators and the final consumers of energy". Nevertheless, the programme granted only a marginal role to non-state actors—a reflection of the energy sector's pre-liberalisation state. Consequently, this Decision did not shift focus from state to non-state actors, though it did refer to individuals as "final consumers" (see Article 3), without defining the term. Dupont (2016, p. 64) notes that the Altener programme represented a "weak policy instrument" due in large part to limited funding.

Three years later, the first Electricity Directive 1996/92/EC established "common rules for the internal market in electricity". According to its preamble, this Directive aimed to "constitute a further phase of liberalisation [...] of the internal market in electricity". It marked foundational milestones by unbundling electricity activities such as generation, transmission, and distribution. The Directive also organised access to the electricity system, enabling new actors to utilise existing infrastructures.

This Directive, in Article 2, provides definitions related to non-state actors (European Parliament and Council of the European Union, 1996):

- "producer" *shall mean a natural or legal person generating electricity;*
- "autoproducer" *shall mean a natural or legal person generating electricity essentially for their own use;*
- "independent producer" *shall mean a producer who does not carry out electricity transmission or distribution functions in the territory covered by the system where they are established;*

- "in Member States in which vertically integrated undertakings do not exist and where a tendering procedure is used, a producer corresponding of point (a), who may not be exclusively subject to the economic precedence of the interconnected system;
  - "customers" *shall mean wholesale or final customers of electricity and distribution companies*;
  - "final customer" *shall mean a customer buying electricity for their own use*;

In this Directive, individuals are recognised either as consumers of electricity or as market participants, whether as producers or wholesale customers. The definition of an autoproducer or independent producer hints at what we now consider "active consumers" or "prosumers", as these individuals could carry out transmission or distribution functions beyond the established electricity grid. However, the terms "autoproducer" and "independent producer" were later omitted from the 2003 Electricity Directive. In addition to opening the sector to new actors, this Directive emphasised protecting "against dominant position mechanisms" (Article 22, 1996/92/EC), highlighting the EU's focus on price stability and competition. Thus, consumer protection was already a priority in this policy.

In response to persistent challenges in market access and the dominance of large incumbents, the EU introduced a second legislative package in 2003, with the third package following in 2008–2009 to consolidate and expand market liberalisation (Dupont, 2016; Dutton, 2015). According to Nies (2018, p. 137), this third package "led to an explosion of groups and institutions" stemming from liberalisation and small-scale renewable developments, which spurred the entry of more companies and new structures into the energy sector.

A significant policy shift occurred in 2014 when the European Commission adapted its state aid guidelines, favouring a competitive, auction-based system for renewable energy support (Leiren and Reimer, 2018). Unlike feed-in tariffs or green certificate schemes, auction systems tend to favour large-scale renewable generation and benefit mainstream actors with access to financial capital, presenting challenges for smaller, community-based energy actors (Fitch-Roy et al., 2019; Hall et al., 2016).

These developments illustrate the EU's central role in energy governance and in incorporating non-state actors into its energy policy. The fourth energy package, however, marked the most significant step for RECs. In 2016, the European Commission launched the CEP, a core component of the Energy Union and one of the Juncker Commission's top priorities, to accelerate the clean energy transition, growth, and job creation in alignment with the Paris Agreement. The CEP introduced "legislative proposals and facilitating measures" designed to modernise the economy and attract investments in clean energy sectors (European Commission, 2016, p. 3). A revised Renewable Energy Directive (RED II), part of the CEP, provided a formal definition of RECs and required Member States to establish an "enabling framework", aimed at facilitating the emergence and scaling of RECs (European Parliament and Council of the European Union, 2018). However, the degree to which this framework advances REC development is left to the discretion of individual Member States (Cobut, 2021).

The European Green Deal, introduced by the von der Leyen Commission in December 2019, aims to make Europe the first climate-neutral continent by 2050, proposing new economic opportunities through a holistic approach to climate governance (European Commission, 2019; Dupont et al., 2023). The 2021 "Fit for 55" package further proposes measures to reduce net greenhouse gas emissions by at least 55% by 2030, covering all economic sectors under legally binding climate targets (European Commission, 2023). Although the Green Deal has expanded EU climate governance, it gives less direct attention to RECs compared to the CEP. This may reflect the von der Leyen Commission's broader focus on transforming all sectors. It is worth noting that the most recent revision of the RED facilitates the expansion of REC activities, including energy sharing, supply, and heating and cooling. It also introduces fast-track permitting for wind turbines and simplified registration processes, which may support RECs in engaging with both onshore and offshore projects.Within the Green Deal, the European Climate Pact aims to build on the CEP, encouraging the establishment and expansion of RECs. The Pact allows RECs to register climate pledges, which can inspire similar initiatives across Europe. It also promotes citizen participation through dialogues and assemblies, encouraging inclusion of diverse perspectives. (European Commission, 2020; Cobut, 2021). The Pact also highlights the challenge of addressing energy poverty and supporting vulnerable consumers, particularly through RECs, which should be a central focus of the forthcoming Citizens Energy Package (European Commission, 2024).The year 2024 has seen the release of the Draghi and Letta reports at the EU level. While the Draghi report emphasises competitiveness and growth, the Letta report proposes a "Single Market narrative" focused on locally embedded citizens, aligning with REC projects as locally based beneficiaries of EU support.

An examination of Rescoop.eu's position papers reveals concerns regarding the transposition of REC provisions from the RED II. Rescoop.eu's focus is on ensuring that national enabling frameworks effectively support REC development (Rescoop.eu, 2024a). In this chapter, we examine the implementation of these provisions in CEE, in order to better assess the EU's ambition to promote local action through the energy sector, how it varies across CEE Member States, and how these provisions enhance polycentric energy governance in EU energy policy.

## An analysis of energy community provisions transposition

### *Polycentric energy governance*

The decentralisation of renewable energy provision, the liberalisation of energy markets, and the inclusion of non-state actors have contributed to a polycentric energy governance system within the EU (Petrovics et al., 2022). This system allows multiple actors across various scales (from local to supranational) and sectors (from markets and state to communities) to influence RES deployment and use. Actors such as RECs, municipalities, citizens, and established market players have differing degrees of decision-making power. This evolving landscape introduces potential risks related to collective decision-making, diverse interest representation, and citizens' entry into energy markets.

In navigating these risks, Member States play a crucial role in transposing REC provisions into national legislation, with considerable flexibility in (re) interpreting EU Directives to suit their contexts. Large discrepancies among Member States highlight challenges for a cohesive polycentric governance system. Jordan et al. (2018b) describe central characteristics of polycentric governance, including overarching rules, local action, trust, mutual adjustment among actors, and system-wide learning. In the following, we examine the role played by the EU in steering polycentric energy governance systems across selected Member States.

### *Case selection*

Rescoop.eu's transposition tracker monitors the progress of REC provisions being incorporated into national legislation across all Member States. It assesses how each Member State has integrated the provisions outlined in Table 4.1 into its legal framework. By examining the content of each piece of national legislation in detail, the tracker serves as a key source of information for this analysis. Additionally, the tracker enhances transparency in the transposition process by documenting and providing access to new national legislation related to REC implementation. This transparency ensures the tracker's reliability, making it a central source for our analysis.

As of late 2024, Member States are at varying stages in this process, with notable differences in their approaches. For instance, Sweden has yet to transpose EU Directives into national law, while other countries may be classified into three progress groups. The well-advanced group includes Belgium, France, Italy, Denmark, or Czechia; a partially advanced group includes Lithuania, Bulgaria, or Croatia; and the least advanced group includes Slovenia, Latvia, or Poland. A majority of Member States have only partially or minimally transposed REC provisions.

**TABLE 4.1** EU Provisions for RECs

| | |
|---|---|
| *Autonomy principle* | The RED II mandates that RECs maintain autonomy from individual members, traditional market actors, and external shareholders. Autonomy is essential to preserve cooperative values and decision-making integrity, protecting RECs from undue influence by incumbents. Our analysis evaluates whether national legislation supports REC autonomy, identifying potential risks where this principle is lacking. |
| *Governance principle (open and voluntary participation, effective control)* | Governance principles are central to citizen empowerment through RECs. Open participation means that anyone living near a REC can join voluntarily, while effective control grants members influence over the REC's operations and decision-making. |
| *Local proximity* | Local proximity requires that members and shareholders are embedded in the geographic proximity of the REC. This is to ensure that issues such as value capture by (physically distant) investors are averted. However the exact boundaries of this locality are not uniform. |
| *Citizen participation* | Citizen participation is key to alleviate the risk of cooptation by incumbent actors. However, in certain countries such as Germany for example, there are requirements set as to how many natural persons should be included in the REC to safeguard against this. |
| *ICA principles* | The ICA principles set the broad framing for any type of cooperative. The observance of these principles can set a strong foundation for the organisational model of RECs. Nevertheless, these principles should also be considered with local needs and contexts in mind. The ICA principles include: voluntary membership, democratic control, economic participation by members, autonomy and independence, education and training, cooperation of cooperatives and a community-focus (ICA, 2024). These principles are reflected in the Directive, but the latter does not promote the cooperative model per se. The choice of cooperatives as the legal form for a REC depends on the national legal framework. When this form exists at the national level, the ICA principles are incorporated based on how they are applied within the country's legislation. |
| *Purpose (non for profit, sharing dividends and use of funds)* | The primary purpose of RECs should be other than generating a financial return. Specifically this means that a REC should be funded with environmental or social value creation aims at the forefront. |

(*Continued*)

**TABLE 4.1** (Continued)

| | |
|---|---|
| *Controlling authority* | A controlling authority is necessary to both ensure the proper legislative grounding of individual RECs and to also monitor the development of RECs overall. Such authorities then can also assist in policy- and decision-making around grid governance and grid services as well as the enabling frameworks needed for RECs to take off. |
| *Enabling Frameworks* | EU energy community provisions also call for enabling frameworks, which include removing unjustified barriers to REC participation, fair market access, close collaboration with Distribution System Operators (DSOs), support for vulnerable groups, and equitable treatment of RECs alongside other market actors (Article 22, RED II). |

This analysis focuses on RECs due to their exclusive focus on renewable energy technologies. Unlike Citizen Energy Communities (CECs) which were defined in the revised electricity directive and can focus on any energy-related activity that benefits its members regardless of locality, RECs represent citizen-driven initiatives that operate at the local level and exclusively work with RES. We selected three Member States from CEE—post-socialist countries that joined the EU after 2004—for a more diverse perspective on REC development. These Member States also often lag in climate and energy policies, making them valuable cases for analysis (Ćetković and Buzogány, 2019).

Using the latest data from Rescoop.eu, we selected one Member State from each progress group with regard to the implementation process: one well advanced (Czechia), one partially advanced (Lithuania), and one poorly advanced (Poland).[1] As already mentioned above, CEE countries only count 12 RECs. Three of them are located in Czechia, one in Poland while Lithuania does not have any (Rescoop.eu, 2025).

The Rescoop transposition tracker serves as the primary data source, supplemented with literature where applicable. Before presenting the analysis, the elements used from the Rescoop transposition tracker are defined. They represent the main issues related to the transposition of the REC definition and related enabling frameworks and derive directly from RED II.

### *REC provisions*

The table below presents the EU provisions for RECs. Each Member State is tasked with advancing its implementation. Using the Rescoop transposition tracker, we assess the progress made by each Member State, highlighting varying levels of advancement, as illustrated by the Member States we have selected.

## *Case analysis*

### *Czechia (well-advanced group)*

Czechia illustrates the group of well-advanced Member States. It is the only Member State from CEE in this group. Several contextual elements help to explain why Czechia is well advanced in its implementation process. One of them underlines that the country's energy sector "has been going through a rapid transition in recent years, and effective business models in support of the decarbonization process have been envisaged. One of those mechanisms is represented by [REC] projects" (Pechancová et al., 2022). Czechia benefits from a network of municipalities that steer renewable energy projects with the aim of enhancing energy efficiency. It is also marked by the persistence of collective ownership models, even after 1989, which makes a fertile ground for REC projects (Pechancová et al., 2022; Malý et al., 2019).

Czechia has published legislation in January 2024 which ensures the autonomy and effective control by members of their renewable energy projects. The governance principles are covered in detail and align with EU criteria, ensuring amongst others, local proximity and voluntary participation. The same applies to the purpose of RECs, which aligns with EU criteria. Another effective measure is the establishment of a regulatory authority managed by the Energy Regulatory Office, which oversees registration processes, data updates, and publishes the public registry online (Rescoop.eu, 2024b).

There are, however, no provisions pertaining to citizen participation. While Czechia's legislation refers to some of the International Cooperative Alliance (ICA) principles present in the EU provisions, it has not linked RECs to the cooperative model in its national legislation. Is it up to RECs to choose the cooperative legal form, in which case the ICA principles will be naturally integrated according to how they are implemented under Czech legislation? At last, Czechia has poorly transposed provisions on the enabling framework. The existing national framework is not properly designed as it lacks an assessment of specific barriers. It lacks proper support mechanisms that assist RECs in accessing information, expertise. However, financing tools have been made accessible for RECs and reforms are being made to enhance cooperation with DSOs to facilitate energy sharing (Rescoop.eu, 2024b).

### *Lithuania (partially advanced group)*

Lithuania is a part of the group of partially advanced Member States. As one of the three Baltic States, Lithuania stands out from Czechia and Poland. Baltic States have inherited a particular historical context as their electricity grid was connected to that of the former Soviet Union (Fang et al., 2024). In parallel to the 2004 enlargement process, the Baltic States viewed their synchronous interconnection with Russia and Belarus as both a threat to their sovereignty and a potential risk to their electricity supply (Westphal et al., 2022). In light of this specific geopolitical context, Lithuania has placed RECs at the margin of its energy transition strategy, instead focusing on developing its independence from the Russian electricity grid. Over the past decade, Lithuania has concentrated on expanding cross-border electricity trade and synchronising its electricity grid with continental Europe (Fang et al., 2024) and advancing nuclear energy initiatives in partnership with foreign investors (Ball, 2015).

In this context, the role of local-level actors in the energy transition has not been the topic of political discussion (Ball, 2015). Yet, Lithuania has published a law on RES in 2011 in which it already defined RECs, ensuring their autonomy and governance principles in detail. The purpose of REC projects had also already been made clear and found itself in line with EU criteria. ICA values and principles are reflected in this law and align with the EU definition but without making specific reference to the cooperative model. An effective controlling authority is already in place and citizen participation is ensured.

Lithuania's main weakness stems from the approach it has taken to transpose provisions related to the enabling framework. Currently, the country has yet to establish a comprehensive, detailed plan for fully implementing these provisions, with only preliminary steps in place. While progress is ongoing, the lack of a clear, structured roadmap creates uncertainty regarding the timeline and effectiveness of the enabling framework's rollout. It should be noted, however, that RECs may benefit from financing tools set up to enhance renewable energy deployment (Rescoop.eu, 2024c).

### *Poland (poorly advanced group)*

Poland illustrates the group of poorly advanced Member States. Poland is usually pictured as the most reluctant Member States of the Visegrad group when it comes to EU climate and energy policies. It is the country that voted the most against energy legislation between 2013 and 2018 (Ćetković and Buzogány, 2019). Poland amended its Law on Energy in July 2023 to integrate REC provisions. One main issue is that RECs are not singularly defined and are presented as a subset of CECs, meaning that EU criteria are not well transposed (Rescoop.eu, 2024d).

A pre-existing definition of the concept of energy cooperative is still present creating a legally ambiguous situation for RECs. The autonomy of RECs is not ensured as other members than citizens, such as SMEs or local authorities may become members, which creates the risk of cooptation of REC projects. The participation of citizens is not guaranteed by any provisions in the law. The purpose of RECs is not entirely in line with EU criteria: it ensures that the purpose has to meet environmental, social, and economic benefits, but the law is written in a way that it excludes RECs to provide service or carry activities for non-members. The reference to the ICA principles is present but is merely a copy-paste from the EU Directives and does not mention the cooperative model and fails to take into consideration the national context to ensure its operationality. The participation of citizens is not guaranteed by any provisions in the law. At last, a controlling authority is present and only requires CECs to register. The situation for RECs regarding this register is unclear, leading to even more confusion (Rescoop.eu, 2024d).

Poland has designed a few rules to support energy cooperatives that produce renewable energy locally for their members' consumption needs. There is, however, no specific framework that helps RECs access information, expertise, and finance. Poland has not yet assessed obstacles and potential for the emergence and upscaling of RECs. It may be noted that the relationship between energy cooperatives and traditional electricity suppliers and DSOs is budding, yet unjustified regulatory and administrative barriers still persist (Rescoop.eu, 2024d).

## Conclusion and discussion

Focusing on CEE Member States reveals that each national transposition occurs within a specific historical, cultural, socio-political, and national context. Czechia, Lithuania, and Poland each follow distinct paths in implementing EU REC provisions, highlighting a heterogeneity of approaches to developing local and community-led energy projects.

Czechia shows that Member States that are well advanced in the transposition of EU REC provisions are not only from Western and Northern Europe and that having an important number of existing energy communities on its territory is not a prerequisite to a detailed implementation of REC provisions. Czechia indicates that implementing these provisions is not merely about copy-pasting words from EU Directives; it is about paying attention to its own national context.

Lithuania's Soviet legacy drives this country to prioritise energy security, namely by synchronising its electricity grid with continental Europe. With the ongoing war in Ukraine, Lithuania highlights how energy policy adapts during conflicts. In this context, RECs play a strategic role in energy geopolitics, emerging as essential tools for securing independence from Russian energy. Lithuania also illustrates that a future challenge in the implementation process is to design proper enabling frameworks that will effectively support REC projects.

Poland reveals that some Member States still lag behind in the transposition process and currently offer an ambiguous environment for RECs to emerge and upscale. In Poland, a partial transposition of EU provisions is delivered as three different definitions of energy communities overlap. This creates a barrier to new projects as it leads to situations of uncertainty and ambiguity. In Poland and Czechia, however, citizen participation is not ensured. We consider this a worrying development considering the citizen empowerment potential of RECs and the cooperative nature of these undertakings.

In a broader manner, this analysis of REC provisions illustrates an evolution of EU energy policy. Referring back to the concept of polycentric governance helps to realise that a variety of transpositions of REC provisions mirrors the variety of polycentric governance models that may (not) exist throughout the EU. As mentioned, building a polycentric energy governance system is the result of a multi-level and multi-actor process. This chapter shows the role played by the EU (and therein higher-level/supranational actors) when it comes to drawing the lines of the so-called "overarching rules" of polycentric governance. While drawing these rules is in part in the hands of the EU legislator, it appears that the transposition of these rules leads to a situation where the same rules (EU Directives) do not create the same level-playing field for RECs, from one Member State to another.

This development highlights that a polycentric energy governance system can emerge in multiple forms. As part of the EU harmonisation process, some partially fulfil the normative promises set out by polycentric governance thinking, while others illustrate a nominal transition to polycentric energy governance, albeit at the risk of diluting the cooperative and community-based models of RECs. What is visible in the transposition of REC provisions is a gap between some Member States that are willing to advance the agenda of energy decentralisation and energy democracy parallel, and others that lag behind. This produces a situation of mismatched heterogeneity across Member States rather than one of harmonisation.

This also questions the way overarching rules are designed. It appears that overarching rules set out by the EU do not lead to polycentric energy governance systems that fulfil the intended goals of REC provisions. This seems to be highly dependent on Member States' contexts and political priorities. Analysing the transposition process helps shed light on paradoxical dynamics, marked by a process of EU harmonisation in diverse polycentric energy governance systems. The primary paradox we see is a process of (market) liberalisation and decoupling rooted in neoliberal economic and political thought being met with a new playing field where alternative organisational forms (RECs) also become an element of the energy system. Here we see a tension between market actors interested in generating a profit, community actors aiming to benefit their local communities, and states navigating legal and regulatory frameworks between these interests. This also illuminates the role of power in polycentric governance thinking; namely how local-scale, citizen-led initiatives relate to political decision-makers and incumbent actors such as electricity utilities or DSOs. The varied transposition realities presented in this chapter highlight that past formulating overarching rules, effective polycentric governance systems also hinge on the illustration of viable local-scale action and political willingness of decision-makers.

Considering the story of European energy policy and markets is meant to be one of harmonisation, in the process of REC transpositions we see one that risks to result in fragmentation (at least in terms of how and what RECs will look like in different national jurisdictions). Another key paradox stands out: by recognising RECs in the RED II, as a way to facilitate the emergence and development of further RECs, the EU has also drafted a precarious definition of the REC model, which has become more prone to business capture, eventually putting the autonomy of individual initiatives at stake. Ensuring that the internal market becomes a place of opportunities for both traditional energy actors and new entrants is a considerable challenge for the EU, as reminded by the Letta report that also underlined the social need to further integrate local citizens into the EU energy market.

To conclude, one evident limit of this work is that the transposition process is still underway, which means that our observations are still at an early stage and can shed light on potential risks and opportunities in the transposition processes rather than prescribe a state of being. Future research needs to monitor the evolution of the transposition process.

## Note

1 The case of Czechia was last updated in April 2024, Lithuania in December 2023, and Poland in March 2024.

## References

Abbott, K. W., Genschel, P., Snidal, D., & Zangl, B. (2015). *International Organizations as Orchestrators*. Cambridge University Press, 430 p.

Ball, C. (2015). The transformation of the energy sector and "Citizen energy": Insights from Germany with comments on the Lithuanian context. *Social Transformations in Contemporary Society*, *3*, 6–13.

Bauwens, T. (2017). Polycentric Governance Approaches for a Low-Carbon Transition: The Roles of Community-Based Energy Initiatives in Enhancing the Resilience of Future Energy Systems. In Labanca, N. (ed.), *Complex Systems and Social Practices in Energy Transitions*. Green Energy and Technology (pp. 119–145). Springer. https://doi.org/10.1007/978-3-319-33753-1_6

Bauwens, T., Gotchev, B., & Holstenkamp, L. (2016). What drives the development of community energy in Europe? The case of wind power cooperatives. *Energy Research & Social Science*, *13*, 136–147. https://doi.org/10.1016/j.erss.2015.12.016

Bauwens, T., Schraven, D., Drewing, E., Radtke, J., Holstenkamp, L., Gotchev, B., & Yildiz, Ö. (2022). Conceptualizing community in energy systems: A systematic review of 183 definitions. *Renewable and Sustainable Energy Reviews*, *156*, 111999. https://doi.org/10.1016/j.rser.2021.111999

Brisbois, M. C. (2019). Powershifts: A framework for assessing the growing impact of decentralized ownership of energy transitions on political decision-making. *Energy Research & Social Science*, 50, 151–161. https://doi.org/10.1016/j.erss.2018.12.003

Buchner, B. K., Oliver, P., Wang, X. Carswell, C. Meattle, C., & Mazza, F. (2017). *Global Landscape of Climate Finance 2017*. Climate Policy Initiative.

Burke, M. J., & Stephens, J. C. (2017). Energy democracy: Goals and policy instruments for sociotechnical transitions. *Energy Research and Social Science*, *33*, 35–48. https://doi.org/10.1016/j.erss.2017.09.024

Capellán-Pérez, I., Johanisova, N., Young, J., & Kunze, C. (2020). Is community energy really non-existent in post-socialist Europe? Examining recent trends in 16 countries. *Energy Research & Social Science*, *61*, 101348. https://doi.org/10.1016/j.erss.2019.101348

Ćetković, S., & Buzogány, A. (2019). The political economy of EU climate and energy policies in Central and Eastern Europe revisited: Shifting coalitions and prospects for clean energy transitions. *Politics and Governance*, *7*, 124–138. https://doi.org/10.17645/pag.v7i1.1786

Cobut, L. (2021). *Renewable Energy Communities in the European Energy Policy: An Evidence of Clean Energy for All, Not by All.* Egmont - The Royal Institute for International Relations, Paper 111.

Council of the European Union. (1993). Council Decision of 13 September 1993 Concerning the Promotion of Renewable Energy Sources in the Community (Altener programme) (93/500/EEC). No L235/41.

Dupont, C. (2016). *Climate Policy Integration into EU Energy Policy.* Routledge, Earthscan, 193 p.

Dupont, C., Moore, B., Boasson, E. L., Gravey, V., Jordan, A., Kivimaa, P., Kulovesi, K., Kuzemko, C., Oberthür, S., Panchuk, D., Rosamond, J., Torney, D., Tosun, J., & von Homeyer, I. (2023). Three decades of EU climate policy: Racing toward climate neutrality? *Wiley Interdisciplinary Reviews: Climate Change*, *15*(1), e863. https://doi.org/10.1002/wcc.863

Dutton, J. (2015). *EU Energy Policy and the Third Package Energy Policy Group Working Paper 1505*. University of Exeter. Available at: https://ukerc8.dl.ac.uk/UCAT/PUBLICATIONS/EU_energy_policy_and_the_third_pac kage.pdf [Accessed 2 October 2022]

European Commission. (2016). *Clean Energy for All Europeans*. Brussels. COM (2016) 860 final. Available at: https://eur-lex.europa.eu/resource.html?uri=cellar:fa6ea15b-b7b0–11e6–9e3c-01aa75ed71a1.0001.02/DOC_1&format=PDF [Accessed 20 October 2024].

European Commission (2019). *Communication on The European Green Deal.* COM(2019) 640 final. Available at: https://eur-lex.europa.eu/legal-content/EN/TXT/?qid=1588580774040&uri=CELEX%3A52019DC0640 [Accessed 3 November 2024].

European Commission. (2020) *Communication [on the] European Climate Pact.* COM(2020) 788 final. Available at: https://ec.europa.eu/commission/presscorner/detail/en/ip_20_2323 [Accessed 3 November 2024].

European Commission (2023). *The European Green Deal. Timeline.* Available at: https://commission.europa.eu/strategy-and-policy/priorities-2019-2024/european-green-deal_en [Accessed 3 November 2024].

European Commission (2024). Work begins on the Citizens Energy Package to ensure a fair and inclusive energy transition. Available at: https://energy.ec.europa.eu/news/work-begins-citizens-energy-package-ensure-fair-and-inclusive-energy-transition-2024-12-19_en [Accessed 3 June 2025].

European Parliament and Council of the European Union. (1996). Directive 96/92/EC of 19 December 1996 concerning common rules for the internal market in electricity. *Official Journal of the European Communities*, L 27, 10 p.

European Parliament and Council of the European Union. (2018). Directive (EU) 2018/2001 of 11 December 2018 on the promotion of the use of energy from renewable sources (recast). *Official Journal of the European Union*, L 328, 128 p.

Fang, S., Jaffe, A. M., Loch-Temzelides, T., & lo Prete, C. (2024). Electricity grids and geopolitics: A game-theoretic analysis of the synchronization of the Baltic States' electricity networks with Continental Europe. *Energy Policy*, *188*, 114068. https://doi.org/10.1016/j.enpol.2024.114068

Fitch-Roy, O., Benson, D., & Woodman, B. (2019). Policy instrument supply and demand: How the renewable electricity auction took over the world. *Politics and Governance*, 7(1), 81–91. https://doi.org/10.17645/pag.v7i1.1581

Hall, S., Foxon, T. J., & Bolton, R. (2016), Financing the civic energy sector: How financial institutions affect ownership models in Germany and the United Kingdom. *Energy Research and Social Science*, *12*, 5–15. https://doi.org/10.1016/j.erss.2015.11.004

International Cooperative Alliance (ICA). (2024). *Cooperative Identity, Values and Principles*. Available at: https://ica.coop/fr/coop%c3%a9ratives/identite-cooperative [Accessed 3 November 2024].

Johnstone, P., Stirling, A., & Sovacool, B. (2017). Policy mixes for incumbency: Exploring the destructive recreation of renewable energy, shale gas 'fracking,' and nuclear power in the United Kingdom. *Energy Research and Social Science*, *33*, 147–162. https://doi.org/10.1016/j.erss.2017.09.005

Jordan, A., Huitema, D., Van Asselt, H., & Forster, J. (2018a). *Governing Climate Change. Polycentricity in Action?* Cambridge University Press.

Jordan, A., Huitema, D., Van Asselt, H., & Forster, J. (2018b). *Preface of Governing Climate Change.* Polycentricity in Action? Cambridge University Press.

Leiren, M. D., & Reimer, I. (2018). Historical institutionalist perspective on the shift from feed-in tariffs towards auctioning in German renewable energy policy. *Energy Research & Social Science*, *43*, 33–40. https://doi.org/10.1016/j.erss.2018.05.022

Malý, V., Šafařík, M., & Matoušek, R. (2019). Consumer (Co-)Ownership in Renewables in the Czech Republic. In Lowitzsch, J. (ed.), *Energy Transition* (pp. 201–222). Springer International Publishing. https://doi.org/10.1007/978-3-319-93518-8_10

Nies, S. (2018). An Introductory Overview on Institutional Change. In Nies, S., & Delors, J. (eds.), *The European Energy Transition. Actors, Factors, Sectors.* (pp. 44–48) Claeys & Casteels, European Energy Studies.

Pechancová, V., Pavelková, D., & Saha, P. (2022). Community renewable energy in the Czech Republic: Value proposition perspective. *Frontiers in Energy Research*, *10*, 14 p.https://doi.org/10.3389/fenrg.2022.821706

Petrovics, D., Huitema, D., & Jordan, A. (2022). Polycentric energy governance: Under what conditions do energy communities scale? *Environmental Policy and Governance*, *32*(5), 438–449. https://doi.org/10.1002/eet.1989

Rescoop.eu. (2024a). *Transposition Tracker*. Available at: https://www.rescoop.eu/policy/transposition-tracker [Accessed 3 November 2024].

Rescoop.eu. (2024b). *Transposition Tracker, Czechia*. Available at: https://www.rescoop.eu/policy/transposition-tracker/rec-cec-definitions/czech-republic; https://www.rescoop.eu/policy/transposition-tracker/enabling-frameworks-support-schemes/czech-republic-2 [Accessed 3 November 2024].

Rescoop.eu. (2024c). *Transposition Tracker, Lithuania*. Available at: https://www.rescoop.eu/policy/transposition-tracker/rec-cec-definitions/lithuania; https://www.rescoop.eu/policy/transposition-tracker/enabling-frameworks-support-schemes/lithuania-2 [Accessed 3 November 2024].

Rescoop.eu. (2024d). *Transposition Tracker, Poland*. Available at: https://www.rescoop.eu/policy/transposition-tracker/rec-cec-definitions/poland-rec-cec-definitions; https://www.rescoop.eu/policy/transposition-tracker/enabling-frameworks-support-schemes/poland [Accessed 3 Nov. 2024].

Rescoop.eu. (2025). *Map*. Available at: https://www.rescoop.eu/network/map/ [Accessed 14 December 2025].

Smith, A., Hargreaves, T., Hielscher, S., Martiskainen, M., & Seyfang, G. (2016) Making the most of community energies: Three perspectives on grassroots innovation. *Environment and Planning*, *48*(2), 407–432.

Stirling, A. (2019). How deep is incumbency? A 'configuring fields' approach to redistributing and reorienting power in socio-material change. *Energy Research and Social Science*, *58*, 101239. https://doi.org/10.1016/j.erss.2019.101239

Westphal, K., Pastukhova, M., & Pepe, J. M. (2022). *Geopolitics of Electricity: Grids, Space and (Political) Power*. Stiftung Wissenschaft und Politik (SWP). Research Paper 6. https://www.swp-berlin.org/10.18449/2022RP06/.

Zindulková, K. (2023). *Czech Community Energy on the Crossroad: Drivers, Challenges and Justice Implications in Shifting Energy Landscapes*. Master's Thesis, Charles University of Prague.

# PART 3
# Agriculture and biodiversity

# 5

# AGRICULTURE AND BIODIVERSITY IN EUROPE

## The faded Green Deal – reversing the biodiversity collapse in the European farming sector

*Ariel Brunner and Marilda Dhaskali*

*In memoriam of: Harriet Bradley*

### Europe's biodiversity crisis

The Intergovernmental Platform for Biodiversity and Ecosystem Services' (IPBES) recent report (2019) paints a dramatic picture of the collapse of global biodiversity. Over a million species are at imminent risk of extinction, and extinction rates are many orders of magnitude higher than they have been for tens of millions of years. Indeed, scientists believe we are already in the midst of a sixth great mass extinction, one on the same scale as the catastrophe that wiped out the dinosaurs at the end of the Cretaceous. Extinction of species is just the tip of the iceberg. Sixty percent of global animal populations have been wiped out since 1970 (WWF, 2018), and whole food chains and ecosystems are disintegrating. Several international environmental organisations including BirdLife International (see Box 5.1) are relaying scientific information for governments to react.

**BOX 5.1 BIRDLIFE INTERNATIONAL**

BirdLife International, created in 1993, is a global partnership of national grassroots conservation organisations that strives to conserve birds, their habitats and global biodiversity, working with people towards sustainability in the use of natural resources. It is the world's largest partnership of conservation organisations, with over 120 partner organisations around the world. BirdLife Europe and Central Asia, a regional division of BirdLife International, works with various stakeholders to conserve birds and biodiversity by focusing on species,

DOI: 10.4324/9781003479260-9

sites and habitats to ensure ecological sustainability and to engage people on these issues. It coordinates the activities of 46 national partner organisations throughout Europe and Central Asia, including all current European Union (EU) countries. BirdLife takes an active role in the development and implementation of the EU nature and biodiversity policy and the effective integration of biodiversity protection into other policy areas. The BirdLife European and Central Asian Partnership represents over 2,000,000 citizens, has over 4,000 staff and owns or manages more than 6,000 nature reserves, covering over 320,000 hectares.

Source: Authors' own elaboration.

The United Nations (UN) has already warned us about the danger of toying with the potential extinction of our own species (Watts, 2018). The "Planetary Boundaries" framework, which defines the safe operating space for humanity, shows that loss of genetic diversity is at the highest risk zone. With climate change, biosphere integrity is classified as a "core boundary", meaning surpassing it would "drive the Earth System into a new state" (Steffen et al., 2015).

The European situation is reflecting worldwide trends, with biodiversity impacted particularly in agricultural areas. Farmland birds have more than halved across the EU since 1980 (Rigal et al., 2023). Birds are doing worse now on farmlands than other habitats in France (Geffroy, 2018). Insects have collapsed by 76% in Germany (within protected habitats) over 27 years (Hallmann et al., 2017), while 64% of grassland species (non-birds), 86% of grassland habitats and 70% of cropland species (non-birds) are in an unfavourable state (European Environment Agency, 2015a, see also Figure 5.1). The biodiversity crisis is advancing in lockstep with the climate crisis, with the two reinforcing each other as climate change puts further pressure on wildlife, while ecosystems that are more fragile fail to act as carbon sinks.

### *Agriculture as the main driver of biodiversity loss in the EU*

The collapse of biodiversity is very clearly driven by human activities with agriculture being the primary source of impacts (Tollefson, 2019). This is due to the fact that agriculture is the dominant form of land use in much of the world, occupying about 40% of the land in the EU. Study after study paints the same picture – of a broken agriculture system playing a leading role in the collapse of the living world.

At the heart of the problem lies the simple fact that too many people consume too much of the Earth's resources. Land, fertile soil and water are finite,

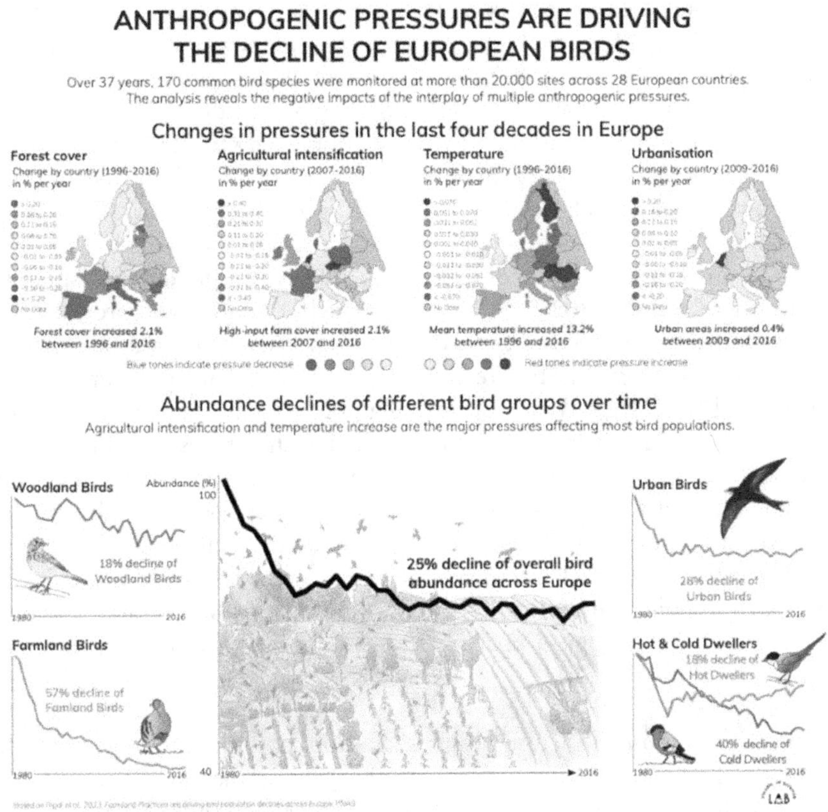


**FIGURE 5.1** Decline of bird populations in Europe 1980–2016, showing that farmland species are doing the worst

and each bit of land is covered by vegetation that transforms the sun's energy into energy available for animals, humans included. Agriculture is at its heart, manipulating ecosystems so that they channel more of the sun's energy captured by photosynthesis into products humans can consume, for example, agriculture replaces a natural forest that produces very few fruits edible to humans with a wheat field or apple orchard that produces a lot more product that humans can consume. The problem is that the amount of photosynthesis a given piece of land can perform is fairly constant. Therefore, agriculture is something of a thermodynamic zero-sum game: the more of the "solar cake" we take, the less we leave for all the remaining organisms. This principle is neatly captured by the "human appropriation of net primary productivity index" (Haberl et al., 2014) which shows how much of the solar energy captured by plants is taken up by humans. In most agricultural regions this is now more than 50% with much of Europe above 70% (Haberl et al., 2014). In large parts of Europe there is hardly anything left unharvested.

This chapter delves into the critical relationship between agriculture and biodiversity loss in Europe, highlighting the role of the Common Agricultural Policy (CAP) as both a driver of environmental degradation and a potential tool for positive change. This chapter begins by addressing the scale of Europe's biodiversity crisis and its primary cause: agricultural expansion and intensification. It then examines the broader environmental impacts of farming practices, exploring how they exacerbate biodiversity loss. This chapter continues by discussing potential solutions within the agricultural sector, including reforms to the CAP and the broader legislative landscape. Attention is given to the 2021–2027 CAP reform, which, despite some progress, has largely failed to reverse the trends of environmental harm. A key focus of this chapter is the European Green Deal and its ambitions to integrate biodiversity and climate goals into EU farming policies, while acknowledging the significant backlash it has faced. The discussion then turns to climate policy as a potential game-changer for agricultural practices, considering how shifts in land use and food consumption can help address both the climate and biodiversity crises. Finally, this chapter examines the challenges of land competition and concludes by emphasising that a comprehensive and coherent approach to policy, alongside the political will for transformation, is crucial for averting the ecological crisis while maintaining food security and supporting rural livelihoods.

This rather theoretical scheme translates in reality into two main processes: land use change and agricultural intensification, both driving large-scale loss in biodiversity and creating a host of other environmental problems, from water pollution and floods to greenhouse gas (GHG) emissions and local air pollution.

The worst and fastest biodiversity loss occurs when natural habitats, particularly large and still relatively intact ecosystems, are cleared to make space for cultivation. Much of Europe was cleared of its native vegetation in previous centuries (and sometimes millennia). European farming still contributes substantially to global clearing through demand for animal feed (mainly soy) (Pendrill et al., 2019) and through indirect effects (e.g. when European oil seeds have been diverted to biofuel production causing the food market to switch to palm oil) (Gawel and Ludwig, 2011). That said, we still see large-scale land use change going on in Europe, most notably the destruction of semi-natural grasslands, traditionally grazed or mown for hay, that are being converted to arable or permanent crops, often with near total loss of their typical biodiversity (BfN, 2017, 2014; Moreira et al., 2012).

However, the main phenomenon driving biodiversity loss in Europe is agriculture intensification (European Ornithologists Union et al., 2019; European Environment Agency, 2015b; Tscharntke et al., 2007; Donald et al., 2006). This takes different forms depending on regions, landscapes and production systems, but the common trend is broadly the same: the replacement of

FIGURE 5.2 Land use change in Europe: satellite images from France, Atlantic Loire, showing the simplification of the same region between 2000 and today

diverse, small-scale and extensive production systems by homogenised, large-scale, high-input-high-output systems. This is driven by a combination of market forces, demographic changes and technological progress but is still massively driven and enabled by public policies and in particular by the EU's CAP. Regional- and farm-level specialisation, the increase in field size and abandonment of complex rotations in favour of monocultures, happening across Europe as illustrated by Figures 5.2 and 5.3, all lead to biodiversity loss through loss of spatial heterogeneity and habitat diversity.

Species may find a nesting place but no food, or food in one season but not in another. Some species can adapt to a certain crop, but others are lost if no other crops are available. Overall, the simplification of the farmed landscape leads to a simplification of the biotic community (Benton et al., 2003). Generalist predators like foxes or crows become much more efficient, for example, in picking up ground-nesting birds, in simplified landscapes that are easy to systematically search. The destruction of non-crop vegetation (hedges, bushes, vegetated field margins, streamside vegetation, rocky outcrops, etc.) has removed most of the native plants from farmed landscapes and hence the insects feeding on them, or birds depending on their seeds (Benton et al., 2003; Newton, 2004). Heavy machinery use causes the destruction of nests and burrowing animals, as well as damage to soil communities. Pesticides cause both direct and indirect poisoning, removing the insects and wild plants that are at the base of the food chain (Geiger et al., 2010). The heavy use of synthetic fertilisers or the spreading of concentrated sludge from intensive livestock operations causes eutrophication of water bodies, kills soil organisms and favours a handful of generalist plants (e.g. nettles) that smother herbaceous communities (Sutton, 2011).

While most biodiversity loss and other environmental harm are caused by agricultural intensification, in some marginal land (e.g. some mountain

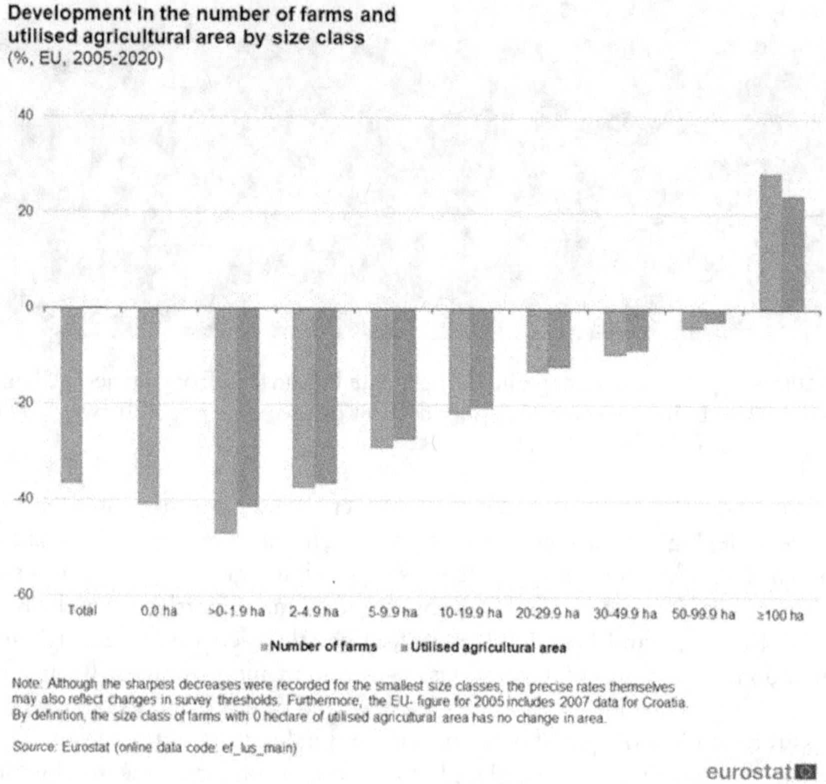


**FIGURE 5.3** Change in the number of farm holdings in Europe 2005–2016, showing that all but the largest farms (over 100 hectares) are declining in Europe

areas, some regions undergoing rapid depopulation) biodiversity is also lost due to the abandonment of traditional farming practices, essentially extensive (especially pastoral) grazing and haymaking for animal fodder (Vickery et al., 2001). Grasslands represent some of the most biodiversity-rich habitats in Europe. Indeed, in Europe, for palaeoecological reasons, grasslands are often biologically richer than forests. Before human colonisation, much of Europe was a dynamic mosaic of forest, open parkland and grassland, maintained by the action of large herbivores, with dynamics similar to those that can be still observed in the few African savannahs that maintain their megafauna. After the extinction of the late Pleistocene megafauna and the progressive impoverishment of European wild herbivore communities through hunting, many species have survived by adapting to secondary habitats maintained by farming, in particular by traditional extensive livestock systems. These labour-intensive systems have been collapsing over the past century, essentially due to the efficiencies brought by the use of fossil fuels which triggered far-reaching technological, demographic and cultural changes (Oppermann et al., 2012).

## Other environmental impacts that reinforce biodiversity loss

Farming is now the biggest source of diffuse water pollution in Europe (European Environment Agency, 2018) with impacts ranging from marine dead zones to closure of drinking water wells. Farming is also the source of 92% of ammonia emissions (European Environment Agency, 2024) that are a key component of air pollution (including inside cities) and cost the lives of tens of thousands of citizens a year (Taddei, 2020). Around 7.2% of EU agricultural land suffers from severe soil erosion (Panagos et al., 2018), and 25% of the EU's territory suffers from soil erosion by water (European Environment Agency, 2015b). Loss of soil organic matter and landscape elements means that the water retention ability of land has been going down relentlessly in agricultural regions, leading to worsening of floods and increased vulnerability to droughts. Pollinator populations have been crashing across Europe (IPBES, 2016) as have the populations of predators of pest species. All of these pose an existential threat to the future of European farming itself, as modern farming is sucked ever deeper into a vicious cycle of environmental degradation and technological responses, such as increased use of irrigation, fertilisers and pesticides, further aggravating the problems. Loss of natural predators forces an ever-growing use of pesticides, which keep eroding natural pest control while breeding resistance in pests. Loss of soil organic matter makes crops ever more dependent on irrigation, just as increased irrigation, climate change and loss of water retention make water resources ever scarcer. This spiralling environmental crisis goes hand in hand with relentless consolidation. Fewer farmers run bigger farms, with the most intensive farmers driving the less intensive ones out of business. In addition to all these environmental problems, EU farming models are fuelling the climate crisis, as agriculture is responsible for 11% of the GHG emissions that is also exacerbating the above-mentioned issues while increasing losses of agriculture production due to natural disasters (European Scientific Advisory Board on Climate Change, 2024). It is worth noting that agriculture's actual contribution is much higher as GHG emissions from feed, fertiliser, fuel, machinery, etc. are attributed to other sectors and countries.

## The potential solutions within agriculture

Modern agronomy, agroecology and decades of agri-environmental experience offer ready-made solutions to most of these problems. Keeping around 10% of a farm under sympathetically managed natural vegetation can provide a habitat for many species, including pollinators and pest predators (Traba and Morales, 2019; Albrecht et al., 2020). Leaving unmowed or ungrazed wide strips on grasslands can massively increase biodiversity, without significantly reducing fodder production. Buffer strips around water courses can both provide habitat and control runoff. Appropriate crop rotation can increase spatial heterogeneity, improve soil fertility, help prevent pest outbreaks and hedge the risks of crop failures and market fluctuation. Species-rich, semi-natural grasslands can produce high-quality meat and dairy with no need for heavy investments, while locking in carbon, slowing runoff and hosting wildlife. Pesticides can be used sparingly following the principles of Integrated Pest Management, where chemical control is used as the last resort or even be entirely replaced by biological control and other non-chemical methods.

For almost any problem, in almost any farming system, solutions are available that have been tested on the ground by farmers and proven to work. Ever more farmers are embracing systemic change, exploring whole new (and sometimes old) ways of producing food. From organic farming to agroforestry, urban farming to high-tech circular economy systems, the last decade has seen a flourishing of new, bold experiments in sustainability. The problem is that best practice remains rare, malpractice remains the norm and the "ecological transition" remains even slower in farming than in other parts of the economy. Structural lock-ins are compounding resistance to change as farmers often have invested in machinery that can only be used following current logics, are tied to supply chains that would only buy current products and get advice often directly or indirectly shaped by the companies selling them inputs.

Ecological and climate sustainability is inherently incompatible with current levels of production and consumption particularly in the livestock sector (see Figure 5.4). Good practice can make a huge difference, in some cases with no trade-off with productivity. For example, there is growing evidence that flower strips in arable farming maintain or even increase yields, even when land is taken out of production to deliver the strips (Pywell et al., 2015). However, in many cases, a more ecologically balanced farming system would need to produce less, in particular have fewer farmed animals as basic thermodynamics dictate that it takes a lot more arable land to produce calories or proteins by feeding crops to animals than by feeding them directly to people.

Europeans consume meat and dairy at levels higher than the ones considered healthy, while bad diets are causes of ill health and the ballooning costs of healthcare (EAT-Lancet Report, 2019; Buckwell and Nadeu, 2018; FAO, 2013). Very substantial reductions in meat and dairy consumption could be compatible with healthy and pleasant diets in line with cultural traditions. Enacting policies in this field is challenging, but from public procurement to labelling, education and fiscal measures, much could be done. Moreover, tens

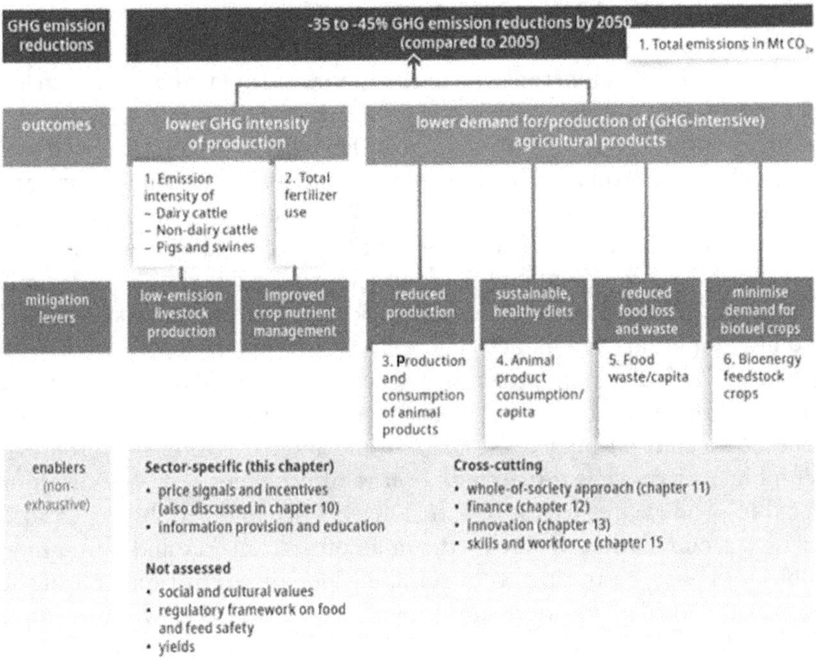


Source: Advisory Board (2024)

**FIGURE 5.4** Assessment framework of the agriculture, food and land-use sector identifying levers to reduce the GHG emissions by 2050

of millions of hectares are needlessly dedicated to the production of bioenergy, which is now well known to be mostly harmful for both the climate and biodiversity (Searchinger, 2022; Hellmann and Verburg, 2010). This demand is subsidies-driven and mostly created, at great expense, for the purposes of creating more demand for the products of intensive agriculture, hence the support of the main farm unions like Copa-Cogeca for bioenergy promotion (Copa-Cogeca, 2013).

The EU has so far done little to curtail its land and climate footprint in terms of agricultural commodities or to improve food chains' sustainability.

## The fundamental driver: the Common Agricultural Policy

Underpinning much of current European farming and land management is the EU's CAP. Established in 1962, it is one of the world's largest subsidies system, spending almost 60 billion Euros per year and absorbing around one-third of the EU budget. The behaviour and choices of most European farmers are dictated as much by CAP incentives as by market forces, demography and environmental conditions.

Originally, the CAP was aimed at boosting food production in a continent still reeling from wartime hunger, at supporting the income of farmers who were then still very numerous and mostly poor, and at facilitating the creation of a common market by buying powerful rural interests into the opening of infra-European borders. The main tools deployed were high external tariffs to protect local producers against imports and guaranteed minimum prices to boost production, with state intervention to buy up any temporary gluts and prevent price slumps. The income support provided by the CAP also led to increase the investment power of farmers who, advised by representatives of manufacturers of agricultural input, increasingly embraced specialisation and high-input farming model.

The early CAP can be probably considered to have succeeded on its own terms. Europe went from food importer to food exporter, food prices have gone down and farming consolidation and modernisation has progressed. With time, success has turned into a series of problems. The one-directional signal to produce, regardless of actual market demand, led, by the 1980s, to massive accumulation of stocks (the infamous wine lakes and butter mountains), runaway costs that were spiralling out of control and raising tensions with other global trading partners. Overproduction was also driving large-scale ecological destruction, and dumping of excess production was devastating for poor farming economies in developing countries. Starting from the end of the 1980s, a long reform cycle that carried on until around 2008 tried to deal with these problems (see Box 5.2).

### BOX 5.2 SPACE FOR NATURE: FROM HAPPY INCIDENT TO FAILED POLICY

In 1988, the EU established a "set aside" (land not in production) obligation to limit the overproduction caused by guaranteed prices (initially 15% of land, reduced to 10% in 1996). Originally intended as an economic measure, recent reviews show that it gave a lifeline to biodiversity by supplying a certain amount of fallow land. This requirement was abolished during the 2008 commodity spike, leading to the loss of farmland wildlife. In the 2021 reform, the enhanced conditionality, requirements to comply with for farmers to receive their direct payments, set a series of good agricultural and environmental conditions (GAEC) amid which one devoted to maintaining non-productive features and land for biodiversity (GAEC 8), rating at best 7% and at worse 3% of land laying fallow. This fell short of the 10% of non-productive elements, including fallow land that is critical for the maintenance and restoration of biodiversity on agricultural land. Moreover, the indicator supposed to measure this instrument is either not used by Member States (half of them have set the indicator to 0% or 2%) or artificially inflated with weighting factors that have no justification. Finally, as a response to the farmer's protest in January 2024, the Commission decided an "emergency" simplification of the CAP, removing the obligation to comply with some of GAECs, notably GAEC 8. It also exempted farmers below 10 Ha from any controls and penalties (thus making all requirements void).

Source: Authors' own elaboration.

The main political driver behind CAP reform was trade liberalisation as the EU tried to open up its economy and chased the economic promises of globalisation, while maintaining on board its powerful farm lobby. Guaranteed prices were progressively abandoned in favour of market prices, with farmers "compensated" through direct payments, originally linked to specific commodities and designed to close the gap between the old guaranteed price and the new, lower, world market price.

At a later stage, "decoupling" broke the link to specific commodities thus freeing the farmers to react to market demands (although some limited coupled support still remained, especially for meat and dairy). Subsidies born as a temporary compensation for expected drop in commodity price had morphed into entitlements to receive CAP payments (that can be bought and sold) that do not have any clear relation to what farmers actually do, nor possess any clear policy objective. A temporary adjustment measure has become a simple privilege bestowed on landowners, an "income support" paid per hectare of land owned, leading to a legitimate debate about fairness. The per-hectare payment logic has entrenched the inequality of the subsidies' distribution where a small portion of beneficiaries capture most of the money. Around 80% of funds go to the top 20%, while the top 1%–2% capture about a third (Apuzzo and Gebrekidan, 2019). Although all reform rounds have claimed to have addressed the issue, current studies show they have had no effect (Alfaro-Navarro and Andrés-Martínez, 2021; Sadłowski et al., 2022) (Figure 5.5).

In parallel to the evolution from price support to current decoupled direct payments, there has been an effort to move part of the CAP budget from farm subsidies to forms of payments for the delivery of public goods such as environmental stewardship and into investments in improvement of life and economic performance of rural areas. The first attempt, starting in the late 1980s, has led to development of agri-environmental schemes, voluntary contracts that allow farmers to be paid for the cost incurred or the income forgone linked to commitments supposed to improve their environmental performance.

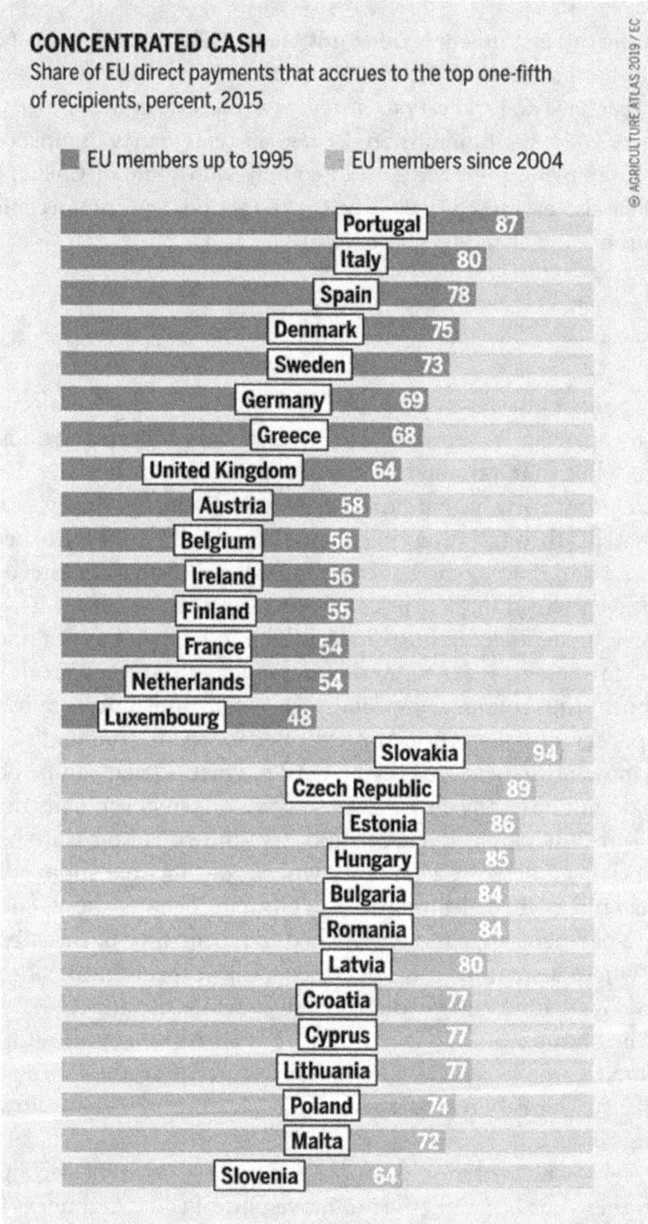


**FIGURE 5.5** Percentage of CAP direct payments going to the top 20% of recipients

The latest reform has also introduced so-called eco-schemes, a reincarnation of the "pillar 1 Greening" mechanism introduced in the previous reform round. These schemes for the climate, the environment and animal welfare are fully funded by the pillar and take the form of yearly payments to farmers who voluntarily enrol. The risk that Member States propose schemes with no added value for biodiversity is still very high, especially due to the high flexibility left to Member States to design their eco-schemes without any binding specifications, as it is the case for agri-environmental measures.

Another element bolted onto the CAP at the beginning of this century has been "Cross Compliance". Essentially, it is a requirement for CAP beneficiaries to respect a number of EU environmental laws and a set of "good agricultural and environmental conditions". Cross compliance has been muddled from the start by competing agendas. Environmental non-governmental organizations (ENGOs) and environmental authorities were looking at a way to get agricultural authorities, which are well equipped to do farm-level controls, involved in environmental enforcement which has always been awfully inadequate on farmland.

Meanwhile, the growing power of Member States and loss of control by a weaker Commission have sent some previous reforms into reverse. Coupled support (i.e. production subsidies) has been increasing over recent years (Matthews, 2015, 2018), mainly favouring the intensive livestock sector (dairy, pigs).

## The 2021–2027 CAP reform

As the environmental crisis gathers speed, and evidence of the failures of the CAP become incontrovertible (ECA special report 13/2020, ECA special report 20/2021 and ECA special report 16/2021), the debate on its future has been raging. The new CAP Plans, implemented as of 1 January 2023 are very unlikely to provide the driving force to a radical transformation of our food systems that is needed in light of the crises we are facing. At the moment, it is not even clear whether they are bringing any small increase in environmental and climate performance. The overall assessment of the CAP subsidies distribution (see Figure 5.6) reveals that 49% of the CAP Strategic Plan budgets are allocated to untargeted decoupled income support payments, 30% of the budgets are devoted to climate and environmental objectives, which does not represent any increase in (supposedly) green spending compared to the previous period and presenting the same concerns in regard to the quality of that spending, while 17.4 billion Euros will go the environmentally harmful "coupled support", of which 70% is flowing to the livestock sector (BirdLife and EEB Report, 2022).

There is evidence that hectare-based income support payments are ineffective for all dimensions of sustainability. They are unevenly distributed,

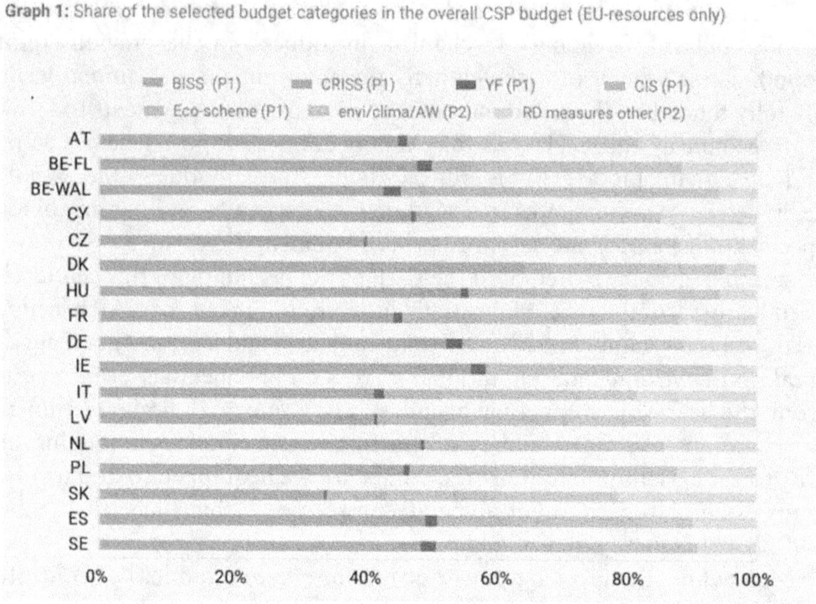


**FIGURE 5.6** Share of the different categories of the CAP Plans budget for 17 MS, for Pillar 1 and 2

leak towards land rents (JRC, 2021) and non-farmers, distort land markets and fail to maintain reasonable incomes or to halt rural employment decline (Pe'er at al., 2019). The fact that half of CSP budgets are allocated to these ineffective instruments calls into question the effectiveness of CAP as a whole to deliver on declared objectives and societal needs.

Regarding Pillar 2, in most countries, the highest share of environment and climate spending was allocated to agri-environment-climate measures but up to 50% of that budget can be spent as support for areas with "natural or other areas-specific constraints". These are de facto more per-hectare payments that do not necessarily provide benefits for biodiversity or climate. On the contrary, due to a lack of environmental safeguards, they may well support agricultural intensification such as livestock overstocking.

The overall CSP budgets are unlikely to support the transition to sustainability and to meaningfully contribute to environmental and climate objectives. Member States have also failed to use the flexibility provided to shift funds between the pillars, to transfer money from untargeted direct payments to the rural development budgets better suited to the achievement of specific objectives. Whatever hypothetical residual gains from the last reform are likely to have been wiped out by the high-speed CAP mini-reform rushed in 2024 ahead of the European Parliament elections and in the wake of the farmer protests wave. Conditionality, the baseline for the CAP's green architecture that was already flawed by very low mandatory requirements has been further watered down (Regulation EU-2024/1468).

ENGOs are extremely worried by the prospect of another lost decade and a return to a race to the bottom on production subsidies driving another wave of intensification and environmental degradation. A consensus among the main ENGOs is around the need for the following elements (BirdLife, EEB and WWF, 2023): (i) invest fully in the transition; (ii) no more untargeted or harmful subsidies; (iii) a just mechanism and fair conditions: reward those delivering for the environment; (iv) a focus on performance and accountability; (v) new sustainable food systems governance and effective enforcement; (vi) from coping to preventing risks; (vii) a more transparent and digital administration.

Reform after reform, the CAP has been a sequel of adjustments at the margins to a policy based on increasing the production, but the very foundation of the policy remained. Farmers are still being paid because they are "farmers" (and that includes corporations and extremely wealthy individuals), independently of how they manage their land. It's hard not to conclude that four decades of attempts at an environmental reform of the CAP have fundamentally failed.

## Beyond the CAP-environmental legislation

Tackling the environmental crisis in farming depends increasingly on other policy tools, notably on environmental legislation. The EU has already some of the most progressive environmental legislation in the world. The Birds and Habitats Directives protect biodiversity and mandate the management of the world's largest coherent network of protected areas – "Natura 2000". The Water Framework Directive protects water bodies. The Nitrates Directive imposes limits on water pollution while the Sustainable Use Directive is supposed to drive a reduction in the use of pesticides though it has spectacularly failed so far. Nevertheless, the implementation of this set of environmental legislation is still very poor. The European Commission has agreed to derogations lasting decades, while Member States have been delivering blanket exemptions from obligations such as appropriate assessments in Natura 2000 sites.

The Dutch example is a perfect case in point. The Dutch government has, over several decades, sought endless derogation and loopholes, and resorted to plain law-breaking, to allow its livestock sector to ramp up production without regard to nitrogen pollution. In 2022, the European Commission ruled to not give any further derogations while national courts have forced the Government to respect the Habitats Directive protection of sensitive sites. This triggered an institutional crisis and a belated emergency plan to buy out dairy farms. This crisis was then manipulated by the farm unions and far-right politicians to blame the EU and its environmental legislation for all the problems farmers are facing. The failure of successive Dutch governments to effectively address nitrogen pollution and progressively reduce livestock densities with tailored financial instruments is a cautionary tale for the future.

On a more positive note, law enforcement on the ground might eventually improve as new technologies and pressure from citizens combine to expose abuse and shame authorities into action. Modern remote sensing can pinpoint land use change in near real time and in great detail, and analysis of past images can pin down the exact moment of conversion so that those responsible cannot hide behind claims that conversion happened before legal protection, or before they took ownership of land, for example.

## A potential game-changer climate policy

The tensions we have seen around CAP reform and enforcement of environmental legislation are likely to grow and the most consequential element is rapidly becoming climate change (see also Chapter 1, this volume). Legally binding climate commitments, which still enjoy a very broad support, cannot be met without radical improvements to farming and land management as foreseen by the Land Use, Land Use Change, and Forestry regulation aiming at capturing 310 Mt of $CO_2$eq by 2030. Farming is one of very few European sectors seeing GHG increasing in recent years. As other sectors decarbonise, the farm lobby's commitment to the status quo will eventually become untenable as the costs of inaction in farming are transferred to other parts of society. All credible climate mitigation scenarios rely on massive restoration of carbon sinks which require both release of farmland for restoration of carbon-rich habitats, such as forests, grasslands and peatlands, and reversal of the loss of carbon from farmland. Methane emissions are a key component of the climate crisis that cannot be tackled without tackling the livestock sector. So far, farming has mostly managed to avoid legally binding changes, and even during the latest communication by the EC on future climate policies, agriculture is the only sector where no reduction has been foreseen (COM/2024/63). It is hard to imagine that this approach can hold (Box 5.3).

### BOX 5.3 THE GREEN DEAL PROMISES ON AGRICULTURE AND FOOD SACRIFICED AT THE ALTAR OF FARMER'S PROTEST

The European Green Deal, the main political project of Ursula von der Leyen's Commission, included some strong strategies to tackle jointly the climate and biodiversity crises. The Farm to Fork Strategy, published in 2020, was meant to lead the EU towards "a fair, healthy and environmentally friendly food system". It has however run into a solid wall of opposition from the farm lobby, ending up with a near complete failure. The Plant protections products regulation, supposed to slash the pesticides use and risk by 50% by 2030, was withdrawn in February 2024 after farmers' protests all around the EU. The long-awaited Sustainable Food System Framework, looking to foster healthy food environments, was simply never proposed. Some other initiatives set out in the Biodiversity Strategy such as the organic production action plan and the better nutrient management action plans never saw the light of day. As an exception, and due to relentless lobbying action by large agrobusiness, the new genomic techniques proposal made it through quite fast and the proposal is at the Council's hands, which needs to find an agreed position. This is in stark contrast to the energy side of the Green Deal which has seen dozens of pieces of legislation approved, already triggered leading to visible speeding up in the rollout of renewable energy, energy efficiency and electrification of transport. The farm lobby anti-Green Deal backlash is currently spilling beyond farming and land use, fuelling a generalised anti-climate and anti-environment drift among centre-right parties, which now extends also to some energy files like the electrification of transport. Even more worrying is the growing alignment between the farm lobby and extreme right ideologies, and its effective use of disinformation, conspiracy theories and violence.

Source: Authors' own elaboration.

## Land competition

Careful land planning is another powerful policy tool to deal with biodiversity loss, restore carbon stocks and simply survive the onslaught of floods, droughts and forest fires due to climate change. Both space and productive land are finite, entailing unavoidable trade-offs between producing food, producing energy, housing people and leaving space for nature. The Green Deal has tackled the latter in the Biodiversity Strategy that led to the Nature Restoration Law, proposed by the Commission in Summer 2022. Agriculture is the key land use, even in the densely populated EU, with 40% of the land devoted to farming, of which 71% is used to produce animal proteins (Green Alliance Report, 2024). In a world where there is a misleading narrative of food security, a race to increase even more production is visible. But modern science is very clear that neither continued food production nor safe continuation of human civilisation are likely without large-scale restoration of nature. Fortunately, once the biodiversity crisis is collectively understood as an emergency, we do have solutions that can enable us to free up land for nature. Firstly, the EU needs policies that redirect both consumption and production towards more plant-based diets and reduced numbers of livestock. Secondly, renewables such as wind, solar and geothermal are way more efficient in producing energy per-hectare than biofuels or burning biomass (BirdLife Report, 2021, Burn or restore); fortunately, their costs are plummeting. Thirdly, by implementing a circular economy and a hierarchy of usages we can reduce the primary need for raw materials and make much better use of residues and waste, including the by-products from agriculture (Box 5.4).

## BOX 5.4 REWILDING EUROPEAN LANDSCAPES

Arnaud Gane UCLouvain Saint-Louis – Bruxelles

At a time when 81% of protected habitats in the EU are in poor condition (European Environment Agency, 2021), the effectiveness of its management of natural protected areas is questionable. It is usually the result of a compromise. For instance, some preserved areas can be maintained in their anthropised form, in particular by maintaining agroforestry or agro-pastoral activities. Rewilding offers a response to these difficulties by proposing to rethink conservation from a perspective that aims the emergence of large-scale self-sufficiency ecosystem through minimal human intervention, sometimes through initial restoration and reintroduction of key species. This movement was born in North America, when activist Dave Foreman and his organisation *Earth First!* came together with the radical philosophy of deep ecology expressed by Arne Næss and others, and with the new ideas of conservationist biologists such as Michael Soulé, who were trying to reform their discipline. Rewilding was new, but related to the old controversy between conservation and preservation, between a vision of protected areas for the benefit of humans or of nature itself.

Rewilding proposes to decentralise our relationship with nature by thinking of conservation projects in the interests of nature, i.e., as wild as possible. In Europe, this new approach was first adopted by practitioners and experts in the field, who conducted the first experiments in the 1980s, notably in Scotland with the *Trees for Life* association. In the early 2000s, several major continental organisations such as Rewilding Europe, Wild Europe, Large Herbivore Network and Wildland Research Institute emerged. As of 2025, they continue to shape the European rewilding movement, which is experimenting with a growing number of local experiments. These experiments are concentrated in isolated regions on a continental scale, such as the north of Scandinavia, areas in Eastern Europe and the mountainous part of the Iberian Peninsula. Several projects are also being implemented in the British Isles particularly by private landowners.

To spread rewilding on a larger European scale in the future, priority is given to abandoned farmland in areas that are unsuitable for mechanised farming. This land abandonment could represent up to 4% of the EU's total surface area by 2030. But there is also a growing movement to rewild the oceans. The Dutch

NGO ARK Rewilding Nederland, one of the most active national rewilding organisations in Europe, has carried out a rewilding project in the estuaries and coastal areas of The Netherlands. In particular, the project focused on the active reintroduction of key species such as flat oysters and shellfish near offshore wind turbines.

Rewilding raises growing interest among the public and academia. In England, the journalist Georges Monbiot popularised the concept by pointing to the aesthetic and meaningful dimension of wild nature; and showing how reconnecting with wild nature could help make sense of the modern world. Oxford professor Paul Jepson has led numerous leading research and advocacy campaigns on rewilding and is on the board of the largest continental organisation, Rewilding Europe. Other continental organisations are very dynamic, including the European Wilderness Institute, the Wildland Research Institute, which produces guidance documents and mapping, and Wild Europe, which raises public awareness through films.

Existing projects explore various national legislations to find space to experiment with rewilding. In some regions, such as Scotland, private landowners can play a leading role in the implementation of projects. In other countries, such as Belgium or France, experiments are being carried out by the State within the framework of national parks' policy, with the concept of "buffer zone". However, at the European level, rewilding is absent from official legislation, except for a 2009 parliamentary resolution calling for the development of "wilderness areas". This does not mean the EU is inactive. It is a key partner, particularly in terms of financial support of rewilding organisations. One example is the LIFE funding programme, which has supported the reintroduction of many large herbivores and of which Rewilding Europe has been one of the partners. More recently, Rewilding Europe has defended the European "Nature Restoration Law" alongside many other leading biodiversity NGOs in a public statement. This law, which was finally adopted in June 2024 after several months of difficult negotiations, could help to create a favourable context for rewilding, while not making any explicit reference to it. Rewilding and restoration are highly complementary depending on the areas considered, and rewilding could be a tool for achieving the objective of restoring 20% of the EU's terrestrial and marine areas by 2030, as set out in this new text. While rewilding and ecological restoration share common goals, rewilding is distinguished by its emphasis on process-led recovery, often involving the reintroduction of keystone species and reduced human management.According to Jepson and Schepers (2016), rewilding faces three main challenges in Europe: gaining scientific and socio-cultural legitimacy, integrating into EU policies as a complementary tool, especially in the CAP policy, and securing sustained financial and political support.

To achieve these objectives, rewilding must also deal with internal tensions that may limit its ability to attract a wider audience. First, rewilding suffers from a lack of clarity as it incorporates very disparate practices under the same label. For example, some rewilding projects are based on the reintroduction of large herbivores, while others are based on radical non-intervention, or "letting go", in the rewilded areas. Secondly, some rewilding projects have experienced difficulties to get accepted by local populations. In England, rewilding has been accused of accompanying rural gentrification, by favouring the vision of nature of the upper urban classes to the detriment of that of the inhabitants of rural areas. Some projects like Oostvaardersplassen (Netherlands) or Vercors Vie Sauvage (France) have led to conflicts with the local population over what constitutes a "good" nature and a "good relationship" with nature. In the wider context of what has been called "post-nature", in reference to the disappearance of the great separation between nature and society, rewilding is certainly a useful tool. But it would need to articulate a social vision with its radical interpretation of nature as autonomous.

Cite this box as: Gane, A. (2025). Rewilding European landscapes. In A. Orsini & E. Kavvatha (Eds.), *EU environmental governance: Current and future challenges* (Box, pp. 113–114). Routledge.

Sources: European Environment Agency (2021); Jepson and Schepers (2016); Pereira et al. (2024); Perpiña Castillo et al. (2018).

## Conclusion

Agricultural policy will make or break our ability to tackle the biodiversity and climate crises in the coming decade. Up to now, the EU agriculture system and the CAP have been major obstacles for protecting biodiversity and have been responsible for the constant erosion of Europe's natural capital. The climate crisis requires all sectors to decarbonise, and land use is also the key to adaptation. It is increasingly apparent that the current food system is massively unsustainable and that the window of opportunity for non-traumatic change is rapidly closing. Reasonable change could unlock huge amounts of land for nature and carbon sink restoration and for more benign forms of farming. Averting the ecological crisis while providing Europe with sufficient food and rural communities with decent livelihoods is still possible. But it would require a coherent set of policies around land planning, just transition in farming, rewarding landowners for allocating space for nature and reorientation of consumption away from meat and dairy towards plant-based diets. Such policies remain at best patchy and at worse absent.

## References


Albrecht M., et al. 2020. The effectiveness of flower strips and hedgerows on pest control, pollination services and crop yield: a quantitative synthesis. Ecology Letters. doi: 10.1111/ele.13576

Alfaro-Navarro, J.-L., & Andrés-Martínez, M.-E. (2021). A Longitudinal and Cross-sectional Analysis of the Distribution of Common Agricultural Policy aids in European Countries. *Agricultural Economics*, 67(9), 351–362. DOI: 10.17221/87/2021-agricecon

Apuzzo, M., & Gebrekidan, S. (2019). Who Keeps Europe's Farm Billions Flowing? Often, Those Who Benefit. *The New York Times*, 11 December, sec. World. Available at: www.nytimes.com/2019/12/11/world/europe/eu-farm-subsidy-lobbying.html (accessed December 2019).

Benton, T. G., Vickery, J. A., & Wilson, J. D. (2003). Farmland Biodiversity: Is Habitat Heterogeneity the Key? *Trends in Ecology & Evolution*, *18*(4), 182–188. DOI: 10.1016/s0169–5347(03)00011-9

BfN. (2014). Grünland-Report: Alles Im Grünen Bereich? Bonn, Federal Agency for Nature Conservation (BfN). Available at: www.bfn.de/fileadmin/MDB/documents/presse/2014/PK_Gruenlandpapier_30.06.2014_final_layout_barrierefrei.pdf (accessed December 2019).

BfN. (2017). Agriculture Report 2017: Biodiversity in Agricultural Landscapes. Bonn, Federal Agency for Nature Conservation (BfN). Available at: www.bfn.de/fileadmin/BfN/landwirtschaft/Dokumente/Agriculture_Report_2017.pdf (accessed December 2019).

BirdLife Europe and Central Asia. (2021). Burn or Restore: Meeting Competing Demands for Land in the Best Way for Nature, the Climate, and Human Needs. Available at: birdlife-europe-report-burn-restore-forests-land-use-bioenergy-may2021.pdf (accessed November 2024).

BirdLife Europe and Central Asia, and European Environmental Bureau. (2022). New CAP Unpacked and Unfit – Joint Report on CAP Strategic Plans. Available at: https://www.birdlife.org/wp-content/uploads/2023/06/CAP-Strategic-Plans-unpacked.and-unfit_Final.pdf(accessed November 2024).

BirdLife Europe and Central Asia, European Environmental Bureau and WWF. (2023). A Brighter Future for EU Food and Farming. Available at: https://www.birdlife.org/wp-content/uploads/2023/09/Joint-vision-for-EU-food-and-farming.pdf (accessed November 2024).

Buckwell, A. & Nadeu, E. (2018). What Is the Safe Operating Space for EU Livestock? Brussels, RISE Foundation. Available at: https://risefoundation.eu/wp-content/uploads/2020/07/2018_RISE_Livestock_Full.pdf (accessed December 2019).

Copa-Cogeca. (2013). Copa-Cogeca's Position on the EU's Biofuels Policy. Available at: https://www.euractiv.com/section/eet/news/biofuels-debate-continues-despite-eu-agreement/

Donald, P. F., Sanderson, F. J., Burfield, I. J., & van Bommel, F. P. J. (2006). Further Evidence of Continent-Wide Impacts of Agricultural Intensification on European Farmland Birds, 1990–2000. *Agriculture, Ecosystems & Environment*, *116*(3–4), 189–196. DOI: 10.1016/j.agee.2006.02.007

EEA, 2024, Air pollution in Europe: 2024 reporting status under the National Emission reduction Commitments Directive, Briefing No 07/2024

European Commission Joint Research Centre. (2021). Joint Research Centre Annual Report 2021. Available at: (accessed November 2024).

European Court of Auditors 16. (2021). Special Report. Common Agricultural Policy and Climate: Half of EU Climate Spending but Farm Emissions Are Not Decreasing. Available at: https://www.eca.europa.eu/en/publications/SR21_16(accessed November 2024).

European Court of Auditors 20. (2021). Special Report. Sustainable Water Use in Agriculture: CAP Funds More Likely to Promote Greater Rather than More Efficient Water Use. Available at: https://www.eca.europa.eu/en/publications/SR21_20(accessed November 2024).

European Environment Agency. (2015a). State of Nature in the EU: Results from Reporting Under the Nature Directives 2007–2012. Luxembourg, EUR-OP. Available at: https://bookshop.europa.eu/uri?target=EUB:NOTICE:THAK15002:EN (accessed March 2020).

European Environment Agency. (2015b). The European Environment: State and Outlook 2015: Synthesis Report. Luxembourg, Publications Office. Available at: https://bookshop.europa.eu/uri?target=EUB:NOTICE:TH0115001:EN:HTML (accessed March 2020).

European Environment Agency. (2018). *European Waters: Assessment of Status and Pressures 2018*. Luxembourg: European Environment Agency. DOI: 10.2800/303664

European Environment Agency. (2021). Conservation Status of Habitats under the EU Habitats Directive. https://www.eea.europa.eu/en/analysis/indicators/conservation-status-of-habitats-under

European Ornithologists Union, European Mammal Foundation, Societas Europaea Herpetologica, Societas Europaea Lepidopterologica, and Butterfly Conservation Europe. (2019). Reform the CAP: Harmful Agriculture Is Destroying Nature. 5 November. Available at: www.birdlife.org/europe-and-central-asia/open-letter-reform-cap (accessed March 2020).

European Scientific Advisory Board on Climate Change. (2024). Towards EU Climate Neutrality: Progress, Policy Gaps and Opportunities. Available at: https://climate-advisory-board.europa.eu/reports-and-publications/towards-climate-neutral-and-resilient-energy-networks-across-europe-advice-on-draft-scenarios-under-the-eu-regulation-on-trans-european-energy-networks(accessedNovember 2024).

Food and Agriculture Organization. (2013). *The State of Food and Agriculture 2013*. Rome, Italy: Food & Agriculture Organization of the United Nations (FAO).

Gawel, E., & Ludwig, G. (2011). The iLUC Dilemma: How to Deal with Indirect Land Use Changes When Governing Energy Crops? *Land Use Policy*, *28*(4), 846–856. DOI: 10.1016/j.landusepol.2011.03.003

Gebrekidan, S., Apuzzo, M., & Novak, B. (2019). The Money Farmers: How Oligarchs and Populists Milk the E.U. for Millions. *The New York Times*, 3 November, sec. World. Available at: www.nytimes.com/2019/11/03/world/europe/eu-farm-subsidy-hungary.html (accessed March 2020).

Geffroy, L. (2018). Où sont passés les oiseaux des champs? *CNRS Le journal*, March 20. Available at: https://lejournal.cnrs.fr/articles/ou-sont-passes-les-oiseaux-des-champs (accessed March 2020).

Geiger, F., Bengtsson, J., Berendse, F., et al. (2010). Persistent Negative Effects of Pesticides on Biodiversity and Biological Control Potential on European Farmland. *Basic and Applied Ecology*, *11*(2), 97–105. DOI: 10.1016/j.baae.2009.12.001

Green Alliance. (2024). A New Land Divided: The Opportunity of Alternative Proteins in Europe. Available at: https://green-alliance.org.uk/publication/a-new-land-dividend-the-opportunity-of-alternative-proteins-in-europe/(accessed December 2024).

Haberl, H., Erb, K.-H., & Krausmann, F. (2014). Human Appropriation of Net Primary Production: Patterns, Trends, and Planetary Boundaries. *Annual Review of Environment and Resources*, *39*(1), 363–391. DOI: 10.1146/annurev-environ-121912-094620

Hallmann, C. A., Sorg, M., Jongejans, E., et al. (2017). More than 75 Percent Decline over 27 Years in Total Flying Insect Biomass in Protected Areas. *PloS One*, *12*(10), e0185809. DOI: 10.1371/journal.pone.0185809

Hellmann, F., & Verburg, P. H. (2010). Impact Assessment of the European Biofuel Directive on Land Use and Biodiversity. *Journal of Environmental Management*, *91*(6), 1389–1396. DOI: 10.1016/j.jenvman.2010.02.022

IPBES. (2016). The Assessment Report of the Intergovernmental Science-Policy Platform on Biodiversity and Ecosystem Services on Pollinators, Pollination and Food Production. Bonn, Germany, Secretariat of the Intergovernmental Science-Policy Platform on Biodiversity and Ecosystem Services. Available at: https://naturalsciences.ch/uuid/d265532d-3189-57fb-9bf7-cd09e6d1670e?r=20190807115818_1565135779_e2ff4c34-10a9-5950-a77c-28d88118cb67 (accessed March 2020).

IPBES. (2019). Summary for Policymakers of the Global Assessment Report on Biodiversity and Ecosystem Services of the Intergovernmental Science-Policy Platform on Biodiversity and Ecosystem Services. Available at: www.ipbes.net/global-assessment-report-biodiversity-ecosystem-services (accessed March 2020).

Jepson, P., & Schepers, F. (2016). *Making Space for Rewilding: Creating an Enabling Policy Environment*. Oxford: University of Oxford. DOI: 10.13140/RG.2.1.1783.1287

JRC, 2021. The capitalisation of CAP subsidies into land rents and land values in the EU. https://publications.jrc.ec.europa.eu/repository/bitstream/JRC125220/jrc125220_land_capitalization.pdf

Matthews, A. (2015). Two Steps Forward, One Step Back: Coupled Payments in the CAP – CAP Reform, 16 April. Available at: https://capreform.eu/two-steps-forward-one-step-back-coupled-payments-in-the-cap/

Matthews, A. (2018). Brakes Removed from Voluntary Coupled Support – CAP Reform, 19 February. Available at: http://capreform.eu/brakes-removed-from-voluntary-coupled-support/ (accessed March 2020).

Moreira, F., Silva, J. P., Estanque, B., et al. (2012). Mosaic-Level Inference of the Impact of Land Cover Changes in Agricultural Landscapes on Biodiversity: A Case-Study with a Threatened Grassland Bird. *PloS One*, *7*(6), e38876. DOI: 10.1371/journal.pone.0038876

Newton, I. (2004). The Recent Declines of Farmland Bird Populations in Britain: An Appraisal of Causal Factors and Conservation Actions. *The Ibis*, *146*(4), 579–600. DOI: 10.1111/j.1474-919x.2004.00375.x

Oppermann, R., Beaufoy, G., & Jones, G. (eds.). (2012). High Nature Value Farming in Europe. Ubstadt-Weiher, Verlag regionalkultur. Available at: https://verlag-regionalkultur.de/media/pdf/a1/3c/b2/bib_657_3.pdf (accessed March 2020).

Panagos, P., Standardi, G., Borrelli, P., Lugato, E., Montanarella, L., & Bosello, F. (2018). Cost of Agricultural Productivity Loss Due to Soil Erosion in the European Union: From Direct Cost Evaluation Approaches to the Use of Macroeconomic Models. *Land Degradation and Development*, *29*(3), 471–484. DOI: 10.1002/ldr.2879

Pe'er, G., Zinngrebe, Y., Moreira, F., et al. (2019). A Greener Path for the EU Common Agricultural Policy. *Science*, *365*(6452), 449–451. DOI: 10.1126/science.aax3146

Pendrill, F., Persson, U. M., Godar, J., Kastner, T., Moran, D., Schmidt, S., & Wood, R. (2019). Agricultural and Forestry Trade Drives Large Share of Tropical Deforestation Emissions. *Global Environmental Change: Human and Policy Dimensions*, *56*, 1–10. DOI: 10.1016/j.gloenvcha.2019.03.002

Pereira, H. M., Hines, J., & Fernández, N. (2024). Conservation: Meeting Biodiversity Targets through Rewilding. *Current Biology*, *34*(20), R918–R921. DOI: 10.1016/j.cub.2024.08.052

Perpiña Castillo, C., Kavalov, B., Diogo, V., Jacobs-Crisioni, C., Batista e Silva, F., & Lavalle, C. (2018). *Agricultural Land Abandonment in the EU within 2015–2030*. European Commission, Ispra, 2018, JRC113718.

Pywell, R. F., Heard, M. S., Woodcock, B. A., Hinsley, S., Ridding, L., Nowakowski, M., & Bullock, J. M. (2015). Wildlife-friendly Farming Increases Crop Yield: Evidence for Ecological Intensification. *Proceedings. Biological Sciences*, *282*(1816), 20151740. DOI: 10.1098/rspb.2015.1740

Rigal, S., Dakos, V., Alonso, H., & Devictor, V. (2023). Farmland Practices Are Driving Bird Population Decline across Europe. *PNAS*, *121*(21), e2216573120.

Sadłowski, A., Beluhova-Uzunova, R., Popp, J., Atanasov, D., Ivanova, B., Shishkova, M., & Hristov, K. (2022). Direct Payments Distribution between Farmers in Selected New EU Member States. *Agris On-Line Papers in Economics and Informatics*, *14*(4), 97–107. DOI: 10.7160/aol.2022.140408

Searchinger, T., James, O., Dumas, P., Kastner, T., & Wisrenius, S. (2022). EU Climate Plan Sacrifices Carbon Storage and Biodiversity for Bioenergy. *Nature*, *612*, 27–30.

Steffen, W., Richardson, K., Rockström, J., et al. (2015). Planetary Boundaries: Guiding Human Development on a Changing Planet. *Science*, *347*(6223), 1259855. DOI: 10.1126/science.1259855

Sutton, M. A., Howard, C. M., Erisman, J.-W., et al. (eds.). (2011). *The European Nitrogen Assessment*. Cambridge: Cambridge University Press. DOI: 10.1017/cbo9780511976988

Taddei, 2020. Case C-723/17 *Craeynest*: New Developments for the Right to Clean Air in the EU - 24 Hours access. *Journal of Environmental Law*, Volume 32, Issue 1, March 2020, Pages 151–160, https://doi.org/10.1093/jel/eqaa001

Tollefson, J. (2019). Humans Are Driving One Million Species to Extinction. *Nature*, *569*(7755), 171. DOI: 10.1038/d41586-019-01448-4

Traba, J., & Morales, M. B. (2019). The Decline of Farmland Birds in Spain Is Strongly Associated to the Loss of Fallowland. *Scientific Reports*, *9*(1), 9473. DOI: 10.1038/s41598-019-45854-0

Tscharntke, T., Tylianakis, J. M., Wade, M. R., Wratten, S. D., Bengtsson, J., & Kleijn, D. (2007). Insect Conservation in Agricultural Landscapes. In Stewart, A. J. A., New, O. R., & Lewis, O. T. (eds.), *Insect Conservation Biology. The 23rd Symposium of the Royal Entomological Society* (pp. 383–404). Wallingford: CABI. DOI: 10.1079/9781845932541.0383

Vickery, J. A., Tallowin, J. R., Feber, R. E., Asteraki, E. J., Atkinson, P. W., Fuller, R. J., & Brown, V. K. (2001). The Management of Lowland Neutral Grasslands in Britain:

Effects of Agricultural Practices on Birds and their Food Resources. *The Journal of Applied Ecology*, *38*(3), 647–664. DOI: 10.1046/j.1365-2664.2001.00626.x

Watts, J. (2018). Stop Biodiversity Loss or We Could Face Our Own Extinction, Warns UN. *The Guardian*, 6 November, sec. Environment. Available at: www.theguardian.com/environment/2018/nov/03/stop-biodiversity-loss-or-we-could-face-our-own-extinction-warns-un (accessed March 2020).

WWF (World Wildlife Fund). (2018). *Living Planet Report*. Gland, Switzerland: World Wildlife Fund.

# 6

# THE EUROPEAN UNION'S INTERNATIONAL OCEAN GOVERNANCE FRAMEWORK, A PATH TOWARD A TURQUOISE DIPLOMACY

*Tiago Pires Da Cruz*

## Introduction

In 2016, the European Commission brought forward the sustainable development of deep-seabed mining as an important element of its ocean governance strategy (European Commission, 2016f). Six years later, the Commission instead chose to promote a science-based precautionary approach for deep-seabed mining. This chapter looks at the governance of seabed and ocean governance at the level of the European Union (EU) and investigates who gets to decide on this specific subfield of ocean policy. It more particularly analyzes how the EU went from the support of a "blue" growth paradigm – a will to better exploit marine spaces and resources – to an advocate of a "green" position through a precautionary approach (Singh et al., 2024). This shift from blue to green is actually nuanced, which leads this chapter to suggest a new in-between color to qualify the EU's strategy: turquoise.

Analytically, this chapter uses the analytical tools of critical geography to grasp discourses and the collective geographical understandings they crystallize. We mobilize John Hannigan's four major narratives to account for the spatialization of ocean space and mineral resource policies as they provide a framework for understanding the political dynamics of the seabed. These narratives produce and crystallize collective understandings of the deep ocean as either, or simultaneously, (1) a common heritage of mankind to be managed and governed collectively, (2) a subject of tensions linked to state interests in controlling resources, (3) a set of resources to be exploited or (4) a set of environments in need of protection (Hannigan, 2016).

DOI: 10.4324/9781003479260-10

At the EU level, such narratives are carried out by different institutions and actors who clash and change positions. To bring these processes to light, this chapter relies on process tracing and focuses on extractive seabed issues in EU-related institutions. Documents were completed by interviews with stakeholders and direct observation to better identify causal links and trends. This chapter starts with the presentation of the ocean environmental issue the EU is dealing with. It then focuses on how the EU has been developing policies on the subject, describing its "turquoise" stance. Finally, it analyzes what is still at stake and the main challenges to come.

## The environmental issues of the ocean and the challenge of its sustainable governance

Since the 1990s, considering environmental issues on a planetary scale has been gaining important traction at multiple policy levels (Compagnon, 2013). While the ocean sector had been structured since the 1970s as a distinct area of global governance, the coming into force of global environmental concerns has raised academic questions about the issue of its coherence in light of raising environmental concerns (Friedheim, 1993; Mendenhall, 2019; Tanaka, 2019).

In fact, since 2015, a momentum toward international ocean governance has been gaining formal traction at the international level. Diplomats and international organizations' personnel came to see ocean and climate challenges as part of the same nexus. In parallel, the issue rose at the United Nations (UN) level: the ocean was increasingly seen as a global common, and, therefore, its environmental problems came to be seen as global too (Jouffray et al., 2020). These efforts have led to the emergence of a discourse to promote a dual need to both *save* and *govern* the Ocean.

In parallel, in the 2010s, the UN hosted ocean negotiations on a new international legally binding instrument. Developed under the framework of the 1982 UN Convention on the Law of Sea, it aimed at the conservation and sustainable use of marine biological diversity, focusing mostly on areas beyond national jurisdiction. Adopted in June 2023, the treaty, known as the "Biodiversity Beyond National Jurisdiction" (BBNJ) Treaty, would create, once ratified, a new set of rules and instruments for the environmental governance of the high seas, a space formerly unregulated regarding scientific and economic activities. To position itself in these negotiations, the EU developed its own international ocean governance approach in the second half of the 2010s. This was an important step in the negotiations, since the EU Member States account for an important percentage of the total marine spaces at the global level, as displayed in Figure 6.1.

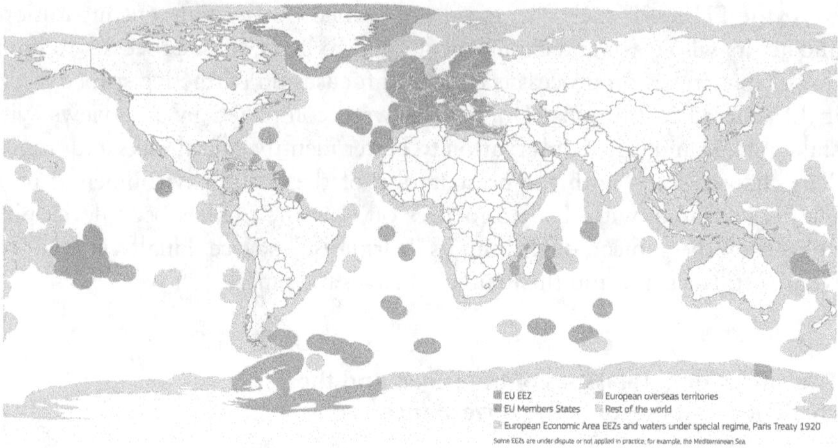


**FIGURE 6.1** EU's Exclusive Economic Zone and overseas territories (European Commission 2016b)

## The turquoise reality of the EU's international ocean governance approach

The sustainability discourse, which posits the possibility that marine activities' development and coexistence be possible through time and compatible with good ecological conditions, has been developed as a keystone of EU marine environmental governance. However, its practical developments and effects must be systematically assessed to better understand how these promises fare. In this sense, the intrinsically blurred coexistence between a "blue" economic growth and a "green" environmental protection that this discourse has often been associated with can be described as "turquoise". The EU's international ocean governance framework focuses on the high seas beyond the direct authority of its Member States, mainly under the work of the Commission's Directorate-General for Maritime Affairs and Fisheries (DG MARE). The framework must be understood in light of the Commission's efforts to develop a marine governance framework directed at its states' jurisdictions.

### *The development of an EU international ocean governance framework*

In 2015, European Commission President mandated its Commissioner for the Environment, Maritime Affairs and Fisheries, to participate in the development of an international ocean governance strategy. This demand was linked to the official beginning of the start of the BBNJ's negotiations (European Commission, 2016c) and DG MARE was given responsibility for a consultation on international ocean governance (European Commission, 2016a).

To justify its involvement on the issue, the Commission first stated its commitment to developing good governance practices to foster common goals such as economic development and sustainability and stated that it could claim to have authority over the largest combined Exclusive Economic Zone in the world. To develop its first coherent policy document on the issue, DG MARE started in 2016 by gathering the opinions of private and public actors, including intergovernmental organizations, public authorities (26%), concerned citizens (19%), NGOs (17%) and businesses (17%). The results summarized in one document stated that most actors considered the contemporary ocean governance framework to be ineffective in managing the oceans sustainably, due to increasing levels of pollution, overexploitation, climate change and acidification. According to the document, the most important factor was ineffective implementation and insufficient coordination, as well as a lack of transparency in decision-making processes. As a conclusion, DG MARE unequivocally presented its public consultation results to be reflective of a general demand for an international legal instrument for marine biodiversity outside national jurisdictions. In parallel, the European Commissioner launched an international "listening tour" to set out the Commission's framework and gather additional stakeholders' views (European Commission, 2016c).

As a result, the European Commission released its first international ocean governance agenda in 2016. The document identified the architecture of ocean governance as being the major issue, because of its lack of coordination. The Commission, through DG MARE, announced it would be working toward the adoption, ratification and implementation of what would become the BBNJ Agreement with its Member States and international partners (European Commission, 2016b). In a joint press release, the European Commission and the High Representative for Foreign Affairs announced their alignment on the issue. The basis for a "turquoise" approach was set, as the oceans' spaces and resources were both deemed to offer considerable economic potential, while the improvement of the governance framework would enable to pursue the fight against climate change, poverty and food security (European Commission, 2016d). From a narrative perspective, the 2016 Communication articulates both the ideas of "saving" and "governing" the Ocean, while also adding the will of "harvesting" its resources. This approach was then rolled out among the main global environmental institutions where the Commission tried to reinforce its agenda: the BBNJ Agreement, but also the UN's Sustainable Development Goals, the Convention on International Trade in Endangered Species of Wild Fauna and Flora or the Convention on Biological Diversity (European Commission, 2019).

While not a formal regulatory development, the 2016 Communication led to the activation of other EU institutions. The issue was, for instance, worked upon in 2017 by the European Parliament, which framed it as an environmental issue (European Parliament, 2018a). Most interestingly, the Recitals of the final 2018 Resolution, concluding a series of declarations, mention the importance of recognizing the transboundary nature of the ocean. Its interconnected ecological nature was viewed to impose a need for collaboration between governments to ensure the sustainability of its shared resources. In a similar narrative as the 2016 Communication, it recognized the importance of investing in a sustainable blue economy, while also conserving marine resources. The Parliament called for the development of a marine economy that would free Europe from dependence on finite resources and lean toward sustainability (European Parliament, 2018b).

### *The Blue reality of the EU's policies*

Between 2012 and up until its revision in 2021, the "blue growth" paradigm has been systematically brought forward by the EU Commission, through DG MARE, as the mantra for ocean policy. Before that, EU institutions had already been investing time and resources in that sense, focusing mainly on fisheries developments since the 1970s (da Conceiçao-Heldt, 2004). This "blue" side of EU policies is also deployed at the international level, as the EU is a member of the 1982 UN Convention on the Law of the Sea and developed efforts to become an important player in fisheries diplomacy. These included diplomatic presence in fora beyond its own waters and in the high seas, as well as in regional seas conventions on pollution (Singh et al., 2024).

Since the Treaty of Lisbon of 2007, the EU has "exclusive competence" in the area of "the conservation of marine biological resources under the common fisheries policy" (European Commission, 2007). This power onto fisheries led, as early as 2008, to a Marine Strategy Framework Directive which instigated the Member States duty to cooperate at the international level and "use existing regional institutional cooperation structures" (European Commission, 2008b). According to legal academic Annie Cudennec, European multilateral marine policy should be seen as operating on two levels: on the one hand, it works "in order to contribute to the definition of the major international principles of sustainable ocean management". On the other hand, it strives to see them "implemented and specified at regional level" (Translated – Cudennec, 2019, p. 257). The EU therefore appeared in the 2010s as an important actor of ocean governance, asserting its competences within Regional Fisheries Management Organizations all over the globe (European Commission, 2024).

The EU stance on marine policies therefore leans toward a willingness to "harvest" marine resources for economic development purposes, preferably in a sustainable manner. After the 2007 Economic crisis, as economic growth and employment opportunities were being sought after actively, marine spaces and resources appeared as a natural avenue for such a search. As a result, the European Commission published the 2012

Communication: "Blue Growth – Opportunities for marine and maritime sustainable growth" (European Commission, 2012). It was the first step in an exercise developed around four major "blue growth focus areas": blue energy; aquaculture; tourism and marine mineral resources. While fisheries were in crisis because of reduced fish populations, these appeared as new frontiers for the EU's economy. While the development of these activities was paramount, it was nonetheless accompanied by a sustainability discourse, in line with the international legal framework.

### *The greening of the EU Governance framework*

The inclusion of sustainability principles in growth-oriented marine policies has been described in the field of social sciences has a process of "greening the blue". It consists in producing "green" sustainable horizon discourses that allow "blue" marine activities to be presented as a better and inevitable future (Childs, 2019). At EU level, the developments of the European Common Fisheries Policy are, according to legal expert Annie Cudennec, an example of the changing approach of European institutions toward marine and environmental issues. She describes a shift from a productivist paradigm to one of transition. Since 1983, with the adoption of the first regulations for the conservation of marine living resources, the policy has gradually incorporated an ecological transition approach (Cudennec & Lazuech, 2018). These dynamics have also been observed within the EU's marine diplomacy (Cudennec & Lazuech, 2018).

Outside of fisheries the "green" and the "blue" have cohabited too. In 1992, the Habitats Directive led legal experts to consider that the EU has since asserted itself as a "player in the sustainable management of the oceans" (Translated – Cudennec, 2019, p. 262). This Directive provided a basis for area-based management via Marine Protected Areas. It was supplemented by the 2008 Directive aiming "to achieve or maintain good environmental status in the marine environment by the year 2020 at the latest" (European Commission, 2008b). In a sense, while the "blue" remained the keystone, it was backed by a sustainable "green" narrative aiming at increased environmental coherence. An example of that is the 2014 Maritime spatial planning Directive to "organize human activities in marine areas to achieve ecological, economic and social objectives" (European Commission, 2014). Systematically, the way in which activities are governed is identified as the issue needing political efforts, rather than the activities themselves.

At this stage, "blue" economic development therefore was the priority, while "green" policies were viewed as essential, but subsidiary with respect to the former. However, after the initial 2016 international ocean governance communication, in the aftermath of the European Green Deal and the Biodiversity strategies, the original blue growth Communication was identified by the Commission, in 2021, as in need of revision toward a "green recovery for the blue economy". From the outset, this led the Commission to assume its commitment to strengthen its language and promote "a sustainable use of aquatic and marine resources while contributing to the recovery of the EU economy" (European Commission, 2021a). The Commission defined a number of economic and environmental development objectives, rather than, as previously, defining precise actions on a limited number of high-potential sectors to be developed (Transforming the EU's Blue Economy, 2021). While the aforementioned prioritization of economic development was still present, the importance of the sustainability discourse was nevertheless on the rise.

Following these developments, in 2021, the European Parliament announced its will to produce a Resolution, which could ultimately be taken up by the European Commission as a binding text, replacing the idea of blue growth with a more nuanced blue economy in line with sustainability principles (Carvalhais, 2022). In June 2022, the European Commission applied this rationale to its communication on international ocean governance entitled "setting the course for a sustainable blue planet", just prior to the second UN Ocean Conference in Lisbon, Portugal. Among various objectives, it promoted the negotiations toward the Agreement and the future Convention on Biological Diversity negotiations as well as a new large-scale MPAs in the Southern Ocean and further measures toward sustainable fisheries and aquaculture, the fight against climate change and marine pollution (DG MARE, 2022).In conclusion, the greening of the domestic EU marine governance framework was reflected in its global diplomacy. However, while sustainability principles and instruments have been steadily promoted, growth and development questions still appeared as paramount. The "turquoise" nature of the EU marine governance describes this in-between state, still leaning toward the blue.

## The turquoise nature of the EU's deep-sea mining engagements

Since 2016, the sector where the Commission's international ocean governance strategy has been knowing the greatest shift is deep-sea mining. Understanding the changes related to this shift can help illustrate the limits of the turquoise approach described in the previous sections.

### *The blue growth rationale for deep-seabed mining*

Deep-seabed mining is an activity which consists in extracting minerals from or resting on the seabed at depths below 200 meters. Although it had not started prior to both the Commission's 2016 and 2022 international ocean governance strategies, mineral deposits have been found in EU Member States' marine jurisdictions. Commercial interests focus on Pacific and Polynesian spaces, as well as in areas beyond national jurisdictions, as shown in Figure 6.2.

In the high seas, the extraction of seabed minerals is regulated by the International Seabed Authority (ISA), which entered into force in 1994,

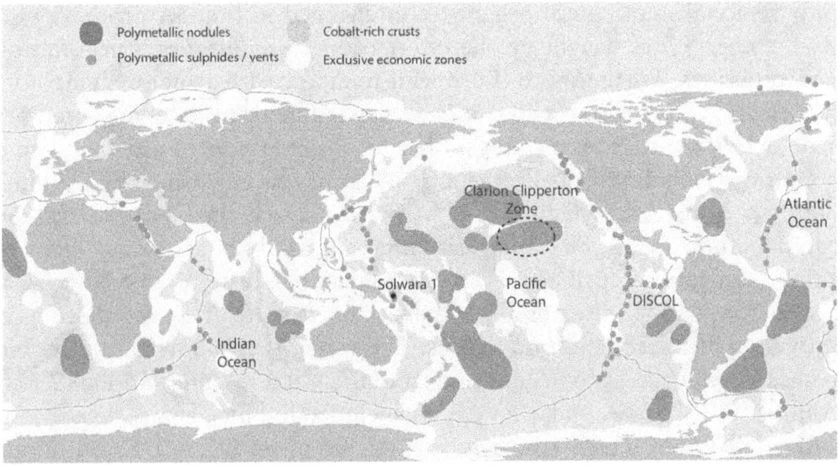


**FIGURE 6.2** Major occurrences of deep-sea metal-bearing minerals within a plate tectonic context, Exclusive Economic Zones in white

*Source:* www.frontiersin.org/journals/marine-science/articles/10.3389/fmars.2017.00418/full

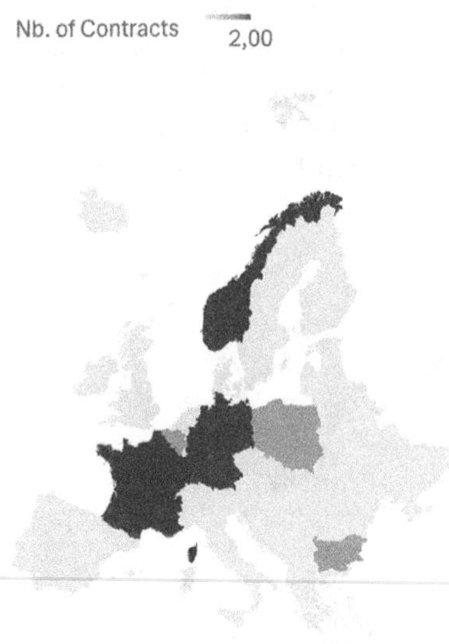


**FIGURE 6.3** Number of ISA exploration contracts related to European countries (excluding Russia) (2024.04). Designed by the author

pursuant to the 1982 UN Convention on the Law of the Sea. The EU acceded to the Convention as full member in 1998; it is therefore also a member of the ISA to the same extent as other State parties, including its members (Singh et al., 2024). At the ISA, European countries play an important role, possessing several exploration contracts: seven out of the 30 existing in 2024, as shown in Figure 6.3. The EU appears as the only regional multilateral organization engaged in the ISA process, while its Member States weight on the international organization's negotiations and processes. Furthermore, European mining and marine dredging companies and contractors (who can be either public or private) are often present at ISA negotiation meetings, offering their services to a wide range of countries, including outside the EU. Their status regarding EU Member States is displayed in Figure 6.4. At the ISA level, their political activity can take three forms. First, they can intervene as observers, where membership organizations such as the French Maritime Cluster or the Norwegian Forum for Marine Minerals allow them to participate in the negotiations by providing them accreditation badges. This has notably allowed service companies to gain access to negotiations and their actors (International Seabed Authority, 2024). Second, they can be included in official State delegations, such as Belgium which regularly invites one of its Belgium-based mining company representatives to join the delegation. Finally, mining companies can be contractors to the ISA, through State sponsorship. It allows them to obtain exploration contracts and, *in fine,* might open the possibility to acquire exploitation rights to areas beyond national jurisdictions (ISA Delegations, 2023). This is the case of two of the most prominent companies of the 2010s at the ISA level: "The Metals Company" from Canada and "DEME-GSR" from Belgium (Casson, 2020). Finally, the importance of the EU at the ISA, in 2024, is also linked to five of the 36 members of the ISA's (executive) Council being EU Member States.

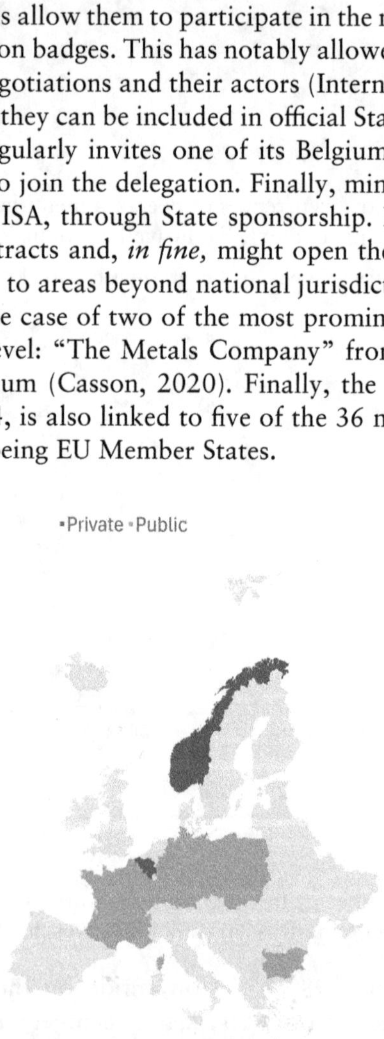


**FIGURE 6.4** Status of the European companies holding ISA exploration contracts (excluding Russia) (April 2024). Designed by the author

Even before the adoption of an international ocean governance strategy by the EU in 2016, the Union already had leverage and recognition in deep-sea mining governance, although it lacked the mandate to speak on behalf of its Member States. The first EU-level efforts to position itself on the issue of deep-sea mining did not come from DG MARE. In fact, in the late 2000s, efforts to securitize mineral resources at the EU level rose. As China was beginning to make political use of its position in the global mineral value chain, Western States, including the United States, became fearful of their supply chains. The question of the security of supply of non-energy and non-agricultural raw materials became a major concern, according to a DG GROW (Directorate-General for Internal Market, Industry, Entrepreneurship and Small and Medium Enterprises) Director (Cozigou, 2016). Seabed environments came to be included in these reflections as a mining frontier to "harvest" to avoid further dependence on third countries, as would China, Russia, India or the United States (Frozel Barros, 2021; Kang & Liu, 2021).

As a result, in 2008, the European Commission introduced its new "raw materials initiative". Among the proposed developments, the need to develop scientific and technical knowledge to enable the "sustainable exploitation of the seabed" was brought forward (European Commission, 2008, p. 11). In following communications, the European Commission identified marine spaces and resources as a future sector to decouple the "economic growth from the use of resources" (Europe, 2010, 2020). The efforts of DG GROW first concluded in 2011, when a list of priority critical raw materials to identify and monitor was published. A few months later, a roadmap was issued, clearly stating the need to position on the issues of "marine resources". As the effects of the financial and economic crisis were still being felt, supply and employment opportunities were paramount. Seabed extraction, therefore, appeared in most documents as an issue to be carried forward, though technicalities remained to be solved (European Commission, 2011). But, in 2011, the issue was not under the remits of DG GROW anymore and had already been taken up by DG MARE.

Shifting DG had no impact on the willingness to open this new extractive blue frontier. The work of DG MARE consisted, first, in contracting a consultancy in 2010. Its final report, in 2012, focused on the issue of "future potential", consisting in "innovation, competitiveness, employment, policy relevance, impact/synergies and sustainability" (Ecorys et al., 2012, p. 42). The consortium proposed a timetable, based on interviews with private actors that had interests in mining activities. They evaluated that the sector would start thriving in the next ten years and that EU's help was needed in the meantime. To develop the sector, the Commission announced a stakeholder consultation plan with a "blue" orientation framing, aiming at "assessing how European industry can become competitive in extracting minerals from the seafloor and how best to ensure that this activity does not prevent future generations from benefiting from hitherto untouched ecosystems" (European Commission, 2012). As a result, in 2012, the EU defined the blue economy and included seabed mining as a sector of economic development: marine mineral resources were associated with vast promises: "global annual turnover of marine mineral mining can be expected to grow from virtually nothing to €5 billion in the next ten years and up to €10 billion by 2030" (European Commission, 2012).

This policy gave momentum to finance the production and dissemination of scientific and technical knowledge through EU institutions. Rather than investing in potentially controversial political or diplomatic avenues, the Commission financed research programs (at around €350 million annually, in 2014). These studies were conducted in specific mining-oriented frameworks, as they "are intended to guide national authorities, the EU and the [ISA] in making future decisions [...], all aimed at the development of activities aimed at the economic growth of the European Union" (Trocan, 2023, p. 24). These efforts appeared as a first EU step to develop a network of like-minded actors and lower the entry cost for new actors.

However, the EU Member States remained divided regarding the exact role which should be attributed to the EU on this issue. In 2021 pro-mining countries in the short term, such as Poland, opposed others who would tend toward a lower-paced approach. The array of Member States positions fluctuated over time, appearing linked to factors such as, but not limited to, prior investments in the development of science and technology, the presence of important country-based industrial actors that allow to hope for economic spinoffs, tensions with other States, advocacy efforts from the ISA Secretariat or other international bodies, the mobilization of the civil society against these developments ... Finally, the Commission's efforts to assert its role to act on behalf of its Members at the ISA fell short (European Commission, 2021b). This suggests that views were still too polarized and, above all, that States were willing to keep on carrying their own voices at the ISA (Singh et al., 2024, p. 310).

### *Deep-seabed mining, a paramount turquoise example of the EU international ocean governance strategy*

On seabed mining, environmental issues appeared as a second thought, to which scientific progress would find answers to enable a "coherent and planned management" (European Commission, 2011). This attitude was a common feature of DG GROW and DG MARE. Furthermore, since the 1960s, scientific efforts have always been mostly linked to commercial mining interests (Hannigan, 2016; Childs, 2020; Le Meur & Muni Toke, 2022; Martínez-Rius, 2022) and military purposes (Oreskes, 2021). Because oceanographic campaigns are extremely expensive, seabed policies have always been conceived and developed in a scarcity of knowledge. Therefore, the first NGOs opposing seabed extractive activities were conscious of this reality and saw the development of environmental sciences, as mainly financed by, and directed toward, mining interests (Interview with an NGO representative, June 2023). Although a few campaigns were designed specifically toward environmental goals, they were often complementary to prospective and regulatory mining aims (Pires da Cruz, 2023).

Yet, as presented earlier, EU institutions have been progressively committing to a sustainable international (and internal) ocean governance agenda. Since 2015, the Commission's stance toward mining has been shifting and reframed as an international governance topic. As the "blue growth" framing became less mobilized, the idea of a sustainable governance of the high seas grew. Seabed institutions did not escape from this new framing at DG MARE, although the DG remained aligned with the idea of making seabed mining sustainable in the short term (European Commission, 2016). The goal was to assert itself as having a crucial role to play in improving ocean governance, including in the field of deep-seabed mining. It did so by developing diplomatic activities in international fora, by sharing its expertise, by financing scientific research and cooperating with non-member partners (European Commission, 2016). A key player asking for more sustainability has been NGOs (see Box 6.1: Case study).

### BOX 6.1 CASE STUDY: THE ROLE OF EUROPEAN NGOS IN THE CAMPAIGN AGAINST THE BLUE GROWTH PARADIGM

Reacting to the 2012 EU growth-oriented narrative, a coalition of 22 NGOs came together in October. Led by organizations such as Seas At Risk, The Mediterranean Information Office for Environment, Culture and Sustainable Development and the Clean Baltic Coalition they produced a joint position paper on European marine policies. The message was clear from its title, "The limits of growth". It multiplied references to the precautionary principle, an approach recognized by EU institutions. The message conveyed was focused on recognizing that most marine ecosystems were already overexploited. Therefore, there was a need to focus on developing sustainable economic activities according to the needs of the EU and its populations (Seas At Risk et al., 2012).

This united call opposed the concept of "blue growth", a concept that was first endorsed and developed by the European Commission and developed as the fisheries crisis grew. This led the European Commission's actors to dismiss the fisheries sector as a promising area of the blue growth strategy and seek new avenues for economic growth. Facing these developments, a discourse questioning the blue growth strategy as promoted by European institutions emerged in the academic field, stressing that the strategy was likely to "reinforce the growth paradigm without the appreciation that there are natural limits to growth" (Hadjimichael, 2018, p. 163). In 2020, a new coalition was built around a "Blue Manifesto for 2030", where the blue growth agenda would, again, be criticized (Seas At Risk, 2021).

Deep-sea mining would not be forgotten in these processes. NGOs such as Seas At Risk, at the European level, or the Deep Sea Conservation Coalition, at the global level, invested increasing time and resources to specifically address the seabed issues. Through time, they managed to create a cohesive coalition around the idea of a deep-sea mining "moratorium", the idea of bringing forth a precautionary approach. As NGOs managed to get access to European parliamentarians, the seabed mining issue was brought on the agenda in the wider international ocean governance process. At that time, the issue did not appear as particularly contentious, as most debates focused on other questions, such as the role of oil and gas. The final 2018 Resolution encouraged States and the Commission to stop subsidizing mining licenses in areas beyond national jurisdiction and to defend the importance of transparency and stakeholder participation in ocean governance, including at the ISA. As a result, since 2018, the European Parliament took up the "moratorium" concept, dear to the NGOs:

> 40. Urges the Commission to call on Member States to stop sponsoring deep-sea mining exploration and exploitation licenses in Areas Beyond National Jurisdiction and on and not to issue permits for deep-sea mining on Member States' continental shelf;
>
> 42. Calls on the Commission and the Member States to support an international moratorium on commercial deep-sea mining exploitation licences until such time as the effects of deep-sea mining on the marine environment, biodiversity and human activities at sea have been studied and researched sufficiently and all possible risks are understood.

Sources: European Parliament (2018a); Seas At Risk (2021).

In parallel with its internal policy efforts, the Commission's work and the ongoing BBNJ Agreement negotiations contributed to bringing ocean issues under the spotlight. Following the adoption of the European Green Deal, the European Commission worked on an "EU Biodiversity Strategy for 2030". It included the following main objectives: "The negative impacts on sensitive species and habitats, including on the seabed through fishing and extraction activities, are substantially reduced to achieve good environmental status". The Strategy also identified external means of "promoting the EU's ambition", including the issue of international ocean governance (European

Commission, 2020). When mobilized again on deep-seabed mining, the Parliament reiterated its position. In its final Resolution, the Parliament proposed a second version of its Resolution on international ocean governance, which also brought forward the concept of "moratorium" (European Parliament, 2022).

In this context, the Commission developed its second communication on international ocean governance. To justify a new position on deep-seabed mining that would fit with the promotion of sustainable governance of marine spaces and resources in the high seas, it based its argumentation on an explicit bibliography. Two articles on the need for scientific actors and processes to obtain more time before the start of industrial-scale activities were mentioned (Amon et al., 2022; Miller et al., 2018). It was completed by an article from the World Economic Forum on the place that marine mineral resources should have in supply chains in the building of responsible materials sourcing (World Economic Forum, 2022). This allowed justifying its new precautionary approach as based on a scientific and political consensus, quoting Resolution 122 of the September 2021 International Union for the Conservation of Nature World Congress which pleaded for a "moratorium". This approach relies on the idea that knowledge is still insufficient for the commercial development of extractive activities (European Commission, 2022).

The EU's Commission renewed its commitment to this precautionary approach after the 2024 European Parliament elections (DG GROW, 2024). However, concerns remain regarding the growing pressures exerted by the demand for raw materials linked to the energy transition and the fears associated with a shortage of minerals (Deberdt & Billon, 2023; Draghi, 2024, p. 54). The controversy among EU Members States might still continue, despite EU sustainability-driven policies:

> while there seems to be a broad desire within the EU to secure marine minerals as a resource for future exploitation activities, an overarching conflict appears to be looming between the exploitation interests of certain [Members States] and the wider interest of the EU in ensuring the protection of the marine environment.
>
> *(Singh et al., 2024, p. 291)*

Furthermore, these EU precautionary-oriented developments remain mostly declarative in nature. These ratchet decision-making mechanisms oriented toward the conservation and precautionary approach are reversible. An example is the process led by DG GROW toward developing a critical raw materials Regulation to face the ever-increasing minerals demand linked to the so-called "energy transition" (Breton, 2022). The Commission's 2023 text opened the possibility for deep-sea mining by defining "extraction" as: "the primary extraction of ores, minerals and plant products from their original source, including from a mineral occurrence underground, mineral occurrence under water, sea brine and trees" (European Commission, 2023). In this process, NGOs had less favorable access to EU and diplomatic actors and only secured last-minute improvements with mixed results (Interview with an NGO representative, April 2024).

Such developments keep NGO representatives alert regarding the possibility to obtain new gains through the EU. Many therefore fight to secure the 2022 Commission and Parliament precautionary declarations, and use them as a political resource in other international fora, such as the ISA, regional sea conventions or even when advocating against Norwegian marine mining projects (Norway's DSM Position, 2024).

## Conclusion

What does European environmental governance mean in the case of the new policy subfield of international ocean governance? Looking at the wider temporal policy processes allows us to understand that the question was taken amidst of wider political tensions. As a result, it cannot be reduced to a mere calculation of pros and cons leading to a coherent diplomatic position but, rather, follows a process of EU interests' formation on the issue and beyond. This helps explain how EU institutions have been promoting a greener narrative to their initial blue growth approach.

However, the agenda, carried out by European institutions, to make "sovereignty games", "harvesting competitions", "common governance efforts" and "Saving the ocean narratives" compatible does not always work. For instance, this "turquoise" approach of the governance of marine spaces and resources, for instance, proved to be a circle-squaring process in the case of deep-sea mining, consistently favoring economic priorities, until a precautionary approach was adopted against the backdrop of a new Agreement on the sustainable development of human activities in the high seas where the EU strived to appear as an ocean-leader.

## Bibliography

Amon, D. J., Gollner, S., Morato, T., Smith, C. R., Chen, C., Christiansen, S., Currie, B., Drazen, J. C., Fukushima, T., Gianni, M., Gjerde, K. M., Gooday, A. J., Grillo, G. G., Haeckel, M., Joyini, T., Ju, S.-J., Levin, L. A., Metaxas, A., Mianowicz, K., … Pickens, C. (2022). Assessment of Scientific Gaps Related to the Effective Environmental Management of Deep-Seabed Mining. *Marine Policy*, *138*(105006), 1–22. https://doi.org/10.1016/j.marpol.2022.105006

Breton, T. (2022, Septembre 14). *Critical Raw Materials Act: Securing the New Gas & Oil at the Heart of Our Economy—Blog of Commissioner Thierry Breton*. European Commission. https://ec.europa.eu/commission/presscorner/detail/en/STATEMENT_22_5523

Carvalhais, I. (2022, avril 5). *Report on Toward a Sustainable Blue Economy in the EU: The Role of the Fisheries and Aquaculture Sectors – A9–0089/2022*. European Parliament. https://www.europarl.europa.eu/doceo/document/A-9-2022-0089_EN.html

Casson, L. (2020, décembre 9). *Deep Trouble—The Murky World of the Deep Sea Mining Industry*. J. Alexander, K. Miller, D. Santillo, D. Currie, C. Söderström, G. Duncan-Jones, & S. Losada (Eds.), Greenpeace International. https://www.greenpeace.org/static/planet4-international-stateless/c86ff110-pto-deep-trouble-report-final-1.pdf

Childs, J. (2019). Greening the Blue? Corporate Strategies for Legitimising Deep Sea Mining. *Political Geography*, *74*, 102060. https://doi.org/10.1016/j.polgeo.2019.102060

Childs, J. (2020). Extraction in Four Dimensions: Time, Space and the Emerging Geo(-)politics of Deep-Sea Mining. *Geopolitics*, *25*(1), 189–213. https://doi.org/10.1080/14650045.2018.1465041

Communication from the Commission – Europe 2020: A Strategy for Smart, Sustainable and Inclusive Growth, COM(2010) 2020 final (2010). https://eur-lex.europa.eu/LexUriServ/LexUriServ.do?uri=COM:2010:2020:FIN:EN:PDF

Communication from the Commission to the European Parliament, the Council, the European Economic and Social Committee and the Committee of the Regions - Roadmap to a Resource Efficient Europe, COM(2011) 571 final (2011). https://eur-lex.europa.eu/legal-content/EN/TXT/?uri=CELEX%3A52011DC0571

Communication from the Commission to the European Parliament, the Council, the European Economic and Social Committee and the Committee of the regions – Blue Growth: Opportunities for Marine and Maritime Sustainable Growth (Text with EEA relevance), COM(2012) 494 final (2012). https://eur-lex.europa.eu/LexUriServ/LexUriServ.do?uri=COM:2012:0494:FIN:EN:PDF

Communication from the Commission to the European Parliament and the Council – The Raw Materials Initiative, Meeting Our Critical Needs for Growth and Jobs in Europe {SEC(2008) 2741}, COM(2008) 699 final (2008). https://eur-lex.europa.eu/legal-content/EN/TXT/?uri=celex%3A52008DC0699

Communication from the Commission to the European Parliament, the Council, the European Economic and Social Committee and the Committee of the Regions – EU Biodiversity Strategy for 2030 Bringing Nature Back into Our Lives, COM(2020) 380 final (2020). https://eur-lex.europa.eu/legal-content/EN/TXT/?uri=celex%3A52020DC0380

Communication from the Commission to the European Parliament, the Council, the European Economic and Social Committee and the Committee of the Regions – Roadmap to a Resource Efficient Europe, COM(2011) 571 final (2011). https://eur-lex.europa.eu/legal-content/EN/TXT/?uri=CELEX%3A52011DC0571

Communication from the Commission to the European Parliament, the Council, the European Economic and Social Committee and the Committee of the Regions on a New Approach for a Sustainable Blue Economy in the EU Transforming the EU's Blue Economy for a Sustainable Future, COM/2021/240 final (2021). https://eur-lex.europa.eu/legal-content/EN/TXT/?uri=celex%3A52021DC0240

Compagnon, D. (2013). Chapitre 38. L'environnement dans les RI. In *Traité de relations internationales* (pp. 1019–1052). Presses de Sciences Po; Cairn.info. https://doi.org/10.3917/scpo.balza.2013.01.1019

Consolidated Versions of the Treaty on European Union and the Treaty on the Functioning of the European Union Consolidated Version of the Treaty on European Union Consolidated Version of the Treaty on the Functioning of the European Union Protocols Annexes to the Treaty on the Functioning of the European Union Declarations Annexed to the Final Act of the Intergovernmental Conference Which Adopted the Treaty of Lisbon, Signed on 13 December 2007 Tables of Equivalences (2007). https://eur-lex.europa.eu/legal-content/EN/TXT/?uri=celex%3A12016ME%2FTXT

Cozigou, G. (2016). La stratégie européenne des matières premières. *Annales des Mines – Responsabilité et environnement*, *82*(2), 24–28. https://doi.org/10.3917/re1.082.0024

Cudennec, A. (2019). L'Union européenne, acteur de la gestion durable des océans. *Revue juridique de l'environnement*, *44*(2), 255–274.

Cudennec, A., & Lazuech, G. (2018). Politique Commune de la Pêche. In A. Van Lang, F. C. Dutilleul, & V. Pironon (Éds.), *Dictionnaire juridique des transitions* écologiques (Vol. 18, pp. 619–622). Institut Universitaire Varenne.

da Conceiçao-Heldt, E. (2004). *The Common Fisheries Policy in the European Union: A Study in Integrative and Distributive Bargaining*. Routledge.

Deberdt, R., & Billon, P. L. (2023). Green Transition Mineral Supply Risks: Comparing Artisanal and Deep-Sea Cobalt Mining in a Time of Climate Crisis. *The Extractive Industries and Society*, *14*, 1–10. https://doi.org/10.1016/j.exis.2023.101232

DG GROW. (2024, octobre 9). *NGOs' Concerns Regarding References to Deep Sea Mining in Mario Draghi's Report : The Future of European Competitiveness*. European Commission. https://ejfoundation.org/resources/downloads/Reply_Concerns-regarding-references-to-deep-sea-mining_FINAL_2024-10-10-154645_kyqh.pdf

DG MARE. (2022, juin 24). *Setting the Course for a Sustainable Blue Planet—Joint Communication on the EU's International Ocean Governance Agenda*. European Commission. https://oceans-and-fisheries.ec.europa.eu/publications/setting-course-sustainable-blue-planet-joint-communication-eus-international-ocean-governance-agenda_en

Directive 2008/56/EC of the European Parliament and of the Council of 17 June 2008 Establishing a Framework for Community Action in the Field of Marine Environmental Policy (Marine Strategy Framework Directive) (Text with EEA Relevance), 2008/56/EC, EP, CONSIL, 164 OJ L (2008). https://data.europa.eu/eli/dir/2008/56/oj/eng

Directive 2014/89/EU of the European Parliament and of the Council of 23 July 2014 Establishing a Framework for Maritime Spatial Planning, 2014/89/EU, EP, CONSIL, 257 OJ L (2014). https://data.europa.eu/eli/dir/2014/89/oj/eng

Draghi, M. (2024). *The Future of European Competitiveness Part A | A Competitiveness Strategy for Europe* (A; Draghi Reports, p. 67). European Commission. https://commission.europa.eu/document/download/97e481fd-2dc3-412d-be4c-f152a8232961_en?filename=The%20future%20of%20European%20competitiveness%20_%20A%20competitiveness%20strategy%20for%20Europe.pdf

Ecorys, D., & Océanic Développement. (2012, August 13). *Blue Growth – Scenarios and drivers for Sustainable Growth from the Oceans, Seas and Coasts—Final Report – Call for Tenders No.* MARE/2010/01*- Client: European Commission, DG MARE – Rotterdam/Brussels, 13th August 2012*. Ecorys.

https://maritime-spatial-planning.ec.europa.eu/sites/default/files/blue_growth_final_report_130920122.pdf

European Commission. (2007). Consolidated Versions of the Treaty on European Union and the Treaty on the Functioning of the European Union Consolidated Version of the Treaty on European Union Consolidated Version of the Treaty on the Functioning of the European Union Protocols Annexes to the Treaty on the Functioning of the European Union Declarations Annexed to the Final Act of the Intergovernmental Conference Which Adopted the Treaty of Lisbon, Signed on 13 December 2007 Tables of Equivalences (2007). https://eur-lex.europa.eu/legal-content/EN/TXT/?uri=celex%3A12016ME%2FTXT

European Commission. (2008a). Communication from the Commission to the European Parliament and the Council - The raw materials initiative, meeting our critical needs for growth and jobs in Europe {SEC(2008) 2741}, COM(2008) 699 final (2008). https://eur-lex.europa.eu/legal-content/EN/TXT/?uri=celex%3A52008DC0699

European Commission. (2008b). Directive 2008/56/EC of the European Parliament and of the Council of 17 June 2008 Establishing a Framework for Community Action in the Field of Marine Environmental Policy (Marine Strategy Framework Directive) (Text with EEA Relevance), 2008/56/EC, EP, CONSIL, 164 OJ L (2008). http://data.europa.eu/eli/dir/2008/56/oj/eng

European Commission. (2010). COM(2010) 2020 final- Communication from the Commission - Europe 2020 A strategy for smart, sustainable and inclusive growth, COM(2010) 2020 final (2010). https://eur-lex.europa.eu/LexUriServ/LexUriServ.do?uri=COM:2010:2020:FIN:EN:PDF

European Commission. (2011). Communication from the Commission to the European Parliament, the Council, the European Economic and Social Committee and the Committee of the Regions - Roadmap to a Resource Efficient Europe, COM(2011) 571 final (2011). https://eur-lex.europa.eu/legal-content/EN/TXT/?uri=CELEX%3A52011DC0571

European Commission. (2012). COM(2012) 494 final - Communication from the Commission to the European Parliament, the Council, the European Economic and Social Committee and the Committee of the regions - Blue Growth - Opportunities for marine and maritime sustainable growth - (Text with EEA relevance), COM(2012) 494 final (2012). https://eur-lex.europa.eu/LexUriServ/LexUriServ.do?uri=COM:2012:0494:FIN:EN:PDF

European Commission. (2014). Directive 2014/89/EU of the European Parliament and of the Council of 23 July 2014 Establishing a Framework for Maritime Spatial Planning, 2014/89/EU, EP, CONSIL, 257 OJ L (2014). http://data.europa.eu/eli/dir/2014/89/oj/eng

European Commission. (2016a). *Contributions Received*. European Commission. https://web.archive.org/web/20170124051516/https://ec.europa.eu/dgs/maritimeaffairs_fisheries/consultations/ocean-governance/contributions/index_en.htm

European Commission. (2016b). *International Ocean Governance: An Agenda for the Future of Our Oceans—List of Actions*. European Commission. https://www.eesc.europa.eu/sites/default/files/resources/docs/list-of-actions_en.pdf

European Commission. (2016c). *Summary of the Results of the Public Consultation on International Ocean Governance—Public Consultation from 4.6.2015 to*

*15.10.2015*. European Commission. https://commission.europa.eu/document/download/1d45ddbe-ac0c-44ac-8c6a-2bc5998b63be_en?filename=consultation-ocean-governance-summary_en.pdf&prefLang=fr

European Commission. (2016d, novembre 10). *International Ocean Governance: EU's Contribution for Safe, Secure, Clean and Sustainably Managed Oceans*. European Commission. https://www.eesc.europa.eu/sites/default/files/resources/docs/ip-16-3619_en.pdf

European Commission. (2016e). Joint Staff Working Document - Synopsis of the Outcome of the Consultation on International Ocean Governance Accompanying the Document - Joint Communication to the European Parliament, the Council, the European Economic and Social Committee and the Commisttee of the Regions - International Ocean Governance: An Agenda for the Future of Our Oceans, SWD(2016) 352 final (2016). https://eur-lex.europa.eu/legal-content/EN/TXT/?uri=CELEX%3A52016SC0352&qid=1709541986717

European Commission. (2016f). Joint Staff Working Document - Synopsis of the Outcome of the Consultation on International Ocean Governance Accompanying the Document - Joint Communication to the European Parliament, the Council, the European Economic and Social Committee and the Commisttee of the Regions - International Ocean Governance: An Agenda for the Future of Our Oceans, SWD(2016) 352 final (2016). https://eur-lex.europa.eu/legal-content/EN/TXT/?uri=CELEX%3A52016SC0352&qid=1709541986717

European Parliament. (2018b). Texts Adopted - International Ocean Governance: An Agenda for the Future of Our Oceans in the Context of the 2030 Sustainable Development Goals - Tuesday, 16 January 2018, P8_TA(2018)0004 (2018). https://www.europarl.europa.eu/doceo/document/TA-8-2018-0004_EN.html

European Commission. (2019). Joint Report to the European Parliament and the Council - Improving International Ocean Governance - Two Years of Progress, Pub. L. No. SWD/2019/104 final (2019). https://eur-lex.europa.eu/legal-content/EN/TXT/?uri=CELEX%3A52019SC0104&qid=1709541986717

European Commission. (2020). Communication from the Commission to the European Parliament, the Council, the European Economic and Social Committee and the Committee of the Regions - EU Biodiversity Strategy for 2030 Bringing Nature Back into Our Lives, COM(2020) 380 final (2020). https://eur-lex.europa.eu/legal-content/EN/TXT/?uri=celex%3A52020DC0380

European Commission. (2021a). *Green Recovery for the Blue Economy EUR-Lex—52021DC0240*. European Commission. https://eur-lex.europa.eu/legal-content/EN/PIN/?uri=COM%3A2021%3A240%3AFIN

European Commission. (2021b). Proposal for a Council Decision on the Position to Be Taken on Behalf of the European Union at the Meetings of the International Seabed Authority Council and Assembly, COM(2021) 1 final (2021). https://eur-lex.europa.eu/legal-content/EN/TXT/?uri=CELEX%3A52021PC0001&qid=1709541986717

European Commission. (2022). Join Communication to the European Parliament, the Council, the European Economic and Social Committee and the Committee of the Regions - Setting the Course for a Sustainable Blue Planet - Joint Communication on the EU's International Ocean Governance Agenda, JOIN/2022/28 final (2022). https://eur-lex.europa.eu/legal-content/EN/TXT/?uri=CELEX%3A52022JC0028&qid=1709541986717

European Commission. (2023). Proposal for a Regulation of the European Parliament and of the Council Establishing a Framework for Ensuring a Secure and Sustainable Supply of Critical Raw Materials and Amending Regulations (EU) 168/2013, (EU) 2018/858, 2018/1724 and (EU) 2019/1020, COM(2023) 160 final-2023/0079 (COD) (2023). https://eur-lex.europa.eu/legal-content/EN/TXT/?uri=COM:2023:160:FIN

European Commission. (2024, mars 21). *Responsible and Sustainable Fishing*. European Commission. https://oceans-and-fisheries.ec.europa.eu/facts-and-figures/facts-and-figures-common-fisheries-policy/responsible-and-sustainable-fishing_en

European Parliament. (2018). *2017/2055(INI)—International Ocean Governance: An Agenda for the Future of Our Oceans in the Context of the 2030 SDGs* (A8-0399/2017, p. 5). European Parliament. https://oeil.secure.europarl.europa.eu/oeil/popups/ficheprocedure.do?lang=en&reference=2017/2055(INI)

European Parliament. (2022, octobre 6). *Momentum for the Ocean: Strengthening Ocean Governance and Biodiversity—Thursday,* 6 October 2022. European Parliament. https://www.europarl.europa.eu/doceo/document/TA-9-2022-0356_EN.html

Friedheim, R. L. (1993). *Negotiating the New Ocean Regime, Volume 1992*. University of South Carolina Press.

Frozel Barros, N. (2021). Un objet diplomatique: L'occupation industrielle des fonds marins à travers l'analyse des décisions politiques. *VertigO – la revue électronique en sciences de l'environnement, Hors-série 33*, Article Hors-série 33. https://doi.org/10.4000/vertigo.29686

Hadjimichael, M. (2018). A Call for a Blue Degrowth: Unravelling the European Union's Fisheries and Maritime Policies. *Marine Policy*, *94*, 158–164. https://doi.org/10.1016/j.marpol.2018.05.007

Hannigan, J. (2016). *The Geopolitics of Deep Oceans*. Polity.

International Seabed Authority. (2024). *Observers*. International Seabed Authority. https://www.isa.org.jm/observers/

Join Communication to the European Parliament, the Council, the European Economic and Social Committee and the Committee of the Regions – Setting the Course for a Sustainable Blue Planet – Joint Communication on the EU's International Ocean Governance Agenda, JOIN/2022/28 final (2022). https://eur-lex.europa.eu/legal-content/EN/TXT/?uri=CELEX%3A52022JC0028&qid=1709541986717

Joint Report to the European Parliament and the Council – Improving International Ocean Governance – Two Years of Progress, Pub. L. No. SWD/2019/104 final (2019). https://eur-lex.europa.eu/legal-content/EN/TXT/?uri=CELEX%3A52019SC0104&qid=1709541986717

Joint Staff Working Document – Synopsis of the Outcome of the Consultation on International Ocean Governance Accompanying the Document – Joint Communication to the European Parliament, the Council, the European Economic and Social Committee and the Commisttee of the Regions – International Ocean Governance: An Agenda for the Future of Our Oceans, SWD(2016) 352 final (2016). https://eur-lex.europa.eu/legal-content/EN/TXT/?uri=CELEX%3A52016SC0352&qid=1709541986717

Joint Staff Working Document – Joint Communication to the European Parliament, the Council, the European Economic and Social Committee and the Committee of the Regions – International Ocean Governance: An Agenda for the Future of Our

Oceans, JOIN(2016) 49 final (2016). https://eur-lex.europa.eu/legal-content/EN/TXT/?uri=CELEX%3A52016JC0049&qid=1709541986717

Jouffray, J.-B., Blasiak, R., Norström, A. V., Österblom, H., & Nyström, M. (2020). The Blue Acceleration : The Trajectory of Human Expansion into the Ocean. *One Earth*, 2(1), 43–54. https://doi.org/10.1016/j.oneear.2019.12.016

Kang, Y., & Liu, S. (2021). The Development History and Latest Progress of Deep-Sea Polymetallic Nodule Mining Technology. *Minerals*, *11*(10), Article 10. https://doi.org/10.3390/min11101132

Le Meur, P.-Y., & Muni Toke, V. (2022). Chapitre 12 –Le « renouveau minier » français vu des abysses ultramarins. In S. Chailleux, S. L. Berre, & Y. Gunzburger (Éds.), *Ressources minérales et transitions* (pp. 353–380). Peter Lang. https://www.peterlang.com/document/1280839

Lists of Delegations – Assembly of the International Seabed Authority, Twenty-Eighth Session, ISBA/28/A/INF/10, Twenty-Eighth Session (2023). https://www.isa.org.jm/wp-content/uploads/2023/07/ISBA_28_A_INF_10.pdf

Martínez-Rius, B. (2022). *Making the Seafloor. French Geologists, Marine Resources, and New Deep Territories (1945–1975)* [These de doctorat]. Sorbonne université.

Mendenhall, E. (2019). The Ocean Governance Regime : International Conventions and Institutions. In P. G. Harris (Éd.), *Climate Change and Ocean Governance* (1re éd., pp. 27–42). Cambridge University Press. https://doi.org/10.1017/9781108502238.002

Miller, K. A., Thompson, K. F., Johnston, P., & Santillo, D. (2018). An Overview of Seabed Mining Including the Current State of Development, Environmental Impacts, and Knowledge Gaps. *Frontiers in Marine Science*, *4*, 418. https://doi.org/10.3389/fmars.2017.00418

Oreskes, N. (2021). *Science on a Mission: How Military Funding Shaped What We Do and Don't Know about the Ocean*. University of Chicago Press. https://doi.org/10.7208/chicago/9780226732411.001.0001

Pires da Cruz, T. (2023). Les aires marines protégées de l'Autorité internationale des fonds marins: La négociation environnementale et l'extraction minière: *Négociations*, *37*(1), 115–137. https://doi.org/10.3917/neg.037.0115

Proposal for a Council Decision on the Position to Be Taken on Behalf of the European Union at the Meetings of the International Seabed Authority Council and Assembly, COM(2021) 1 final (2021). https://eur-lex.europa.eu/legal-content/EN/TXT/?uri=CELEX%3A52021PC0001&qid=1709541986717

Proposal for a Regulation of the European Parliament and of the Council Establishing a Framework for Ensuring a Secure and Sustainable Supply of Critical Raw Materials and Amending Regulations (EU) 168/2013, (EU) 2018/858, 2018/1724 and (EU) 2019/1020, COM(2023) 160 final-2023/0079 (COD) (2023). https://eur-lex.europa.eu/legal-content/EN/TXT/?uri=COM:2023:160:FIN

Seas At Risk. (2021). *2020 Progress Assessment Blue Manifesto – The Roadmap to a Healthy Ocean in 2030*. Seas At Risk. https://seas-at-risk.org/wp-content/uploads/2021/06/BLUE-MANIFESTO-Background-FINAL.pdf

Seas At Risk, European Seas Environmental Cooperation, Coalition Clean Baltic, The Mediterranean Information Office for Environment, Culture and Sustainable Development (MIOECSDE), Coalition Clean Baltic (CCB), Black Sea NGO Network (BSNN), Oceana, BirdLife Europe, Friends of the Earth Europe, DPPVN (Society of bird research and nature protection – Slovenia), Mediterranean S.O.S (Greece), U Marinu – CPIE Bastia Golo Méditerranée, Mediterrania – Centre d'Initiatives Ecologiques, MEDASSET (Mediterranean Association to Save the Sea Turtles), The North Sea Foundation, European Environmental Bureau (EEB), Danish Society for a Living Sea, Marine Conservation Society, Sea First, & EcoVitae (2012, octobre). *Limits to Blue Growth—Joint NGO Position Paper*. Seas At Risk. https://mio-ecsde.org/wp-content/uploads/2012/10/En-Joint-Position-Paper-on-the-Limits-to-Blue-Growth.pdf

Singh, P. A., Campanella, V. T., & Maes, F. (2024). Chapter VI.1.4 The European Union and Seabed Mining. In V. T. Campanella (Éd.), *Routledge Handbook of Seabed Mining and the Law of the Sea* (pp. 290–313). Routledge.

Tanaka, Y. (2019). *The International Law of the Sea* (Third edition). Cambridge University Press.

Texts Adopted – International Ocean Governance: An Agenda for the Future of Our Oceans in the Context of the 2030 Sustainable Development Goals – Tuesday, 16 January 2018, P8_TA(2018)0004 (2018). https://www.europarl.europa.eu/doceo/document/TA-8-2018-0004_EN.html

Texts Adopted – Norway's Recent Decision to Advance Seabed Mining in the Arctic – Wednesday, 7 February 2024, 2024/2520(RSP), B9–0095/2024 (2024). https://www.europarl.europa.eu/doceo/document/TA-9-2024-0068_EN.html

Trocan, L. M. (2023). The Interest of the European Union in the Exploration and Exploitation of the Blue Continent' Resources. *International Investment Law Journal*, *3*(1), 17–28.

World Economic Forum. (2022). *Decision-Making on Deep-Sea Mineral Stewardship : A Supply Chain Perspective – White Paper* (p. 71). World Economic Forum. https://www.weforum.org/whitepapers/decision-making-on-deep-sea-mineral-stewardship-a-supply-chain-perspective/

**PART 4**

# Waste and health

# 7

# THE SUSTAINABLE BATTERIES REGULATION

*Cándido García Molyneux*[1]

## Introduction

On July 12, 2023, the European Parliament and Council adopted Regulation (EU) 2023/1542 on Batteries and Waste Batteries ("Sustainable Batteries Regulation" – "SBR"). The SBR was one of the first legislative acts that the European Union (EU) adopted among the ambitious environmental measures that the European Commission announced in its European Green Deal Communication in December 2019. This "deal" for Europe proposed a

> new growth strategy aimed to transform the EU into a fair and prosperous society, with a modern, resource-efficient and competitive economy where there are no net emissions of greenhouse gases in 2050 and where economic growth is decoupled from resource use.

Although the Commission's Green Deal Communication already reflected significant changes in Europe's perception of the world, too much has happened since December 2019, including a worldwide health crisis, an aggression and invasion of European territory and the re-election of a new U.S. Presidency that has made clear its dislike for environmental legislation. Many Europeans now feel more vulnerable and more fearful of a world outside Europe that they do not understand, and they can no longer trust. These developments have strongly influenced the EU's legislative implementation of the European Green Deal and are now leading to a postponement and watering down of already adopted Green Deal legislation.

However, this chapter reviews the SBR from a different perspective. It explains the SBR as reflecting an evolution of EU waste management policies,

DOI: 10.4324/9781003479260-12

particularly the legislation of waste batteries and similar equipment. An evolution from focusing on producer responsibility, to life-cycle analysis and circular economy, to industrial policy, and to trying to influence the batteries' global supply chain. An evolution from basic producer responsibility principles to trying to establish the rules of the global game. This chapter argues that, as a consequence of such evolution, the SBR has become a Frankenstein regulation. It is a patch of different and sometimes conflicting regulatory techniques and concepts that may confuse law students and practitioners.

This chapter shares reflections on the SBR from the point of view of a practitioner of EU environmental law. It analyzes the evolution of waste legislation, and in particular waste battery legislation, and different regulatory techniques, by focusing on what, from a practitioner's point of view, are recurring questions when interpreting and applying EU waste legislation: the "who," the "what," and the "where." Asking the "who" aims at identifying the entity that is responsible for complying with the legislation's requirements. Is it the manufacturer, producer, importer or distributor and is there de facto a difference among them? Or should it be the consumer or the public authority? The "what" is the obligation that is imposed. Is it merely ensuring the sound treatment of waste, or is it paying for the product's waste take back, meeting specific recycling targets, or even ensuring the sustainability of the product's supply chain? The "where" or territorial scope is defining the physical area, in effect the market, that triggers the obligation of the legislation. In a regulatory model that aims for economic and social integration, like the EU, the territorial scope of application of the law becomes paramount. These concepts have varied with the evolution of EU waste policy and are also used with different meanings within the same SBR.

Understanding the EU's waste policy evolution and acknowledging its influence in the different regulatory techniques of the SBR helps better interpret and apply the SBR, and other Green Deal legislation. Law does not necessarily have to be coherent as it has no meaning by itself. However, one should understand its contradictions, and use them for a better purpose.

## The evolution of EU waste batteries legislation from the 1990s to 2024

Council Directive 75/442/EEC on Waste of 1975 was the first piece of waste legislation that the EU adopted.[2] Yet, this chapter looks at the evolution of the EU legislation on waste batteries, and to some extent waste electrical and electronic equipment ("WEEE"), as of the 1990s. The EU adopted the first waste batteries directive in 1991, and it was in the 1990s when a major revolution in EU waste legislation occurred: it transformed from pure waste policy to a product-oriented environmental policy.[3]

The review of the evolution of waste batteries legislation focuses on two parallel tracks (see Figure 7.1). The tracks are not strictly lineal but rather

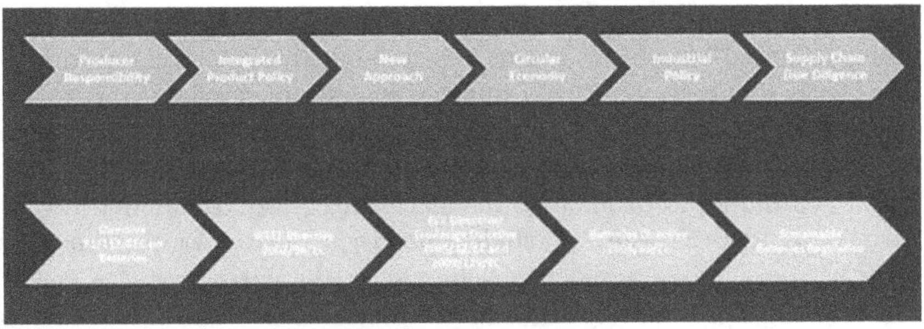

FIGURE 7.1 The evolution of waste batteries legislation

highlight a trend, and in fact the explanations below include detours. One track represents the evolution of general concepts as developed by the European Commission. It starts with the key waste concept of producer responsibility, and thereafter highlights the different conceptual stages of integrated product policy (IPP), the so-called "New Approach" to product regulation, the Commission's communications on circular economy and industrial policy, and the increasing focus on supply chain due diligence principles.

A parallel track looks at the different pieces of legislation matching the evolution of the different concepts and principles, from the first Directive 91/157/EEC on Batteries and Accumulators Containing Certain Dangerous Substances, to Directive 2002/96/EC on WEEE, Directive 2005/32/EC on a Framework for the Setting of Ecodesign Requirements for Energy-Using Products, Directive 2006/66/EC on Batteries and Accumulators and Waste Batteries and Accumulators, and finally the SBR.

### *The first batteries Directive*

The first batteries Directive, Directive 91/157/EEC, is an example of waste legislation prior to the EU's adoption and implementation of the producer responsibility principle. It introduced a limited regulation on batteries and focused mainly on batteries containing hazardous substances and introduced requirements to ensure the sound disposal of "spent" batteries.

The Directive's recitals contained a reference to the "polluter pays" principle and the objectives of environmental policy that are still reflected in Article 191 of the Treaty on the Functioning of the European Union ("TFEU"). The recitals referred to "preventing, reducing and as far as possible eliminating pollution and ensuring sound management of raw materials resources, on the basis also of the 'polluter pays' principle" (Council of the European Communities, 1991, Recital 3).

However, the Directive's objective was solely to "approximate the laws of the Member States on the recovery and controlled disposal of those spent batteries and accumulators containing dangerous substances" (Article 1). It only required Member States to impose limited waste requirements and made clear that they could "not impede, prohibit or restrict the marketing of batteries" covered by the Directive and conforming with its provisions. In particular, the Directive required Member States to prohibit the marketing of specific types of alkaline manganese batteries containing mercury (Article 3); adopt measures, including marking of batteries, to ensure that spent batteries and accumulators were collected separately with a view to recovery (Article 4); and draw programs to reduce metals in batteries, the gradual reduction in household waste of spent batteries and the separate disposal of spent batteries (Article 6). Member States were also to ensure the organization of separate collection (i.e., the creation of the first collective schemes) and were allowed to set up deposit systems where appropriate, and to "introduce measures such as economic instruments in order to encourage recycling" (Article 7). Interestingly, the Directive also imposed a loose requirement on Member States to ensure that batteries could not be incorporated into appliances, other than those excluded, unless they could be readily removed (Article 5).[4]

The Directive also allowed Member States to use economic instruments to encourage recycling, and its recitals emphasized that economic instruments may encourage the separate collection and recycling of spent batteries. However, despite the reference of the recitals of the Directive to the "polluter pays" principle, it did not require producers to take responsibility and pay for the waste take back of their products. Instead, as an example of economic instruments, the Directive's recitals highlighted "deposit systems," which essentially impose the obligation on consumers. Not surprisingly, it was in the late 1980s and early 1990s that the European Court of Justice had to address the discretion of Member States to regulate deposit systems.[5]

In effect, Directive 91/157 only imposed minimum requirements on Member States to encourage the separate collection of waste batteries. The obligations (and discretion) fell on the Member States, not on producers, and the relevant market was therefore that of the national territory.

### *Extended producer responsibility, the Directive on WEEE, and the Directive on batteries and accumulators and waste batteries and accumulators*

The key policy development that revolutionized EU waste law and allowed it to evolve into one also focusing on environmental product regulation and eventually on circularity was the adoption of the then not that obvious approach of extended producer responsibility ("EPR").[6]

In 1996, as part of its review of the Community Strategy for Waste Management, the Commission presented the EPR approach as a derivation of the basic EU environmental principle that the polluter should pay (European

Commission, 1996). The 1996 Communication argued that the product manufacturer has a predominant role when considering the life cycle of a product and preventing the generation of waste. It stated that:

> [t]he manufacturer is the one to take key decisions concerning the waste management potential of his product, such as design, conception, use of specific materials, composition of the product and finally its marketing. The manufacturer is therefore able to provide the means not only to avoid waste by a considered utilization of natural resources, renewable raw materials or non-hazardous materials, but also to conceive products in a way which facilitates proper re-use and recovery.
>
> *(p. 8)*

As the Organisation for Economic Co-operation and Development (OECD) explained a few years later, the EPR approach was intended to extend the producer's responsibility for a product to the waste phase of the product's life cycle in order to (i) shift the responsibility toward the producer and away from the taxpayer or consumer, and (ii) provide incentives to producers to incorporate environmental considerations into the design of their products.

Thus, making producers pay for the waste management of their products was intended to encourage producers to design them in a more sustainable way. Waste policy moved from focusing solely on the sound disposal and treatment of waste, and the producer became responsible for the waste resulting from the products it markets, even if it did not itself generate the waste (i.e., even if it did not itself discard the product). This extended producer responsibility's objective to encourage producers to design their products in a more sustainable way allowed to create a link between waste policy and product-oriented environmental policies, and to transform waste policy from focusing on the end-of-life of products to a more holistic circular economy policy.

### *New legal issues and concepts*

The introduction of this new EPR approach also required addressing new complex and often interrelated legal issues. Who was the producer responsible for the waste phase of the products? Was it the manufacturer no matter where he was located, the importer importing the products into the EU, or the distributor importing the products into the Member State? Should retailers putting their brand names on the products be the responsible producers?

And if the EPR approach was intended to incentivize producers to design their products in a more sustainable way, should the producers' responsibility be collective (i.e., producers as a collective group should share the costs of the waste phase of their products), or should it be individual so that each producer should only pay for the actual costs of the waste phase of their specific products? Would individual responsibility be compatible with producers joining producer responsibility organizations to take care of the waste phase of their products?

Similarly, what did producers have to pay for? For the separate collection of the producer's waste products, or also for the treatment, recycling, recovery and recycling, and sound disposal? And should these costs refer to future waste (i.e., waste resulting from products that were supplied after the entry into force of the legislation) or also historical waste (i.e., waste resulting from products supplied before the entry into force of the requirements). And should the contribution of producers be calculated on the basis of volumes of product marketed, the actual volume of the waste products, or the producer's turnover or profits? Should producers get a discount for the value of the recycled materials from their products? And would producers be required or allowed to show consumers the costs of the waste phase of their products (i.e., visible fee) so that such costs would be a competitive element when marketing products?

Related to these legal issues was also the key issue of the threshold that triggered the producer's responsibility. Should it be any supply of the product? This would arguably mean that all suppliers of the product (e.g., EU manufacturer, importer, distributor, retailer) should contribute. Or should it be the first supply in the EU, or the first supply in the particular Member State? And how would it be possible to identify the producer and ensure that it met its producer responsibility in each particular case?

### *The WEEE directive*

EU institutions, Member States, industry, and other stakeholders were confronted with those legal issues for the first time during the drafting and legislative adoption of the WEEE Directive (García Molyneux, 2001). It was very likely the first time that the EU imposed its EPR on producers of electrical and electronic equipment. Suddenly, the Information Technology (IT) industry was required to pay for the management of the waste resulting from their electronic equipment, not because they discarded the products but because they had marketed them.[7]

The legislative process of adopting the WEEE Directive and later national transposition and application of its requirements revealed the practical and political difficulties of applying the EPR and would lead the EU to apply it only halfway. For example, for WEEE from private households, the Directive stated that for products put on the market later than August 13, 2005, each producer should be responsible for financing the waste operations relating to the waste from "his own products." In that case, producers could choose to fulfill this obligation individually (i.e., creating their own take back schemes) or by joining a collective scheme. In contrast, for WEEE from users other than private households (e.g., B2B), the Directive merely stated that the financing of the costs should "be provided by producers," that Member States could also require that (professional) users also be made partly or totally responsible, and that in any case producers and their professional customers could agree on other financing methods (European Parliament and Council, 2003a, Article 9).

The difficulty of striking the right balance when applying the EPR approach became apparent with respect to WEEE from users other than private households (B2B) during the very last stages of adoption of the Directive. As a result of industry concerns that making them pay for historical waste would impose unreasonable financial burdens on companies, the Parliament, Council, and Commission published a statement promising that should such concerns prove to be founded, the Commission would present a proposal to amend those rules and the Parliament and Council would "undertake to act expeditiously" to adopt it. Less than two years later, the Parliament and Council adopted an amendment to the Directive with new rules on the allocation of the costs for WEEE from users other than private households and historical waste. If the WEEE waste was being replaced by new products or by new products fulfilling the same function, producers should finance the costs of the WEEE, but Member States could also require that professional users also paid totally or partly for such costs. In contrast, where the WEEE was not replaced by new products, users should have to pay the costs. In both cases, and also for future WEEE, producers and professional users could agree on other financing methods.

At the end of the day, the national implementation and application of the WEEE Directive arguably suggests that Member States were more interested in making someone pay than to incentivize producers to design their products in a more sustainable way (García Molyneux, 2008).

### *The second batteries Directive 2006/66*

The new Directive on Batteries and Accumulators and Waste Batteries and Accumulators of 2006 was much more ambitious than its predecessor of 1991. It implemented an EPR approach and also included some elements of circularity. The Directive's objectives were to prevent the marketing of batteries and accumulators containing hazardous substances and introduce specific rules on collection, treatment, recycling and disposal of waste batteries and accumulators to supplement other EU waste legislation and ensure a high level of collection and recycling of waste batteries and accumulators. Its objective was to

> improve the environmental performance of batteries and accumulators and of the activities of all economic operators involved in the life cycle of batteries and accumulators *e.g.*, producers, distributors and end-users

> and, in particular, those operators directly involved in the treatment and recycling of waste batteries and accumulators.
>
> *(European Parliament and Council, 2006a, Article 1)*

The Directive restricted the marketing of batteries containing hazardous substances (Article 4); required Member States to put in place collection schemes for waste portable batteries and accumulators and achieve specific collection targets (Articles 8 and 10); continued to require that manufacturers design appliances in a way that users could readily remove their waste batteries and accumulators (Article 11); required Member States to ensure that producers set up schemes to provide for the treatment and recycling of waste batteries and accumulators (Article 12); required Member States to ensure that users of batteries were informed of the measures available to ensure the separate collection and recycling of batteries and accumulators (Article 20); prohibited the disposal of waste industrial and automotive batteries and accumulators in landfills or by incineration (Article 14); and introduced rules on the export of waste batteries and accumulators to third countries (Article 15).

The new Directive's provisions on EPR addressed the legal issues mentioned above and the lessons learnt from the adoption and implementation of the WEEE Directive. Recital 19 outlined the Directive's EPR approach, as it stated that

> [b]asic principles for financing the management of waste batteries and accumulators should be set at Community level. Financing schemes should help to achieve high collection and recycling rates and to give effect to the principle of producer responsibility. All producers as defined by this Directive should be registered. Producers should finance the costs of collecting, treating and recycling all collected batteries and accumulators minus the profit made by selling the materials recovered.
>
> *(Recital 19)*

In effect, as the WEEE Directive, the Batteries Directive of 2006 resolved most of the relevant legal issues of defining producer, threshold that triggered the requirements and market, by tying them to the territory of the Member States. The Directive defined the producer responsible for compliance with the requirements as

> any person *in a Member State* that, irrespective of the selling technique used, including by means of distance communication as […] *places* batteries or accumulators […] *on the market for the first time within the territory of that Member State* on a professional basis.
>
> *(emphasis added) (Article 3(12))*

The producer was not the entity that manufactured or took responsibility for the manufacture of batteries, but rather that who marketed them. Placing on the market was defined as making available, whether in return for payment or free of charge, to a third party within the Community and included import into the customs territory of the Community. Hence, despite the confusing definition of placing on the market, the producer responsible was an entity in a Member State that brought in or supplied the batteries in the market of a particular Member State for the first time. The producer responsible was the first supplier, and not other distributors and retailers in the supply chain, in a particular Member State.

This clear link with the territory of each Member State was emphasized by the Directive's requirement on Member States to ensure that "each producer is registered [and that] [r]egistration shall be subject to the same procedural requirements in each Member State" (Article 17). Given the practicalities of waste take back, the reality of political convenience and the EU legal structure, EPR was conceived solely within the territorial scope of the Member State. Ultimately, the implementation and application of waste policy and extended producer responsibility was the responsibility of each Member State.

The Directive imposed many of its requirements on "producers." For example, "producers [...] or third parties acting on their behalf" could not "refuse to take back waste industrial batteries and accumulators from end-users, regardless of chemical composition and origin" (Article 9(3)). Similarly, "producers or third parties [were to be required to] set up schemes using best available techniques, in terms of the protection of health and the environment, to provide for the treatment and recycling of waste batteries and accumulators" (Article 12(1)).

Moreover, in line with the EPR approach, Member States were to ensure "that producers, or third parties acting on their behalf, *finance any net costs arising* from" (emphasis added) the collection, treatment, and recycling of all waste batteries (Article 16(1)). The reference to "net costs" indicated that the costs of the waste take back should be "minus the profit made by selling the materials recovered" (Recital 19). These financial obligations applied retroactively to all waste batteries and accumulators, irrespective of the date of their placing on the market (Article 16(6)), although Member States could exempt small producers (Article 18). The Directive also prohibited producers from showing to end-users (i.e., visible fee) the costs of collection, treatment, and recycling at the time of sale of new portable batteries and accumulators (Article 16(4)).

Yet, while Directive 2006/66, established the "[b]asic principles for financing the management of waste batteries and accumulators" at the EU level (Recital 19), unlike the WEEE Directive, it did not address the specific and controversial issue of how producers should pay for the costs of the waste take back, and whether they should pay only for the costs of the waste of their own batteries or whether Member States could introduce other calculation methods that allowed for a collective allocation of the costs. This controversial detail was left for the Member States and their producer responsibility organizations to define.

### *The new approach and integrated product policy*

The environmental regulation of batteries and related electronic equipment outlined until now was specific to waste legislation that focused mostly on the end-of-life of products – not on their manufacture and use phases – with some exceptions, such as chemical restrictions that were mainly intended to address the chemical risks of the end-of-life phase. However, during the 2000s, the EU would also adopt separate legislation that, while it focused on the same categories of products, intended to address their environmental impacts mostly during their manufacture and use.

The Commission announced this distinct regulation in its Communication on Integrated Policy of 2003, where it already acknowledged the importance of life-cycle thinking (European Commission, 2003a). While this separate regulation was also the result of different personalities and competition between different departments of the European Commission it resulted in the use of different legal concepts and techniques. Indeed, the European Commission's Directorate General (DG) for Environment was given the responsibility to regulate the end-of-life phase of products and their waste take back, while other DGs were to focus on legislation regulating the environmental impact of products during their manufacture and use. These other DGs would use legal concepts that were more in line with their policy goal of ensuring a single market and with the product safety legislation for which they were responsible.

#### *Integrated product policy*

In 2003, the European Commission presented its Communication on Integrated Product Policy ("IPP") where it called for a life-cycle and market-based approach to regulating products (European Commission, 2003a). The primary aim of this IPP was "to reduce the environmental impacts from products throughout their life-cycle, harnessing, where possible, a market driven approach, within which competitiveness concerns are integrated." The IPP was based on five key principles: life-cycle thinking; working with the market; stakeholder involvement; continuous improvement; and the use of a combination of a variety of instruments. As explained at the time, life-cycle thinking was intended to consider a product's life-cycle and to reduce the product's environmental impacts from "the cradle to the grave" and to regulate them in an "integrated" manner. This life-cycle thinking and integrated approach would be fundamental to later develop the EU's circular economy approach.

Moreover, the Communication assumed that the producer was responsible for addressing the environmental impacts of the products. For example, it stated that "any product policy needs to ensure that producers and designers become more responsible for ensuring that their products fulfil agreed criteria on health, safety and the environment." Thus, this new environmental product policy assumed the producer's responsibility. It was only possible because the EU had previously developed the EPR approach on the basis of the polluter pays principle enshrined in the TFEU. EPR allowed for the transition from an environmental product legislation that focused mostly on the products' end-of-life to one that regulated products in a more integrated manner, addressing environmental impacts during their entire life-cycle.

### *The new approach legislation*

A step back is necessary. During the 1990s, the DG responsible for internal market and industry developed a so-called "new approach" legislation to regulate the safety requirements of different categories of products that were to be placed on the EU market. The key concepts and objectives of this legislative approach were reflected in the European Commission's Guide to the Implementation of Directives Based on the New Approach and Global Approach (European Commission, 2000a). Examples of this legislation were the Directive on Electromagnetic Compatibility (European Parliament and Council, 2004), the Low Voltage Directive European Parliament and Council, 2006b), and the Directive on Radio Equipment and Telecommunications Terminal Equipment (European Parliament and Council, 1999).

The objective of this so-called "New Approach" legislation was to define basic safety requirements on products to ensure that they could be marketed across all EU Member States. The focus was on the regulation of specific category of products by imposing harmonized safety requirements and allowing for the free trade of the products across Member States. To avoid the political complexities of having Member States agree on detailed technical requirements, the legislation imposed "essential requirements" and left it to European Standardization Bodies, subject to the oversight of the European Commission, to adopt detailed technical standards. Products that complied with the technical standards published by the Commission in the EU official journal (i.e., harmonized standards) were presumed to comply with the different Directives' essential requirements. Each Directive regulating specific categories of products established essentially the same conformity assessment requirements, including preparation of technical documentation, following different conformity assessment procedures, which could include so-called Notified Bodies, Declarations of Conformity, and CE marking.

Because the objective was to encourage an EU single market of the regulated products, the Directives used legal concepts that focused on the EU market rather than the Member States. They required that the products comply with their requirements when they were placed on the market and defined placing on the market "as the *initial action* of making a product available *for the first time on the Community market*, with a view to distribution or use in the Community". A regulated product was to comply with the applicable legislation when it was imported or made available in the EU market for the first time.

The Directives also focused on the concept of "manufacturer" and its supply chain rather than on "producers" that imported or supplied the products in a particular Member State. They typically defined "manufacturer" as "the person who is responsible for designing and manufacturing a product with a view to placing it on the Community market on his own behalf." This manufacturer did not have to be established in the EU market and could appoint an authorized representative to manage some of the formality (rather than substantive) compliance requirements. Other operators, such as importers and distributors, in the EU (rather than per Member State) also had residual responsibilities, including notifying authorities in particular cases of non-compliance, and could be held responsible in full where there was no manufacturer.

### *The energy-using products directive*

The EU would develop the Commission's announced life-cycle thinking and integrated approach by using the same New Approach principles that had been used to adopt prior safety legislation intended to harmonize the regulation of products marketed in the EU, instead of the EPR concepts used under the waste take back legislation. Clear examples of this were a draft for a Directive on the Impact on the Environment of Electrical and Electronic Equipment ("Draft EEE") that the Commission's DG for Enterprise circulated in 2000, and the Directive on Eco-Design for Energy Using Products ("EuPD") that the EU adopted in 2005.

The Draft EEE was the result of a "turf war" between the European Commission's DG for Environment and that for Enterprise (Hunter and García Molyneux, 2000). DG Enterprise perceived that DG Environment's proposals to regulate the waste take back and chemicals in electronic equipment (i.e., the proposed WEEE and RoHS Directives) as tramming into its territory and presented a counterproposal introducing environmental requirements on the equipment itself. It did so by relying on the same New Approach regulatory techniques that it had used to regulate the safety of products.

The Draft EEE was intended to apply to all categories of EEE (European Commission, 2000b). As the New Approach legislation, it defined the manufacturer as the person responsible for compliance, meaning

> the natural or legal person with responsibility for the design and manufacture of a product before it is placed on the market under his own name, regardless of whether these operations are carried out by that person himself or on his behalf.

Also in line with the New Approach techniques, and perhaps most controversial, the draft proposed to define loose essential environmental requirements and to rely on European Standardization Organizations to adopt the necessary technical standards to demonstrate compliance.

The Draft introduced different categories of essential environmental requirements: general category, on design and material selection, manufacture, and use. For example, as a general essential environmental requirement, it stated that "[EEE] shall be designed and manufactured in such a way that, when used under the conditions and for the purposes intended, a high level of environmental performance is achieved throughout their life cycle." As part of the design and material selection, manufacturers were to "take into account," inter alia, life-cycle duration optimization and upgrading of equipment. Materials should be selected taking account of recyclability. In choosing materials, manufacturers would be required to examine renewability of resources, avoid disposal of resources as waste, evaluate adverse impact on the environment for devices or materials that will become disposal waste, and verify availability of supplies. Concerning use, manufacturers were to ensure that energy consumption was minimized during normal use. As with New Approach legislation, producers would be able to comply with these different environmental essential requirements by demonstrating conformity with harmonized standards.

In the end, however, the European Commission did not propose the Draft EEE to the European Parliament and Council for their adoption. Instead, a couple of years after, it decided to present a legislative proposal for a Directive on the Eco-Design of Energy Using Products that would establish a framework for the Commission to adopt specific product regulations imposing eco-design, and mostly energy efficiency, requirements on specific categories of products (European Commission, 2003b). As with the Draft EEE, this new proposal was not the work of DG Environment, but rather of DG Enterprise and DG Energy.

The EuPD was the first EU legislation to incorporate significant environmental requirements in the design phase (development) of products (European Parliament and Council, 2005). It defined criteria on the basis of which the Commission was to adopt regulations for specific categories of producers. It defined manufacturer as the person responsible for compliance. A manufacturer was the person who

> manufactures products covered by the Directive and is responsible for the conformity with this Directive in view of their being placed on the market and/or put into service under the manufacturer's own name or trademark or for the manufacturers' own use.

Like other New Approach legislation, it required manufacturers to ensure that products complied with the applicable requirements at the time they were placed on the market of the EU, and defined placing on the market as the "first time" a product was made available on the "[EU] market."

The EuPD also used the conformity assessment procedures of the New Approach legislation, including the obligation to issue declarations of conformity and to CE mark the products. However, in contrast to the Draft EEE, it relied much less on harmonized standards. While in principle this remained as a theoretical possibility, in practice most of the requirements were detailed by means of Commission regulations. In some cases, the Commission accepted voluntary agreements with industry as a way to regulate specific categories of products under the EuPD.

In 2009, the EuPD was replaced by Directive 2009/125/EC Establishing a Framework for the Setting of Eco-design Requirements for Energy-Related Products, which established a similar framework it provided for an expansion of the covered products and emphasized the need to introduce eco-design, in addition to energy efficiency requirements (European Parliament and Council, 2009). This Directive is being replaced by the new Regulation (EU) 2024/1781 Establishing a Framework for the Setting of Eco-design requirements for Sustainable Products (European Parliament and Council, 2024b).

### *Circular economy*

In its 2014 Communication Towards a Circular Economy: A Zero Waste Program for Europe (European Commission, 2014), the Commission built upon the core concepts of its IPP and EPR approaches and called for a new circular economy approach for its waste management legislation. This was further developed in its 2015 Communication Closing the Loop – An EU Action Plan for the Circular Economy (European Commission, 2015), which called for the EU to adopt measures to make possible an EU economy "where the value of products, materials and resources is maintained in the economy for as long as possible, and the generation of waste minimized" (European Commission, 2015). The plan was intended to be "an essential contribution to the EU's efforts to develop a sustainable, low carbon, resource efficient and competitive economy" and "the opportunity to transform our economy and generate new and sustainable competitive advantages for Europe."

The Commission's plan was intended to create a regulatory framework to develop a circular economy, in the context of the EU single market, and to give "clear signals to economic operators and society at large on the way forward with long term waste targets as well as a concrete, broad and ambitious set of actions" (European Commission, 2015). These actions were intended to focus on the life cycle of products from the design phase to its waste and recycling. It called for the need to adopt measures related to product design (to make products more durable and easier to repair and recycle but also to reduce their energy consumption); efficient use of resources during manufacturing processes; more efficient and sustainable consumption patterns (e.g., labelling, taxation, green procurement, limiting technological obsolescence); waste management (increased re-use, recycling, and recover targets); and boosting the market for secondary raw materials.

This circular economy approach was also emphasized and transitioned to a "sustainable products" approach in the European Green Deal Communication that the European Commission presented in December 2019 (European Commission, 2019). Among other things, the Green Deal Communication announced the Commission's intention to present a new circular economy action plan that would include a "sustainable products" policy to support the circular design of all products based on a common methodology and principles. The objective was to prioritize reducing and using materials before recycling them, set minimum requirements to prevent environmentally harmful products from being placed on the EU market, strengthen extended producer responsibility, and encourage businesses to offer reusable, durable and repairable products. This was reflected in the Communication on A new Circular Economy Action Plan For a cleaner and more competitive Europe that the European Commission presented in March 2020 (European Commission, 2020a).

Specifically for batteries, the new action plan of 2020 announced a new legislative proposal that would reflect the new circular economy and sustainable products approach and introduce rules regulating recycled content, improving the collection and recycling rates of all batteries, ensuring the recovery of valuable materials, addressing non-rechargeable batteries with a view to progressively phasing out; and regulating sustainability and transparency requirements for batteries taking account carbon footprint of battery manufacturing, ethical sourcing of raw materials and security of supply, and facilitating re-use, repurposing, and recycling (European Commission, 2020a, p. 6).

### *Batteries as a strategic sector and the need to ensure the sustainability of the supply chain*

But there were two other developments that would also end up defining the regulation of batteries in the EU. Those were the EU's increasing awareness that batteries were a strategic element to ensure the EU's transition to a net zero carbon economy and of the importance and vulnerability of their supply chains. This sense of the EU's vulnerability was already evident at the time of the Commission's publication of its Green Deal Communication but was emphasized during the legislative process of the SBR. Events like the COVID-19 pandemic and Russia's aggression of Ukraine helped to emphasize this.

Already in 2018, as part of its Communication on Europe on the Move: Sustainable Mobility for Europe: Safe, Connected and Clean, the Commission presented a Strategic Action Plan for Batteries (European Commission, 2018a, Annex 2). The premise of that plan was that "batteries development and production is a strategic imperative for Europe in the context of the clean energy transition and is a key component of the competitiveness of its automotive sector." Therefore, the EU's "immediate challenge [was] to create a competitive and sustainable battery manufacturing" and to "move fast in this global race."

The plan, however, also assumed that the EU's supply chains of the raw materials and components to manufacture batteries were vulnerable. In effect, this sense of vulnerability had already been used to emphasize the need for comprehensive waste management and circular economic policy (European Commission, 2018b). Thus, in the plan the Commission announced that it was

> promoting a cross-border and integrated European approach covering the whole value chain of the batteries ecosystem, and focusing on sustainability, starting with the extraction and processing of raw materials, the design and manufacturing phase of battery cells and battery packs, and their use, second use, recycling and disposal in a circular economy context.

It was supporting "a sustainable battery value chain" that would introduce "requirements for safe and sustainable batteries production – as a key driver for EU competitiveness." That sustainable value chain "should be well-integrated into the circular economy and drive the competitiveness of European products."

Since then, the EU has adopted different measures to support the supply chain of batteries. Examples of this are the Critical Raw Materials Act and the Net Zero Industry Act. However, a particular relevant aspect of the plan was that it emphasized that to ensure the supply chain of batteries, it was necessary to ensure that it was sustainable. This would be done not only by imposing circular economy requirements on batteries to ensure that they are sustainable, but also by imposing "due diligence" sustainability requirements on all economic operators in the supply chain, even if they are outside the EU.

This "due diligence" approach to environmental regulation had already been used in prior EU legislation, such as the Regulation (EU) 995/2010 on the Obligations of Operators Who Place Timber and Timber Products on the Market ("Timber Regulation"). In its legislative proposal for the Regulation, the Commission argued that

> the specific nature of illegal logging requires a new approach that takes into account the structure of the entire timber sector, which aims at a systematic change of behavior in operators and customers and which focuses on promoting sector-specific systems that can prevent illegal timber and timber products entering the Community market.
>
> *(European Commission, 2008)*

Recital 16 of the Regulation further explained that "operators placing timber and timber products for the first time on the internal market should take the appropriate steps in order to ascertain that illegally harvested timber and timber products derived from such timber are not placed on the internal market. To that end, operators should exercise due diligence through a system of measures and procedures to minimise the risk of placing illegally harvested timber and timber products derived from such timber on the internal market." This "due diligence system" included "three elements inherent to risk management: access to information, risk assessment and mitigation of the risk identified." The system should provide

> access to information about the sources and suppliers of the timber and timber products being placed on the internal market for the first time, including relevant information such as compliance with the applicable legislation, the country of harvest, species, quantity, and where applicable sub-national region and concession of harvest.

On the basis of that information, "operators should carry out a risk assessment," and where "risk is identified" they "should mitigate such risk in a manner proportionate to the risk identified."

This due diligence approach entailed a different regulatory technique than that of the EPR or that used to regulate the environmental characteristics of products. It focused on all operators of the supply chain, not merely on the person that first supplied the product in the EU or Member State. It did not impose specific requirements of products or on the management of their waste, but rather to ensure all operators involved in the extraction of the necessary raw materials, manufacture of the components and products, and even those merely supplying the products behaved in a sustainable way, even if they were not operating in the EU. This regulatory approach was also reflected in the SBR that the EU adopted in 2023.

## The sustainable batteries regulation

In July 2023, the EU adopted its new SBR.[8] The SBR replaces the Batteries Directive 2006/66/EC. It applies as of February 2024, but some of its requirements will be phased in (European Parliament and Council, 2023, Article 96).

The SBR applies to all batteries, regardless of their shape, volume, weight, design, material composition, use or purpose (Article 1(3)). It applies to portable batteries, including those contained in marketed devices; electric vehicle batteries; starting, lighting, and ignition batteries; batteries for light means of transport (batteries for electric bicycles); and industrial batteries. Home batteries of more than 5 kg typically qualify as industrial batteries.

As further explained below, while the Regulation tries to regulate the sustainability of batteries and their supply chain in a holistic manner, it continues to use distinct regulatory approaches and concepts for each of its categories of requirements.

### *A holistic environmental regulation of batteries*

The SBR aims to regulate all sustainability aspects of batteries in an integrated manner. It imposes requirements on "sustainability, safety, labelling, marking and information" on batteries marketed in the EU, "minimum requirements for [EPR]" (Article 1(1)), and "due diligence obligations on economic operators placing batteries on the market or putting them into service" (Article 1(2)). To reinforce the sustainability of batteries marketed in the EU, the SBR also imposes green criteria requirements on the public procurement of batteries (Article 85).

First, the SBR imposes extensive and detailed environmental and safety design requirements on batteries (Chapter II). These requirements are supplemented with related labelling, marking, and information requirements (Chapter III). Depending on the type of batteries, the SBR imposes requirements on chemical composition (Article 6), carbon footprint (Article 7), recycled content (Article 8), performance and durability (Articles 9 and 10), removability and replaceability (Article 11),[9] safety (Article 12), an obligation to provide information on the state of health and expective life time of batteries (Article 14), and a batteries passport (Articles 77 and 78).

Second, the SBR imposes broader EPR obligations than those of Directive 2006/66/EC (Chapter VIII). Among other things, it requires Member States to establish registers of producers (Article 55); makes clear that producers have EPR obligations to cover the costs of the take back and management of the waste caused by their batteries (Article 56); regulates the producer responsibility organizations that producers may join to comply with their obligations (Articles 57 and 58); imposes requirements and targets on the collection of waste batteries (Articles 59, 60, 61, 62, 67, 68, and 69); provides for the possibility of regulating deposit and return systems for batteries (Article 63); imposes obligations on the treatment of waste batteries and on operators of treatment facilities (Articles 65 and 70); defines targets for recycling efficiency and recovery of materials of waste batteries (Article 71), and establishes requirements on preparation for re-use or for repurposing of waste batteries (Article 73).

Third, the SBR imposes on economic operators an obligation to adopt supply chain due diligence policies and internal systems with respect to specific critical raw materials of the batteries they supply (Chapter VII). These critical raw materials are cobalt, natural graphite, lithium, nickel, and chemical compounds based on the prior raw materials that are necessary to the manufacture of the active materials of the batteries (Section 1 of Annex X). The policies and associated internal management systems must identify and assess the risks of adverse impacts on the environment (i.e., air, water, soil, biodiversity, hazardous substances, noise and vibration, plant safety, energy use, waste, and residues), human rights, labor rights and industrial relations (including, occupational health and safety, child labor, forced labor, discrimination, trade union freedoms), and community life, including that of indigenous peoples (Section 2 of Annex X).

The due diligence obligations include: requirements to identify and assess the risk of adverse impacts (Article 50(1)(a)); design and implement strategies to prevent, mitigate, and address such impacts (Article 50(1)(b)); the adoption of management systems that, among others, must include a due diligence policy and establish a grievance mechanism (Article 49); and reporting and disclosure obligations (Article 52). The due diligence policies must incorporate international due diligence policy standards (Article 49(1)(b)),[10] be communicated to suppliers and the public (Article 49(1)(a)), and be incorporated into contracts and agreements with suppliers (Article 49(1)(d)). In effect, these due diligence requirements require economic operators to assess the compliance of their suppliers of critical raw materials with international environmental and social agreements. Waste policy has indeed moved from an EPR regulatory framework to trying to impose and enforce environmental and social standards around the world.

### *Different regulatory approaches and concepts*

While the Regulation tries to regulate batteries and their supply chain in a holistic and integrated manner, it uses different regulatory approaches and concepts for its different categories of requirements. Rather than merging into a new regulatory model, the three different regulatory approaches described above (i.e., EPR, new approach, supply chain due diligence requirements) are patched together in a Frankenstein manner in the Regulation. As a result, who is the entity responsible for compliance, what are the conformity obligations, and what is the applicable market and what triggers the requirements will vary depending on the category of obligations. This means that whether a particular person is responsible for the SBR's requirements, and how and when will depend on the different categories of requirements. This may confuse the interpretation of the different requirements, and it is also a reminder that (regulatory) history matters.

### *Sustainability and safety requirements*

For example, the sustainability and safety requirements on batteries are clearly tied to concepts of the new approach intended to regulate the characteristics of products and to ensure a single market. They are subject to conformity procedures that are typical of the EU legislation on products (Chapter IV). Like many other EU product legislation, they introduce a presumption of conformity for batteries that comply with harmonized standards (Article 15), conformity assessment procedures, such as internal production control and quality assurance of production process (Article 17); and Declaration of Conformity and CE marking obligations (Articles 18, 19, and 20).

The Regulation makes the "manufacturer" that places on the market batteries the person principally responsible for compliance with its sustainability and safety requirements (Article 38). The manufacturer is "any natural or legal person who manufactures a battery or has a battery designed or manufactured, and markets that battery under its own name or trademark or puts it into service for its own purposes" (Article 3(33)). This definition of manufacturer does not require it to be established in the EU and a foreign manufacturer may appoint an authorized representative in the EU.

The SBR defines placing on the market as the "first making available of a battery on the Union market" (Article 3(16)), while it defines making available as "means *any supply* of a battery for distribution or use on the Union market in the course of a commercial activity, whether in return for payment or free of charge" (emphasis added) (Article 3(17)). Thus, import of batteries into the EU/European Economic Area (EEA) (i.e., release of into free circulation in the EU/EEA) will normally constitute a placing on the market unless the batteries are put into the warehouse of the importer without further supply to other operators in the supply chain. The focus is on the EU market and on EU harmonization to ensure that batteries complying with the requirements can move freely across the Union.

As other EU product regulation, the SBR also introduces subsidiary roles for other so-called economic operators, such as authorized representatives (Article 40), importers (Article 41), and distributors (Article 42). It also regulates scenarios where importers and distributors may be considered as manufacturers for purposes of the Regulation's battery sustainability and safety requirements, namely where the batteries are placed on the market or put into service under the importer's or distributor's own name or trademark, or where they modify the batteries that have already been marketed (Article 44). In response to the EU's concern with non-EU operators marketing batteries in the EU via online marketplaces or platforms, the Regulation also requires fulfillment service providers to inform and cooperate with national authorities on any action taken to eliminate the risks posed by the batteries (Article 43).

### *EPR and waste management*

In contrast, the EPR and waste management requirements use the same regulatory approach and concepts that were typical of the Batteries Directive 2006/66 and other waste management legislation. They continue to focus on the producer that supply batteries in a particular Member State.

Thus, the producer who has the EPR obligations is the

> manufacturer, importer or distributor or other natural or legal person that, irrespective of the selling technique used, including by means of distance contracts, either: (a) is established in a Member State and manufactures batteries under its own name or trademark, or has batteries designed or manufactured and supplies them for the first time under its own name or trademark, including those incorporated in appliances, light means of transport or other vehicles, within the territory of that Member State; (b) is established in a Member State and resells within the territory of that Member State, under its own name or trademark, batteries including those incorporated in appliances, light means of transport or other vehicles, manufactured by others, on which the name or trademark of those other manufacturers does not appear; (c) is established in a Member State and supplies for the first time in that Member State on a professional basis, batteries, including those incorporated in appliances, light means of transport or other vehicles, from another Member State or from a third country; or (d) sells batteries, including those incorporated in appliances, light means of transport or other vehicles, by means of distance contracts directly to end-users, whether or not they are private households, in a Member State, and is established in another Member State or in a third country.
>
> *(Article 3(47))*

The producer is tied to a Member State. Unless it supplies batteries directly to end-users by means of distance contracts, the producer must be established in an EU Member State. Moreover, it is the entity that supplies batteries in a particular Member State. What triggers the EPR obligations is the supply of the batteries in a particular Member State. This is reinforced by the obligations to register per Member State (Article 55), although arguably producers could join pan-European producer responsibility organizations.

The SBR also imposes some waste take-back obligations on distributors. They must take back waste batteries from end user free of charge without imposing an obligation on the end user to buy or to have bought a new battery, and to hand over the batteries to the producers or product responsibility organizations (Article 62).

In fact, the focus of the EPR obligations on the Member State market creates a dilemma when trying to address the waste of batteries that are supplied from non-EU sellers to EU end-users via online platforms of marketplaces. Article 62(6) tries to solve this by distinguishing between distributors and online platforms and providing that "providers of online platforms [...] that allow consumers to conclude distance contracts with producers" must obtain specific information from producers that offer batteries, such as registration numbers or self-certification of their compliance with the SBR's EPR obligations.

### *Due diligence requirements*

In turn, and in order to ensure the sustainability of the entire supply (or value) chain of batteries, the new due diligence requirements focus on the EU market (Article 48), but extend their reach to all economic operators that market batteries, including

> manufacturer, the authorised representative, the importer, the distributor or the fulfilment service provider or any other natural or legal person who is subject to obligations in relation to the manufacture, preparation for re-use, preparation for repurposing, repurposing or remanufacturing of batteries, the making available or the placing of batteries on the market, including online, or the putting of batteries into service in accordance with this Regulation.
>
> *(Article 3(22))*

In principle, economic operators that are not established in the EU may be subject to the due diligence requirements of the SBR.

### *A recap and some homework*

Thus, the entity responsible and the moment at which compliance will be required will vary depending on the different categories of requirements of the SBR. To understand the impact of this, one may want to consider trying to solve the following illustrative exercise.

Imagine multinational A in the United States with a subsidiary B in China that incorporates into its devices batteries that it purchases from independent contract manufacturer C in Pakistan. Another subsidiary D of the multinational A in Switzerland purchases the devices from the subsidiary B and ships them from that subsidiary B to (i) another subsidiary E in France and (ii) an independent distributor F in Germany. The subsidiary E thereafter ships devices to subsidiaries G in Italy and H in Poland. In this scenario, identify the categories of requirements of the SBR that apply to entities A, B, C, D, E, F, G, and H, and the moment in time when they apply. In a scenario that is very close in time to the applicable deadlines, the general counsel of this multinational might be particularly nervous.

## Final remarks: Why does all this matter?

The thesis of this chapter is rather straightforward: the SBR reflects different regulatory concepts and approaches that derive from different stages of an evolution of EU waste policy and environmental product regulation. Despite the EU's attempt to regulate batteries in a holistic and integrated manner, the regulation is based on a patchwork of different regulatory techniques that were designed to achieve different objectives.

Thus, regulatory history matters. It helps us understand the SBR and interpret and apply its different requirements by keeping in mind that the applicable sources of interpretation (e.g., guidance or concepts of other legislation) may have to vary depending on the requirements of the SBR. For example, one should be cautious when trying to rely too much on the Commission's Blue Guide mentioned above to interpret some of the provisions of the SBR, such as those imposing due diligence requirements through the supply chain.

Moreover, keeping in mind regulatory history and the fact that the EU uses different regulatory approaches is also important to understand other EU environmental product legislation. This legislation may also combine different regulatory approaches, but even when it does not, we should also keep in mind that it may not be correct to rely too much on all types of sources when trying to interpret the rules.

Finally, as a possible line of research, one may want to explore one important question this chapter has not addressed, namely whether the EU's legal and institutional framework would have allowed the EU to regulate batteries in an integrated manner by using a single and coherent regulatory approach. One could argue that the different regulatory approaches of the SBR are not only the result of regulatory history, but also of deeper structural issues, such as the allocation of competences between the EU and Member States. Indeed, if Europeans want to work together to address global challenges, such as transition to a low carbon economy where batteries will be necessary, they may have to significantly change the way they do things.

## Notes

1 To Professor Francis Snyder, who taught me to enjoy the study of law.
2 For an overview of EU waste legislation and in particular on how it applies to the regulation of electronic equipment, see García Molyneux (2021).
3 For a company's perspective of this transformation, see Kautto (2008).
4 This was the first precedent of the current Article 11 of the SBR.
5 Later in 2009, the European Commission, building upon the case law, tried to provide some guidance to Member States. See European Commission (2009).
6 See also Pouikili (2020).
7 For an insider's detailed assessment of the WEEE Directive's requirements, see McLoughlin (2005).
8 The Regulation was based on the legislative proposal that the Commission presented two and a half years earlier in 2020 ( European Commission 2020b).
9 See García Molyneux (2023).
10 The international due diligence policy standards are those established international instruments listed in Section 3 of Annex X to the SBR.

## References

Council of the European Communities. (1975). Directive 75/442/EEC of 15 July 1975 on Waste. O.J.L 191. Retrieved from https://eur-lex.europa.eu/legal-content/EN/TXT/PDF/?uri=CELEX:31975L0442&qid=1751704729070 (accessed December 2024).

Council of the European Communities. (1991). Directive 91/157/EEC of 18 March 1991 on Batteries and Accumulators Containing Certain Dangerous Substances. O.J.L 78. Retrieved from https://eur-lex.europa.eu/legal-content/EN/TXT/PDF/?uri=CELEX:31991L0157&qid=1751704936418 (accessed December 2024).

Court of Justice of the European Union. (1988, September 20). Judgment of September 20, 1988, Commission of the European Communities v Kingdom of Denmark,C-302/86,EU:C:1988:421.Retrievedfromhttps://eur-lex.europa.eu/legal-content/EN/TXT/?uri=CELEX%3A61986CJ0302&qid=1751705246149 (accessed December 2024).

European Commission. (1996). Communication from the Commission to the Council and the European Parliament: Community Strategy for Waste Management, COM [1996] 399 Final. Brussels. Retrieved from https://eur-lex.europa.eu/legal-content/EN/TXT/PDF/?uri=CELEX:51996DC0399&qid=1751696063873 (accessed December 2024).

European Commission. (2000a). Guide to the Implementation of Directives Based on the New Approach and the Global Approach (the "Blue Guide"). Retrieved from https://op.europa.eu/en/publication-detail/-/publication/4f6721ee-8008-4fd7-acf7-9d03448d49e5 (accessed December 2024).

European Commission. (2000b). Second Draft Proposal for a Directive on the Impact on the Environment of Electrical and Electronic Equipment.

European Commission. (2003a). Communication from the Commission to the Council and the European Parliament:Integrated Product Policy - Building on Environmental Life-Cycle Thinking, COM [2003] 302 Final. Brussels. Retrieved from https://

eur-lex.europa.eu/legal-content/EN/TXT/PDF/?uri=CELEX:52003DC0302&qid=1751705459050 (accessed December 2024).

European Commission. (2003b). Proposal for a Directive of the European Parliament and of the Council on Establishing a Framework for the Setting of Eco-design requirements for Energy-Using Products and amending Council Directive 92/42/EEC (COM(2003) 453 Final). Retrieved from https://eur-lex.europa.eu/legal-content/EN/TXT/PDF/?uri=CELEX:52003PC0453&qid=1751705672049 (accessed December 2024).

European Commission. (2008). Proposal for a Regulation of the European Parliament and of the Council Laying Down the Obligations of Operators Who Place Timber and Timber Products on the Market (COM(2008) 644 final). Retrieved from https://eur-lex.europa.eu/legal-content/EN/TXT/PDF/?uri=CELEX:52008PC0644&qid=1751705836074 (accessed December 2024).

European Commission. (2009). Communication from the Commission: Beverage Packaging, Deposit Systems, and Free Movement of Goods, (2009/C 107/01) O.J.C 107. Retrieved from https://eur-lex.europa.eu/legal-content/EN/TXT/PDF/?uri=CELEX:52009XC0509(01) (accessed December 2024).

European Commission. (2014). Communication from the Commission to the European Parliament, the Council, the European Economic and Social Committee and the Committee of the Regions: Towards a Circular Economy: A Zero Waste Program for Europe, (COM(2014) 398 final). Brussels. Retrieved from https://eur-lex.europa.eu/legal-content/EN/TXT/PDF/?uri=CELEX:52014DC0398&qid=1751706079673 (accessed December 2024).

European Commission. (2015). Communication from the Commission to the European Parliament, the Council, the European Economic and Social Committee and the Committee of the Regions: Closing the loop – An EU action plan for the Circular Economy (COM(2015) 614 final). Brussels. Retrieved from https://eur-lex.europa.eu/legal-content/EN/TXT/?uri=CELEX%3A52015DC0614&qid=1751706179668 (accessed December 2024).

European Commission. (2018a). Communication from the Commission to the European Parliament, the Council, the European Economic and Social Committee and the Committee of the Regions: Europe on the Move – Sustainable Mobility for Europe: safe, connected and clean (COM(2018) 293 final). Brussels. Retrieved from https://eur-lex.europa.eu/resource.html?uri=cellar:0e8b694e-59b5-11e8-ab41-01aa75ed71a1.0003.02/DOC_3&format=PDF (accessed December 2024).

European Commission. (2018b). Report on Critical Raw Materials and the Circular Economy. Retrieved from https://op.europa.eu/en/publication-detail/-/publication/d1be1b43-e18f-11e8-b690-01aa75ed71a1 (accessed December 2024).

European Commission. (2019). Communication from the Commission to the European Parliament, the European Council, the Council, the European Economic and Social Committee, and the Committee of the Regions: The European Green Deal, COM (2019) 640 final. Brussels. Retrieved from https://eur-lex.europa.eu/legal-content/EN/TXT/?uri=CELEX%3A52019DC0640&qid=1751706513414 (accessed December 2024).

European Commission. (2020a). Communication from the Commission to the European Parliament, the Council, the European Economic and Social Committee and the Committee of the Regions: A new Circular Economy Action Plan for a cleaner and more competitive Europe, (COM(2020) 98 final). Brussels. Retrieved from https://eur-lex.europa.eu/legal-content/EN/TXT/?uri=CELEX%3A52020DC0098&qid=1751706600471 (accessed December 2024).

European Commission. (2020b). Proposal for a Regulation of the European Parliament and of the Council Concerning Batteries and Waste Batteries, Repealing Directive 2006/66/EC and amending Regulation (EU) No 2019/1020 (COM(2020) 798 final). Retrieved from https://eur-lex.europa.eu/legal-content/EN/TXT/?uri=CELEX%3A52020PC0798&qid=1751706692577 (accessed December 2024).

European Parliament and Council. (1999). Directive 1999/5/EC on Radio Equipment and Telecommunications Terminal Equipment, O.J.L 91. Retrieved from https://eur-lex.europa.eu/legal-content/EN/TXT/PDF/?uri=CELEX:31999L0005&qid=1751706812880 (accessed December 2024).

European Parliament and Council. (2003a). Directive 2002/96/EC of 27 January 2003 on Waste Electrical and Electronic Equipment (WEEE). O.J.L 37. Retrieved from https://eur-lex.europa.eu/legal-content/EN/TXT/?uri=CELEX%3A32002L0096&qid=1751706886843 (accessed December 2024).

European Parliament and Council. (2003b). Directive 2003/108/EC Amending Directive 2002/96/EC on Waste Electrical and Electronic Equipment (WEEE). O.J.L. 345. Retrieved from https://eur-lex.europa.eu/legal-content/EN/TXT/PDF/?uri=CELEX:32003L0108&qid=1751706980638 (accessed December 2024).

European Parliament and Council. (2004). Directive 2004/108/EC of 15 December 2004 on the Approximation of the Laws of the Member States Relating to Electromagnetic Compatibility. O.J.L 390. Retrieved from https://eur-lex.europa.eu/legal-content/EN/TXT/PDF/?uri=CELEX:32004L0108&qid=1751707040165 (accessed December 2024).

European Parliament and Council. (2005). Directive 2005/32/EC of 6 July 2005 Establishing a Framework for the Setting of Ecodesign Requirements for Energy-Using Products. O.J.L 191. Retrieved from https://eur-lex.europa.eu/legal-content/EN/TXT/PDF/?uri=CELEX:32005L0032&qid=1751707125607 (accessed December 2024).

European Parliament and Council. (2006a). Directive 2006/66/EC of 6 September 2006 on Batteries and Accumulators and Waste Batteries and Accumulators. O.J.L 266. Retrieved from https://eur-lex.europa.eu/legal-content/EN/TXT/PDF/?uri=CELEX:32006L0066&qid=1751707194761 (accessed December 2024).

European Parliament and Council. (2006b). Directive 2006/95/EC of 12 December 2006 on the Harmonization of the Laws of the Member States Relating to Electrical Equipment Designed for Use within Certain Voltage Limits, O.J.L 374. Retrieved from https://eur-lex.europa.eu/legal-content/EN/TXT/PDF/?uri=CELEX:32006L0095&qid=1751707250564 (accessed December 2024).

European Parliament and Council. (2009). Directive 2009/125/EC of the European Parliament and of the Council of 21 October 2009 Establishing a Framework for the Setting of Ecodesign Requirements for Energy-Related Products. Retrieved from https://eur-lex.europa.eu/legal-content/EN/TXT/PDF/?uri=CELEX:32009L0125&qid=1751707327500 (accessed December 2024).

European Parliament and Council. (2010). Regulation (EU) No 995/2010 of the European Parliament and of the Council of October 20, 2010 Laying Down the Obligations of Operators Who Place Timber and Timber Products on the Market. O.J.L 2010. Retrieved from https://eur-lex.europa.eu/legal-content/EN/TXT/PDF/?uri=CELEX:32010R0995&qid=1751707398063 (accessed December 2024).

European Parliament and Council. (2023). Regulation (EU) 2023/1542 of the European Parliament and of the Council of July12, 2023 on Batteries and Waste Batteries,

Establishing a Framework for the Sustainable Design and Use of Batteries, and Repealing Regulation (EU) No 1103/2010. O.J.L 2023. Retrieved from https://eur-lex.europa.eu/legal-content/EN/TXT/PDF/?uri=CELEX:32023R1542&qid=1751707471054 (accessed December 2024).

European Parliament and Council. (2024b). Regulation (EU) 2024/1252 of the European Parliament and of the Council of April 11, 2024 Establishing a Framework for Ensuring a Secure and Sustainable Supply of Critical Raw Materials and Amending Regulations (EU) No 168/2013, (EU) 2018/858, (EU) 2018/1724 and (EU) 2019/1020. O.J.L 2024. Retrieved from https://eur-lex.europa.eu/legal-content/EN/TXT/PDF/?uri=OJ:L_202401252&qid=1751707571958 (accessed December 2024).

European Parliament, the Council, and the Commission. (2003). Joint Declaration Relating to Article 9 of Directive 2002/96: Financing in Respect of WEEE from Users Other than Private Households. O.J.L. 37/39. Retrieved from https://eur-lex.europa.eu/resource.html?uri=cellar:ac89e64f-a4a5-4c13-8d96-1fd1d6bcaa49.0004.02/DOC_2&format=PDF (accessed December 2024).

García Molyneux, C. (2001). Important Stage Reached in Drafting of Controversial Electrical and Electronic Equipment Directives. *EuroWatch*, 13(20).

García Molyneux, C. (2008). The EU Environmental Regulation of Electrical and Electronic Equipment. In K. Kritzer et al. (Eds.), *International Contract Manual* (Vol. 2, Chapter 38). Thomson.

García Molyneux, C. (2021). The Dilemma of EU Waste Policy and Circular Economy: Deconstructing the Environmental Regulation of an Electronic Tablet. In A. Orsini & E. Kavvatha (Eds.), *EU Environmental Governance: Current and Future Challenges* (pp. 57–78). Routledge Studies in Environmental Policy.

García Molyneux, C. (2023). New Removability and Replaceability Requirements for Batteries Marketed in the European Union. *Pratt's Energy Law Report*, *23*(5), 179–185.

Hunter, R., & García Molyneux, C. (2000). The Draft EC Directive on Electronic Equipment: An Imprudent and Illegal Proposal. *International Environment Reporter* (BNA), November 22.

Kautto, P. (2008). *Who Holds the Reins in Integrated Product Policy: An Individual Company as a Target of Regulation and as a Policy Maker*. Helsinki School of Economics (HSE Print).

McLoughlin, A. (2005). What Is in a Name? Regulation of Electrical and Electronic Products. *European Environmental Law Review*, 252–264. Retrieved from https://doi.org/10.54648/eelr2005037

Organisation for Economic Co-operation and Development. (2016). Extended Producer Responsibility: A Guidance Manual for Governments. Retrieved from https://www.oecd.org/content/dam/oecd/en/publications/reports/2001/03/extended-producer-responsibility_g1gh2874/9789264189867-en.pdf (accessed December 2024).

Pouikili, K. (2020). Concretising the Role of Extended Producer Responsibility in the European Union Waste Law and Policy through the Lens of the Circular Economy. *ERA Forum*, *20*, 491–508.

# 8

# TOWARD AN INTEGRATED POLICY FRAMEWORK FOR ENVIRONMENTAL GOVERNANCE? GENEALOGIES, DISCOURSES, AND PRACTICES OF "ONE HEALTH"

*David Seke and François Thoreau*

## Introduction

In 2020, a succession of mass lockdowns struck most societies worldwide following the COVID-19 pandemics. Anybody could witness scenes that, not so long ago, would have been unconceivable: classes of teenagers all wearing masks, public services shut down, and anyone's movements dependent on health prevention policies. Fiction somehow overwhelmed reality. Yet fiction was already written. The movie "Contagion" from Steven Soderbergh depicted this reality-to-be already in 2011, nine years ahead of the pandemics, based on the World Health Organization (WHO)'s scenarios of preparedness. The reality followed the script. The only missing element from the movie was the more-than-human question: the virus allegedly circulating from animal species such as pangolins or bats. And yet, this has been one of many more sanitary scandals and public health issues involving animals: from the BSE "mad cow" disease to avian or porcine flus.

Over the last two decades "One Health" (OH) has emerged as an important theme in global health policies to face such predicaments. In a nutshell, OH is a discursive policy tool that contends that the health of humans, animals, and ecosystems are interrelated. Understandably, it took momentum in the wake of the COVID-19 pandemic where the zoonotic risk of disease transmitted from animals to humans became prevalent. Before that, it was promoted by different international institutions, each using different definitions reflecting their own agenda. In the wake of the pandemic, it rose abruptly at the international agenda of health and public health policies which has led to a need to unify the approach.

DOI: 10.4324/9781003479260-13

In this context an expert panel was tasked in 2022 to come up with a unique definition of the term. Known as the One Health High-Level Expert Panel, it has been embraced by a quadripartite alliance of key United Nations institutions[1] (Kaplan et al., 2023). In 2022, this panel came up with the following consensual definition of OH:

> One Health is an integrated, unifying approach that aims to sustainably balance and optimize the health of people, animals and ecosystems. It recognizes the health of humans, domestic and wild animals, plants, and the wider environment (including ecosystems) are closely linked and interdependent. The approach mobilizes multiple sectors, disciplines and communities at varying levels of society to work together to foster well-being and tackle threats to health and ecosystems, while addressing the collective need for clean water, energy and air, safe and nutritious food, taking action on climate change, and contributing to sustainable development.
>
> *(FAO, WHO et al., 2022)*

In this definition, OH remains a rather elusive concept with broad terms such as "integrated, unifying approach" without specifying what is integrated and unified nor by whom "or indeed how they should actually go about it" (Cassidy, 2016). Arguably, somehow by design, a notion like OH must remain vague as its scope has to cover the agenda of such varied institutions as the ones partaking to the high-level group; but by doing so it means it also reaches important regulatory bodies such as the European Commission or the United States (US) Environmental Protection Agency.

And this is what happened. The increasing popularity of the term meant that the European Union (EU) progressively introduced it in its policy and legal discourse as a guiding framework for several European initiatives (Coli & Schebesta, 2023). The language of OH has also been increasingly mobilized by the FAO, WHO, and World Organization for Animal Health (WOAH), "given the influence of these institutions in shaping health research, policy, and practice globally, it is important to understand why OH has had so much traction with these actors" (Cassidy, 2016, p. 214). So far, the EU uses OH mostly as an overarching framework for its pharmaceutical strategy, including the different regulation aimed at limiting antimicrobial resistance and aimed at the prevention and preparedness to zoonotic diseases.[2]

In this chapter, we carry out a literature review and discursive analysis of the OH discourse. To do that, our contribution relies on a state-of-the-art of the scientific literature, including in social sciences and more specifically on the history of science, from which we will show the historic lineage of the OH discourse. (1) We show how this notion now aims at integrating different dimensions that were thus far held separate. Building on grey literature, i.e., reports and policies from international institutions, we then trace back the emergence of this notion as a primary concern for public policies (2) In the last section, we unpack the consequences of the advent of a OH integrated agenda for environmental governance, particularly in the EU setting (3) We finish by discussing some prospective elements.

## From separations to integration

This section offers a condensed history of OH which shows how the health of humans, animals, and ecosystems has come to be thought of as separate. This is important to understand why there is a need to bridge back what has been separated, thus the fundamental context of emergence of OH policies.

### *The great divides*

Genealogies of the OH narrative often insist on it being the product of great man and great visionaries (Evans & Leighton, 2014; Lerner & Berg, 2015). Names such as Rudolf Virchow, Robert Koch, William Osler, John McFadyean, James Steele, and Calvin Schwabe are regularly mentioned in this regard. Works of such key historical figures are presented in a linear fashion, as if a unique path was unavoidably leading to the advent of OH, with the notion possessing some kind of inner historic necessity. Calvin Schwabe, perhaps the most acknowledged of these "founding fathers", published an essay in 1978 in which he demonstrated the close interrelations between medical knowledge and veterinary knowledge. Humans, he contends, learned a great deal of their medical knowledge from the art of taking care of animals. Schwabe offers a very long history of human-animal relationships, going back, for example, to the domestication of sheep and cows.

As valuable these insights may be, Bresalier et al. (2021) point out that it matters to make room for a wider context of emergence that traces back the history of OH to a greater deal of historical figures and moments. Going so far as the pre-modern era, the authors develop a far-reaching historical perspective that acknowledges the historical role of human-animal relations in many civilizations while focusing on ancient Greece. Aristotle, they argue, can be cast as an early adopter of comparative medicine. In his series of books 'Historia Animalium' and through his numerous dissections, he highlighted the common characteristics among different species and the possibility to learn about humans from animals (Bresalier et al., 2021; Evans & Leighton, 2014). Additionally, the physician Hippocrates (460 BCE–367 BCE) wrote about the interdependence between public health and a clean environment, identifying that bodily health was influenced by changes in the environment (Bresalier et al., 2021; Evans & Leighton, 2014). Hippocrates' disciples objected to intentionally cause injuries for the pursuit of knowledge be it on humans or, significantly, on animals (Kirk & Worboys, 2011).

The Aristotelian program of animal research along with the writings of Hippocrates created a medical framework based on humors as a baseline for describing the health of both humans and animals. This framework has been influential in Western thought as far as the 18th century (Ancheta et al., 2021a) when it was thoroughly challenged by the development of modern, scientific medicine. Interestingly, the humorism cut across species boundaries is not mere anthropomorphism but the product of the belief that human and animal share humoral corporality. Assumably, modern medicine and the scientific revolution entrenched a separation between humans and non-humans, between nature and culture, which has led to distinguish, delineate, if not sometimes oppose, humans from animals and ecosystems. In that sense, OH policies can be seen as an attempt to bridge what has been unbridged, to reassemble what has been separated.

### *The advent of the veterinary profession*

The establishment of the first veterinary schools in the second half of the 18th century is a good indicator of that separation. It can be considered as a pivotal moment that distinguished humans from their fellow commensals and environments. This development marked the beginning of a revolution in animal health practices (Bresalier et al., 2021). However, veterinary medicine was not initially a domain apart from human medicine. In fact, many medical men shaped the early development of veterinary medicine by occupying important roles in these schools such as vice presidents, board members, teachers, and examiners (Woods, 2017). The separation of the two disciplines was gradual and followed progressive paths of institutionalization.

The advent of veterinary professionals at the beginning of the 19th century gradually eroded the social, cultural, and epistemological continuity with human medicine in order to gain more disciplinary independence. For instance, in Britain, at the end of the Napoleonic wars in 1815, military demands for veterinary services diminished. Thus, veterinary practitioners needed to enter the competitive market of animal healing in which they would need to coexist with vernacular and other popular healing practices, in such a way that "the veterinarian's superiority was not self-evident" (Woods, 2017, p. 512). Just like medicine (Stengers, 2003), the veterinary profession had to make its own claims to rationality and political legitimacy. Efforts were driven toward establishing "veterinary medicine as an independent sphere, a "sister" profession rather than a child of the medical parent" (Woods, 2017, p. 512).

The history of that institutionalization is needed to understand the very rationale that takes shape throughout the 19th and 20th centuries. The advent of a vet science insisted on what differentiates bodies and anatomical scrutiny from one species to another, rather than what is similar. This specific anatomical science was important to establish a proper veterinarian science on the very grounds that contributed to establish modern medicine. In short, it could be said that human medicine had rather universalist claims over bodies and cure and was seeking to prevail over veterinary medicine, while veterinary reformers, on the contrary, emphasized these distinctions, to establish a distinctive and autonomous identity for their profession (Woods, 2017). This more than anything offers a crucial insight into initiatives such as One Medicine and OH whose very premise should call our attention to various practices through which actors in the field of health negotiate and navigate various disciplinary and institutional boundaries that constitute the field of health.

### *The integrative endeavor of global health: One Medicine*

From the 19th century on, veterinary and medical professions progressively grew apart from one another. Yet some initiatives were developed to try to bring back what had previously been separated. In that respect, initiatives were taken for greater cooperation between the two disciplines. "One Medicine" (OM) is one such initiative. Calvin Schwabe is often considered as the initiator of OM. He was a zoologist and veterinarian who extensively studied parasitology and epidemiology (Woods et al., 2018). When leading a research program on the parasitic tapeworm *E. granulosus*, Schwabe had to study the parasite across many professional boundaries "creating professional and disciplinary connections that brought the tapeworm's movements and exchanges into relief" (Woods et al., 2018, p. 196). His insight gathered into a disciplinary critique of human and veterinary medicine. He argues that the separation between human and veterinary medicine was essentially an artifact from their past histories and rivalries, rather than the product of meaningful differences in their object of study.

Schwabe went on to suggest, in his 700-page textbook *Veterinary Medicine and Human Health* (Kaplan et al., 2023; Schwabe, 1984), a reform of veterinary research, education, and practice using OM as the core organizing principle (Cassidy, 2016, p. 218). OM was born in a context where the status of animals in medical research shifted from a diverse range of varied animals toward a more restricted one, which would consider foremost laboratory animals, i.e., those animals whose form-of-life could be controlled and maintained within the enclosed space of the laboratory[3] (Bresalier et al., 2021). As a result, OM put greater emphasis on biomedical and veterinary research, entertaining a clinical focus which lacked for the most part input from ecology and environmental approaches (Ancheta et al., 2021b).

Although OM was in theory a collaborative effort from the two branches of medicine, "the greater number of medical schools and higher social status of physicians compared to veterinarians discouraged collaboration and gave the biomedical view that medicine had more influence regarding public health issues" (Ancheta et al., 2021b, p. 1132). Nevertheless, between the end of the 1990s and the early 2000s OM gained some traction within international organizations mainly through veterinary public health (Zinsstag et al., 2011). By the beginning of the new millennium "different groups of the ongoing traditions of comparative medicine and veterinary public health came together into a rearticulated vision of 'One Medicine, One Health'" (Bresalier et al., 2021, p. 35). For this reason, OM is often reduced to a direct predecessor to OH. However, both concepts are the results of the convergence of different, albeit overlapping, sets of actors and agendas. Which is why OM is more heavily dominated by clinical research in comparative medicine and veterinary public health (Michalon, 2020).

## The advent of One Health in global health discourses

This section traces back the emergence of the OH discourse in global governance. It details how it has been defined prior to, and after, the 2022 consensual definition. OH used to reflect more clearly the agenda of varied institutions and was enlarged to fit them all, gaining reach and span but arguably losing some traction.

### *First formulations*

After World War II, in international relations "the idea that national interests should be overcome in order to deal with international problems became highly compelling" (Cassidy, 2016, p. 219). It was not before the 1990s that it took shape in international health policy circles. It emerged as a response to diverse health crisis such as HIV and to growing concerns about infectious diseases. This spurred a framing transition from "international" to "global health" (Bresalier et al., 2021; Cassidy, 2016; Leboeuf, 2011). A significant contribution to this evolving field came from the Wildlife Conservation Society (WCS), a U.S.-based NGO composed of experts including, among others, veterinarians, zoologists, and epidemiologists. Their aim was to harness their veterinary and general health expertise to tackle challenges to wildlife conservation (WCS, n.d.).

Following the SARS and avian influenza outbreaks, WCS hosted in 2004 a series of meetings and most importantly the symposium "Building Interdisciplinary Bridges to Health in a Globalized World" in New York. This workshop led to the formulation of the so-called "Manhattan Principles". These detailed a collaborative, transdisciplinary approach. Coined "One World – One Health", these principles were recently updated, in 2019, as the "Berlin principles on One Health" (Gruetzmacher et al., 2021). From these meetings one major contribution of the WCS was to call for more consideration of "ecosystem health". This implies the crucial idea that, just like humans or animals, "ecosystems can also be regarded as patients and evaluated as healthy or otherwise" (Woods et al., 2018, p. 206).

These early formulations are essential to understand that there are, or have been, very much different concerns which feed into the very definition of OH. For instance, even on a central topic such as zoonoses, different *foci* are already apparent. WCS veterinarians, for example, because of the NGO focus on wildlife preservation, tended to tackle infectious diseases through their movements within a variety of hosts across multiple species, in complex loops between wildlife and human livestock among others. Their understanding reflected concerns of biodiversity and insertion of wildlife into humane ecosystems. This stands in sharp contrast with OM and other international agencies approach that tended to consider zoonosis as a mere and rather straightforward transmission of infections from animals to humans (Woods et al., 2018). Such difference matters as these different narratives also trigger different political positions regarding the appropriate policy response for the management of health issues. To understand OH, one must examine the key role that international agencies have played in shaping these narratives.

### *Many One Health agendas*

This section explores the OH agenda through a broad analysis of the documents produced mainly by international agencies from 2008 to 2023 (FAO et al., 2008, 2010, 2017; FAO, WHO, et al., 2022; FAO, WOAH, et al., 2022). By doing so, one is better equipped to grasp how OH exists at the intersection of these various narratives, as well as their importance in shaping the policy response of international agencies.

The first major document in the modern formulation of OH is the Strategic Framework for Reducing Risks of Infectious Diseases at the Animal–Human–Ecosystems Interface (FAO et al., 2008), an interface which included varied international institutions. Indeed, this policy framework laid the foundation for the development of the OH narrative, both with respect to meaning and reach. In particular, the document can be understood as a consequence to the recent avian influenza crisis, said to be in a "logical continuation and expansion of the successful current global response to AHI" (FAO et al., 2008, p. 38). However, the agencies had different matters of concerns when it came to handling the influenza crisis. Referring to "One World, One Health" was a means to unify varied concerns into a comprehensive framework. Drawing on organizational studies as well as Science and Technology Studies, Chien (2013) argues that OH allowed the agencies to resolve their inter-agency conflicts about their differing policy frames of avian influenza in the 2000s. She highlights that, at the time, each agency had its preferred solution to the avian flu crisis, leading to the development of competing policy frames: the technical-biomedical frame, the societal-intervention frame, and the ecological-conservation frame.

For instance, the WHO anchored its response in public health outcomes whereas the FAO and OIE emphasized biosecurity measures in poultry farms and markets (Spencer et al., 2019). This approach can be observed in the FAO's biosecurity campaign slogan on the avian flu which was "Find it fast – Kill it quickly-Stop spreading" (FAO, 2008, p. 13). Other authors (Galaz et al., 2015; Leboeuf, 2011; Scoones & Forster, 2010) make a similar analysis in which these policy responses are accompanied by different albeit overlapping 'outbreaks narratives' each focused on a different framing of the avian flu. This showcases how the tension of articulating heterogenic narratives/perspectives is a central mechanism in the international development of OH. Overall, the policy response to the avian flu has been a crucial step in understanding the development of OH because the period establishes, or at the very least reinforces, the topics that will gain heavy focus and become the backbones of the framework for years to come.

These topics and their varying importance are reflected in "the definitions of One Health advanced by major international agencies which can be assumed to be representative of their respective sectoral interests" (Abbas et al., 2022, p. 387). This was particularly salient before the quadripartite alliance put forth an OH definition in 2022 (FAO WHO et al., 2022). Prior to that integration, the institutional concerns were reflected in the very different matters of concern of different agencies. For instance, the WHO emphasized the threat on human health: "by linking humans, animals and the environment, One Health can help to address the full spectrum of disease control – from prevention to detection, preparedness, response and management – and contribute to global health security" (WHO, 2024). The FAO insisted on food safety as can be observed in its strategic framework, a document that highlights the institutions' priorities. The FAO considers OH a Programme Priority Area which falls under imperatives of 'better production' or to "ensure sustainable consumption and production patterns, through efficient and inclusive food and agriculture supply chains at local, regional and global level (…)" (FAO, 2021). Finally the WOAH, besides the common definition of OH, calls for the detection of zoonosis, in particular if it was shown to have an impact on domestic animal, trade animal, or wildlife (WOAH, 2022, p. 128).

This heterogeneity agenda is a key feature of the integrated OH narrative, leading to its conceptual vagueness. Such vagueness helps to create a functional consensus among actors with dissimilar perspectives and could be called "boundary work", in the sense of "purposeful individual and collective effort to influence the social, symbolic, material or temporal boundaries, demarcations and distinctions affecting groups, occupations and organizations" (Langley et al., 2019, p. 5). The implication is that OH's main strength rests with its ability to be loose enough to accommodate different viewpoints in its very definition, while structuring enough to give direction and allow communication across disciplinary and institutional boundaries (Bresalier et al., 2021; Chien, 2013; Leboeuf, 2011). So far OH policies struggle to equally integrate its three key components, namely human medicine, veterinary medicine, and ecology (Morand et al., 2020). This stake is crucial to the development of the OH narration. Different authors quoted above (Chien, 2013; Galaz et al., 2015) seek to make sense of these boundary dynamics to transform tensions and encourage coordination (Chien, 2013, p. 222), hence attempting to unfold "the challenges associated with bringing together the distinct organizational cultures, priorities, approaches and interests of the Tripartite's members" (Spencer et al., 2019, p. 6).

This perspective invites to investigate the relations between the OH protagonists from multiple angles. For instance, it shows how some specific disciplinary expertise, concerns, and interests, shaped the evolution of various ideas that contributed to the development of global health such as: OM, One World One Health (OWOH), and OH. These efforts are constantly producing tentative versions of OH, shaped by the power dynamics of the actors and institutions involved. More importantly, this genealogy also draws our attention to the respective agendas associated with each actor, set of expertise, and relevant scientific disciplines (Morand et al., 2020; Woods et al., 2018). Each may foster or strengthen its political legitimacy and their own agenda, while absorbing or resorbing their differences, hence their tensions and conflicts, with other institutions. Under the big umbrella of OH and such unifying notions, it is easier for individual bureaucrats, organizations, and experts to reach the consensual idea that we are all, wherever we are, whatever we do or say, somehow complementary. This is why it matters to depart from discourses and try to locate more precisely what OH policies are about in practice and most specifically as they engage with environmental governance.

## The role of environmental governance in One Health

This section details a great discrepancy between discourses of OH and practices. While "environment" is constantly used as a category for public action and legitimizes OH approaches, through examples such as antimicrobial resistance or zoonotic diseases, they remain understudied, and environmental problems underfunded.

### *The prevalence of the environmental agenda setting*

The OH narratives are originally entrenched in infectious diseases, which were already considered a transgression when it emerged on the public agenda.

Considering infectious diseases could lead to "a shift from the primary focus on highly pathogenic avian influenza to (re-)emerging zoonotic diseases at the animal–human–ecosystems interface" (FAO et al., 2008, p. 43). This emphasis on infectious diseases continued in the subsequent statements (FAO et al., 2010, 2017) and up to these days. Environmental issues were considered on the sideline in 2017, when the tripartite alliance determined "[these] three technical topics – **antimicrobial resistance, rabies** and **zoonotic influenza** – to showcase the importance of multi-sectoral collaboration for effective risk management" (FAO et al., 2017, p. 2). These documents show that the primary focus of OH has been mostly on human and animal health issues with ecological considerations remaining notably marginal.

However a closer inspection of the specific topics and disciplines that became important in OH reveals that one frame, the biosecurity frame, has become dominant over the others (Van Patter et al., 2023). For instance, in the first joint statement, climate change is only mentioned once, while infectious diseases have dedicated chapters (FAO et al., 2008). The norms and concerns of a technical-biomedical frame came to dominate the realization of OH (Chien, 2013, p. 222). The result is that biosecurity, i.e., the focus on barrier systems, surveillance, and sanitation, became one of, if not the most developed aspects of the framework. Some critique this biomedical framing of the issue on the grounds that it fosters a conception of disease management whose main criteria for success is the ability to track and prevent the transmission of pathogens by controlling "the flow of undesirables across territories and bodies" (Hinchliffe et al., 2013, p. 4), instead of truly engaging with the heterogeneity of (local) practices "through which health is made" (Hinchliffe, 2015, p. 31). Another consequence is the relative marginality of ecological concerns, until fairly recently. Indeed, the latest OH report recognizes that "the environmental sector (...) is not always routinely incorporated into the OH approach and there has been limited engagement in cross-sectoral initiatives. (...) [and that] there is good potential to integrate environmental considerations more consistently" (FAO, WHO et al., 2022, p. 11).

These evolutions illustrate the significance of framing in understanding how OH operates. Specifically, they reveal three important insights about OH and its adoption by international organizations, including the EU. First, a unifying rhetoric for OH might lead to prevent from studying the intersection of various fields of health. Because OH exists precisely at the intersection of the different frames that constitute it, it can only be understood in the context of the struggles between the different components that coexist within it. Which is why a deeper engagement with social sciences is necessary to clarify these epistemological and ontological positions that are made blurry by the very nature of OH as a construct that crosses and redraws multiple boundaries (Kingsley & Taylor, 2017).

### *The limits to environmentalism in OH policies, discourses, and practices*

In OH policies, environmental concerns are consistently marginalized to the benefit of human (mostly) and animal (partly) problems. This is the case because of rather consistent shortcomings of OH as it is currently formulated. Since it is preoccupied with the health of humans, animals, and ecosystems, OH, by its very definition, is supposed to question the configuration of human and non-human relations. However, as Van Patter et al (2023, p. 2) points out, OH is permeated by "pervasive anthropocentrism". This refers to its general tendency to consider animals and ecosystems only insofar as they impact human health. With respect to the anthropocentric tendencies of OH, Cañada et al. (2022, p. 2) argue that it is "ethically problematic because it does not consider the ecological implications of putting humans first, instead subjugating non-human suffering to human safety." This is a direct result of the dominance of the biomedical concern which is mirrored in the EU's incorporation of OH in its legislature. This approach to OH is in line with what Coghlan and Coghlan (2018) term "One Health instrumentalism", where ethical considerations are minimally integrated to health. This highlights the separation that is often created between OH's "scientific agenda" and its moral or ethical implications. In this instrumentalist perspective OH is "first and foremost a practical and programmatic call for interdisciplinary collaboration" (Verweij & Bovenkerk, 2016, p. 2) and therefore leaves unexamined the "hard ethical" questions such as "how morally to handle trade-offs between human and nonhuman health" (Coghlan & Coghlan, 2018, p. 3).

Furthermore, this anthropocentrism expresses itself in the way OH selectively pursue its 'holistic agenda' (Lindenmayer et al., 2022). In the case of humans, the broad spectrum of factors that comprise human health are more likely to be taken into account. For instance, it is widely accepted that human health goes beyond physical well-being but also includes mental and social well-being (...) not just the absence of disease or infirmity (WHO, 2023). Yet when it comes to other-than-human animals Van Patter et al. (2023, p. 3) observe that "health tends to remain relegated to physical disease and human use factors". Similarly, with regards to ecosystem health, not only is the overall investigation of this dimension of OH research relatively weak (Lindenmayer et al., 2022) but, when included, ecosystem health often remains framed in anthropocentric terms such as making the environment more resilient and sustainable "all with a goal of providing for human societies within the context of perpetual growth and the extractivist logics of capital" (Van Patter et al., 2023, p. 3).

Different authors in social sciences have submitted several propositions in order to move beyond OH anthropocentrism. Drawing on the concept of "expanding circle of moral concern" Coghlan et al. (2021) call for an ethically and practically bolder OH: an expanding circle of moral means to move our ethical considerations from those close by (family, friends, tribe) and extend them toward all humans (Coghlan et al., 2021). This also means that this circle must expand to include non-human animals: "recognising a much higher moral status in nonhuman animals means seriously weighing their interests in moral deliberation" (Coghlan et al., 2021, p. 2). While not overly focused on ecosystems, this concept would nevertheless contribute to covering the blind spots of the instrumentalist approach to OH.

Cañada et al. (2022) address the anthropocentrism of antimicrobial resistance management in OH by drawing on "multispecies ethics in feminist Science and Technology Studies" such as the ethics of care as developed by Puig de la Bellacasa (2017) or Haraway (2013). They argue that this more relational ethics would help to account for the "situatedness of anti-microbial resistance ecologies" and articulate a non-anthropocentric human and non-human relation. It would all the while providing tools that render us more able to address (and maybe go beyond) the potential "flatness emerging from debunking human exceptionalist hierarchies" (Cañada et al., 2022, p. 3). Greater connections with social science make it possible to reflexively examine the ethical implications of OH as a concept. This wider disciplinary engagements suggest a more robust body of research on OH ethics (Capps, 2022; Lindenmayer et al., 2022; Sironi et al., 2022).

## Discussion: the one and the many

John Law describes "one world metaphysics" that essentially operates from the ontological perspective that there exists only "one all-encompassing reality". The implication of that is that any alternative conception of reality is either "merely a belief" or simply mistaken (Law, 2015, p. 127). Hinchliffe (2015) argues that "one world-ism" forms the basis of many OWOH and OH epistemologies. As developed above, this approach is the result of the international framing of the issue, with an emphasis on biosecurity and surveillance. Against this backdrop, Hinchliffe (2015) argues for a broader conception of health (and therefore OH) that accounts for local conditions and peculiar entanglements and, by doing so, moves away from one world-ism to emphasize there are many different contingent worlds. In practice, this could involve placing greater value on, engaging with, and integrating local forms of knowledge into research and disease management. It would imply a greater focus on understanding local health conceptions and policies (Kingsley & Taylor, 2017; Okello et al., 2015).

These considerations suggest that OWOH and OH, in their current formulations, risk "collapsing local diversity into unhelpful 'one-size-fits-all' global policy making" (Kingsley & Taylor, 2017, p. 9). This highlights the tensions that exist when a global agenda travels between contexts, tensions related to what Tsing (2012) calls scalability. Scalability refers to the process by which knowledge and practices are said to expand to different contexts without changing (Ehrenstein & Neyland, 2018; Tsing, 2012). Hinchliffe's critique of the one world-ism in OH can also be interpreted in terms of a relative skepticism in regards to OH's general ambition of scaling up health intervention. This would be in line with Tsing's perspective seeing limits with knowledge and solutions framed as universal. She essentially considers that "scale work is problematic and doomed to fail because it ignores or erases the diversity of our world" (Ehrenstein & Neyland, 2018, p. 5). Ehrenstein and Neyland try to offer an alternative conclusion with regards to scale work that might be valuable to OH research: they argue that by its very nature global health is inevitably confronted with scale work because of its "normative aim [to] provide health... to the greatest number of people" (Ehrenstein & Neyland, 2018, p. 77). This process therefore comes with moral consequences. While they seem to agree with Tsing (2012) that "scalability is never complete" (p. 510) they do not consider it a reason to not attempt it/engage in this process but instead go about it with the upmost care.

## Conclusion

This chapter examined the increasing adoption of OH by international institutions, to address complex global health challenges, such as zoonotic diseases and environmental degradation, by promoting collaborative efforts across various sectors and disciplines. However, this chapter highlighted several critical issues within the current OH framework that necessitate closer scrutiny and reevaluation.

This chapter put forth a genealogy of OH that was careful in highlighting its contextual and contingent nature. It enabled to understand that the OH international agenda is not self-evident nor inevitable. Rather, it has been made through tentative disciplinary and institutional alliances and rivalry. Such insight leads to consider the significant but unavoidable gap between OH's rhetoric of integration and the practical realities of trying to organize diverse and sometimes competing interests and knowledge.

To delve deeper into these tensions, this chapter engaged with various authors of social sciences and their critiques of OH as currently formulated. One of the main critiques of the existing OH paradigm is its inherent anthropocentrism. By prioritizing human health, often at the expense of ecological well-being, the prevailing OH approach tends to marginalize the broader environmental concerns that are equally crucial for sustainable health outcomes. However, more fundamentally, this approach side steps the ethical concerns that dealing in global health comes with.

In conclusion, the adoption and framing of OH by international bodies constitutes a fertile ground to critically examine the way we engage in the promotion and preservation of health. In the case of OH, future research should aim to examine the blind spot put forth in this chapter. An interesting effort in this direction would be to examine OH, under the lens of boundary work. Various authors have made relevant critiques of the concept on the basis of its ethics (Cañada et al., 2022; Van Patter et al., 2023), its epistemology (Hinchliffe, 2015) or the critique of an over-reliance on biomedical knowledge for intervention (Hinchliffe et al., 2013; Kingsley & Taylor, 2017). In essence, what these critiques have in common is their perception of the boundaries of OH as conceptually too narrow and they work to expand them so as to enrich the concept with broader perspectives.

## Notes

1 The Food and Agriculture Organization of the UN (FAO), the WHO, the World Organisation for Animal Health (WOAH, formerly OIE), and the United Nations Environment Program.
2 https://health.ec.europa.eu/one-health_en (last consulted 2024-11-22).
3 For more on this see Anchenta et al. (2021b) who refer in their genealogy of OH to this period as the "Guinea Pig Era".

## Bibliography

Abbas, S. S., Shorten, T., & Rushton, J. (2022). Meanings and Mechanisms of One Health Partnerships: Insights from a Critical Review of Literature on Cross-government Collaborations. *Health Policy and Planning*, *37*(3), 385–399. https://doi.org/10.1093/heapol/czab134

Ancheta, J., Fadaak, R., Anholt, R. M., Julien, D., Barkema, H. W., & Leslie, M. (2021a). The Origins and Lineage of One Health, Part I. *The Canadian Veterinary Journal*, 62(8), 883–885.

Ancheta, J., Fadaak, R., Anholt, R. M., Julien, D., Barkema, H. W., & Leslie, M. (2021b). The Origins and Lineage of One Health, Part II. *The Canadian Veterinary Journal*, 62(10), 1131–1133.

Bresalier, M., Cassidy, A., & Woods, A. (2021). One Health in History. In J. Zinsstag, E. Schelling, L. Crump, M. Whittaker, M. Tanner, & C. Stephen (Eds.), *One Health: The Theory and Practice of Integrated Health Approaches* (Second edition, pp. 19–43). CAB International.

Cañada, J. A., Sariola, S., & Butcher, A. (2022). In Critique of Anthropocentrism: A More-than-Human Ethical Framework for Antimicrobial Resistance. *Medical Humanities*, *48*(4), e16–e16. https://doi.org/10.1136/medhum-2021-012309

Capps, B. (2022). One Health Ethics. *Bioethics*, *36*(4), 348–355. https://doi.org/10.1111/bioe.12984

Cassidy, A. (2016). One Medicine? Advocating (Inter)disciplinarity at the Interfaces of Animal Health, Human Health, and the Environment. In S. Frickel, M. Albert, & B. Prainsack (Eds.), *Investigating Interdisciplinary Collaboration: Theory and Practice across Disciplines* 213–235. Rutgers University Press. https://www.ncbi.nlm.nih.gov/books/NBK395883/

Chien, Y.-J. (2013). How did international agencies perceive the avian influenza problem? The adoption and manufacture of the "One World, One Health" framework. *Sociology of Health & Illness*, *35*(2), 213–226. https://doi.org/10.1111/j.1467-9566.2012.01534.x

Coghlan, S., & Coghlan, B. (2018). One Health, Bioethics, and Nonhuman Ethics. *The American Journal of Bioethics*, *18*(11), 3–5. https://doi.org/10.1080/15265161.2018.1524224

Coghlan, S., Coghlan, B. J., Capon, A., & Singer, P. (2021). A Bolder One Health: Expanding the Moral Circle to Optimize Health for All. *One Health Outlook*, *3*(1), 21. https://doi.org/10.1186/s42522-021-00053-8

Coli, F., & Schebesta, H. (2023). One Health in the EU: The Next Future? *European Papers - A Journal on Law and Integration*, *8*(1), 301–316. https://doi.org/10.15166/2499-8249/652

Ehrenstein, V., & Neyland, D. (2018). On Scale Work: Evidential Practices and Global Health Interventions. *Economy and Society*, *47*(1), 59–82. https://doi.org/10.1080/03085147.2018.1432154

Estebanez, J., & Boireau, P. (2022). One Health: A Social Science Discussion of a Global Agenda. *Parasite*, *29*, 17. https://doi.org/10.1051/parasite/2022014

Evans, B. R., & Leighton, F. A. (2014). A History of One Health. *Revue Scientifique Et Technique (International Office of Epizootics)*, *33*(2), 413–420. https://doi.org/10.20506/rst.33.2.2298

FAO. (2008). *Biosecurity for Highly Pathogenic Avian Influenza: Issues and Options*. Food and Agriculture Organization of the United Nations.

FAO. (2021). *Strategic Framework 2022-31* (p. 39). https://www.fao.org/strategic-framework/en

FAO, WHO, UNEP, & WOAH. (2022). *One Health Joint Plan of Action, 2022–2026*. FAO, UNEP, WHO, WOAH (founded as OIE). https://doi.org/10.4060/cc2289en

FAO, WHO, & WOAH. (2017). *The Tripartite's Commitment Providing Multi-sectoral, Collaborative Leadership in Addressing Health Challenges* (p. 4). https://www.woah.org/app/uploads/2018/05/tripartite-2017.pdf

FAO, WOAH, WHO, & UNEP. (2022). *Strategic Framework for Collaboration on Antimicrobial Resistance* (p. 22).https://www.who.int/publications/i/item/9789240045408

FAO, World Health Oragnization, & World Organization for Animal Health. (2010). *The FAO-OIE-WHO Collaboration. Sharing Responsibilities and Coordinating Global Activities to Address Health Risks at the Animal-Human-Ecosystems Interfaces. A Tripartite Concept Note* (p. 8). https://www.who.int/publications/m/item/the-fao-oie-who-collaboration

FAO, World Health Oragnization, World Organization for Animal Health, UN System Inluenza Coordination, Unicef, & The World Bank. (2008). *Contributing to One World, One Health. A Strategic Framework for Reducing Risks of Infectious Diseases at the Animal-Human-Ecosystems Interface* (p. 68). fao archive. https://www.fao.org/3/aj137e/aj137e00.pdf

Galaz, V., Leach, M., Scoones, I., & Stein, C. (2015). *The Political Economy of One Health Research and Policy* (p. 46). STEPS Centre.

Gruetzmacher, K., Karesh, W. B., Amuasi, J. H., Arshad, A., Farlow, A., Gabrysch, S., Jetzkowitz, J., Lieberman, S., Palmer, C., Winkler, A. S., & Walzer, C. (2021). The Berlin Principles on One Health – Bridging Global Health and Conservation.

*Science of The Total Environment*, *764*, 142919. https://doi.org/10.1016/j.scitotenv.2020.142919

Haraway, D. J. (2013). *When Species Meet*. University of Minnesota Press.

Hinchliffe, S. (2015). More than One World, More than One Health: Re-configuring Interspecies Health. *Social Science & Medicine*, *129*, 28–35. https://doi.org/10.1016/j.socscimed.2014.07.007

Hinchliffe, S., Allen, J., Lavau, S., Bingham, N., & Carter, S. (2013). Biosecurity and the Topologies of Infected Life: From Borderlines to Borderlands. *Transactions of the Institute of British Geographers*, *38*(4), 531–543. https://doi.org/10.1111/j.1475-5661.2012.00538.x

Kaplan, B., Conti, L. A., Carter, C. N., Kahn, L. H., Yuill, T. M., Barrentine, B., Monath, T. P., Woodall, J. P., Chapman, H. J., & Seifman, R. (2023). 'One Medicine–One Health': An Historic Perspective. *World Medical Journal*, *69*(2), 18–27.

Kingsley, P., & Taylor, E. M. (2017). One Health: Competing Perspectives in an Emerging Field. *Parasitology*, *144*(1), 7–14. https://doi.org/10.1017/S0031182015001845

Kirk, R. G. W., & Worboys, M. (2011). Medicine and Species: One Medicine, One History? In M. Jackson (Ed.), *The Oxford Handbook of the History of Medicine* (pp. 1–22). Oxford University Press. https://doi.org/10.1093/oxfordhb/9780199546497.013.0031

Langley, A., Lindberg, K., Mørk, B. E., Nicolini, D., Raviola, E., & Walter, L. (2019). Boundary Work among Groups, Occupations, and Organizations: From Cartography to Process. *Academy of Management Annals*, *13*(2), 704–736. https://doi.org/10.5465/annals.2017.0089

Lapinski, M. K., Funk, J. A., & Moccia, L. T. (2015). Recommendations for the Role of Social Science Research in One Health. *Social Science & Medicine*, *129*, 51–60. https://doi.org/10.1016/j.socscimed.2014.09.048

Law, J. (2015). What's Wrong with a One-World World? *Distinktion: Journal of Social Theory*, *16*(1), 126–139. https://doi.org/10.1080/1600910X.2015.1020066

Leboeuf, A. (2011). *Making Sense of One Health: Cooperating at the Human-Animal-Ecosystem Health Interface* (Etude de l'ifri 7; Health and Environment Reports, p. 91). Institut français des relations internationales.

Lerner, H., & Berg, C. (2015). The Concept of Health in One Health and Some Practical Implications for Research and Education: What Is One Health? *Infection Ecology & Epidemiology*, *5*(1), 25300. https://doi.org/10.3402/iee.v5.25300

Lindenmayer, J. M., Kaufman, G. E., Baker, L., Coghlan, S., Koontz, F. W., Nieuwland, J., Stewart, K. L., & Lynn, W. S. (2022). One Health Ethics: "What Then Must We Do?"*. *CABI One Health*, *2022*, (p. 4). https://doi.org/10.1079/cabionehealth.2022.0011

Michalon, J. (2020). Accounting for One Health: Insights from the Social Sciences. *Parasite*, *27*, 10. https://doi.org/10.1051/parasite/2020056

Morand, S., Guégan, J.-F., & Laurans, Y. (2020). De One Health à Ecohealth, cartographie du chantier inachevé de l'intégration des santés humaine, animale et environnementale. *Décryptage*, *20*(4), 4.

Okello, A., Welburn, S., & Smith, J. (2015). Crossing Institutional Boundaries: Mapping the Policy Process for Improved Control of Endemic and Neglected Zoonoses in sub-Saharan Africa. *Health Policy and Planning*, *30*(6), 804–812. https://doi.org/10.1093/heapol/czu059

Puig de la Bellacasa, M. (2017). *Matters of Care: Speculative Ethics in more than Human Worlds*. University of Minnesota Press.

Schwabe, C. W. (1984). *Veterinary medicine and human health* (3rd ed). Williams & Wilkins.

Scoones, I., & Forster, P. (2010). Unpacking the International Response to Avian Influenza: Actors, Networks and Narratives. In *Avian Influenza*. Routledge. 18–64.

Sironi, V. A., Inglese, S., & Lavazza, A. (2022). The "One Health" Approach in the Face of Covid-19: How Radical Should It Be? *Philosophy, Ethics, and Humanities in Medicine*, *17*(1), 3. https://doi.org/10.1186/s13010-022-00116-2

Spencer, J., McRobie, E., Dar, O., Rahman-Shepherd, A., Hasan, N., Hanefeld, J., & Khan, M. (2019). Is the Current Surge in Political and Financial Attention to One Health Solidifying or Splintering the Movement? *BMJ Global Health*, *4*(1), e001102. https://doi.org/10.1136/bmjgh-2018-001102

Stengers, I. (2003). The Doctor and the Charlatan. *Cultural Studies Review*, *9*(2), Article 2. https://doi.org/10.5130/csr.v9i2.3561

Tsing, A. L. (2012). On Nonscalability. *Common Knowledge*, *18*(3), 505–524. https://doi.org/10.1215/0961754X-1630424

Van Patter, L. E., Linares-Roake, J., & Breen, A. V. (2023). What Does One Health Want? Feminist, Posthuman, and Anti-colonial Possibilities. *One Health Outlook*, *5*(1), 4. https://doi.org/10.1186/s42522-022-00076-9

Verweij, M., & Bovenkerk, B. (2016). Ethical Promises and Pitfalls of OneHealth. *Public Health Ethics*, *9*(1), 1–4. https://doi.org/10.1093/phe/phw003

WCS. (n.d.). *WCS One World One Health Mission statement*. Retrieved September 6, 2023, from https://oneworldonehealth.wcs.org/About-Us/Mission.aspx

WHO. (2023). *Constitution of the World Health Organization*. https://www.who.int/about/governance/constitution

WHO. (2024). One health. https://www.who.int/health-topics/one-health

WOAH. (2022). *Implementation of WOAH standards: The Observatory Annual Report. First Edition, 2022*. O.I.E (World Organisation for Animal Health). https://doi.org/10.20506/obs.3339

Woods, A. (2017). From One Medicine to Two. *Bulletin of the History of Medicine*, *91*(3), 494–523. https://doi.org/10.1353/bhm.2017.0058

Woods, A., Bresalier, M., Cassidy, A., & Mason Dentinger, R. (2018). *Animals and the Shaping of Modern Medicine: One Health and its Histories* (1st ed.). Springer International Publishing; Imprint: Palgrave Macmillan. https://doi.org/10.1007/978-3-319-64337-3

Zinsstag, J., Schelling, E., Waltner-Toews, D., & Tanner, M. (2011). From "One Medicine" to "One Health" and Systemic Approaches to Health and Well-being. *Preventive Veterinary Medicine*, *101*(3), 148–156. https://doi.org/10.1016/j.prevetmed.2010.07.003

# PART 5

# Reconciling environmental concerns with market imperatives

# 9

# EU TRADE AND ENVIRONMENTAL POLICIES

## Between contradiction and complementarity

*Marcus Gustafsson, Victor Crochet and Elyse Kneller**

### Introduction

The European Union (EU) trade and environmental policies complement each other in several respects, but there are also important areas of contradiction and conflict. In the context of a weakened rules-based global order, several governments, including the EU, are relying ever more on unilateral rather than multilateral solutions to achieve their foreign policy goals. In that situation, the EU has been relying on unilateral trade measures to achieve its environmental policies abroad. These include multiple sustainability initiatives, such as the Carbon Border Adjustment Mechanism (CBAM) and regulations on deforestation and corporate due diligence.

The complementarity between trade and enhanced sustainability considerations is also visible in the EU's partnerships with other trading partners. Under pressure from the European Parliament, sustainability chapters in bilateral free trade agreements (FTAs) have consistently grown in importance. However, they have also come to hinder the EU's free trading ambitions by threatening to torpedo certain negotiations, such as with Mercosur. The EU has moreover spearheaded new and innovative partnership models, such as the Coalition of Trade Ministers on Climate, the Raw Materials Club and several Green and Critical Raw Material Partnerships, which seek to use trade and investment measures as a means to further the green transition. Yet in an effort to reach environmental objectives in such innovative ways, these clubs and partnerships also have the potential to bring about new trade-related challenges.

At the same time, the EU's extensive new industrial policies aimed at building a new green industrial base, and which are relying on trade tools to do

DOI: 10.4324/9781003479260-15

so, stand in potential competition with the EU's environmental aims. Green goods and technologies have been incorporated into the EU's plans to boost supply security domestically and supply chain resilience more broadly. As a result, some of those goods will likely be more expensive to procure and may thus drive up the costs of the green transition.

In this chapter, we thus assess how the trade and environmental policy have come to interact and what to expect in the years ahead. We first discuss how unilateral trade measures have come to complement the EU's environmental goals before turning to how FTAs are doing the same. While doing so, we also highlight the tensions that are arising from mixing trade and environmental objectives. In the third section, we turn to how trade measures are being used by the EU to pursue industrial policies and how this might contradict the EU's net-zero goals. We provide some concluding thoughts in the final section.

## Environment and trade policy: complementary unilateralism

Multilateralism is deeply embedded in the EU's DNA. The EU is itself an international organisation that operates on a treaty basis and with conferred powers. The 1990s and early 2000s, when the EU and the United States were together pursuing neoliberal policies and championing multilateral institutions suited the EU well. One notable achievement was the founding of the World Trade Organization (WTO) in 1995. However, it was also the blockage of the appointment of adjudicators to the WTO's Appellate Body – a permanent body of eight adjudicators which could hear appeals from *ad hoc* dispute settlement panels on breaches of WTO law – alongside the United States' withdrawal from the Trans-Pacific Partnership and the British vote for Brexit in 2016, that has come to bookmark the end of that era.

Indeed, in subsequent years, the Biden administration did not seek to return to the negotiation table of trade deals, nor did it devote much effort to reform or revitalise the WTO (Schropp, 2024). Due to its weakened status, rule-following among WTO members has deteriorated and some developing country members, such as India and South Africa, have held back agreement on new WTO rules (Blenkinsop, 2023; Denamiel et al., 2024). This has included one of the WTO's first environmentally focused agreements, the Agreement on Fisheries Subsidies, aimed at preventing subsidies to illegal and unregulated fishing activities. Even though the Agreement was adopted at the 12th Ministerial Conference on 17 June 2022, several matters were left outstanding, and no agreement was reached at the subsequent Ministerial Conference in 2024 (Denamiel et al., 2024). Article 12 of the Agreement stipulates that, unless WTO Members agree otherwise, a failure by Members to agree on further disciplines by the next Ministerial Conference in 2026 will terminate the Agreement in its entirety. The future of this first environmentally focused WTO Agreement – and only second WTO Agreement agreed since the WTO's founding in 1995 – thus still hangs in the balance.

This same trend of increasingly fraught multilateral negotiations can also be observed among the negotiations of environmental agreements. After the success of the Montreal Protocol in 1987 to limit gases injurious to the ozone layer, and the UN Earth Summit in Rio in 1992, which established the UN Framework Convention on Climate Change, successes have been notably fewer and less consequential. Whereas the EU was an enthusiastic supporter of the Kyoto Protocol agreed in 1997 and as such undertook fairly ambitious emissions targets under both its first (2008–2012) and second (2012–2020) commitment periods, several other developed economies lowered their ambitions or declined to participate in the second period. The United States famously never joined. Although UN Members were able to conclude the Paris Agreement in 2015, its Nationally Determined Contributions that set out countries' emissions reductions remain voluntary. It can therefore be argued that, like in the world of international trade, the level of ambition to strike grand multilateral bargains has steadily declined.

This shift to a lower level of ambition at the multilateral level has prompted the EU to rely more on unilateral measures both in the environmental and trade sphere, and to increasingly use trade measures as the means to enforce and propagate its environment measures. We use the term "unilateral" in the sense of the EU taking individual or "autonomous" action, as opposed to taking action based on multilateral or plurilateral negotiations (e.g., under UN conventions, in the WTO or the Organization for Economic Cooperation and Development – OECD), or bilateral negotiations (e.g., as part of bilateral FTAs). Unilateral does not imply that the measures in any way contravene international rules. The rise in unilateralism merely implies that multilateral avenues to affect change are becoming more challenging.

In the first edition of this edited volume, the authors already pointed to the antecedents of this trend. They highlighted the revised Renewable Energy Directive, and the forthcoming CBAM as examples of "a newfound confidence – or perhaps impatience – on the part of the EU", and of "a clear shift from a multilateral to a unilateral approach in EU environmental and trade policy" (Gustafsson & Crochet, 2021). In the intervening years, this trend has continued apace. CBAM became law (in 2023), together with several other major sustainability laws which have an impact on trade. These include: the Corporate Sustainable Reporting Directive (2022); the Batteries Regulation (2023); the Regulation on Deforestation-free Products (2023); the Ecodesign for Sustainable Products Regulation (2024); and the Corporate Sustainability Due Diligence Directive (2024). To different degrees, the above instruments rely on what Anu Bradford has termed the "Brussels Effect", using the EU's first mover advantage in goods and services regulation and its consumer market to induce foreign jurisdictions to adopt similar rules

(Bradford, 2020). And because it is trade policy that conditions access to the EU's inner market, it has become the means by which these rules are often enforced and operationalised. The Corporate Sustainability Directives operate slightly differently from the other instruments by imposing requirements on EU firms rather than on imported products. Yet by conditioning the ability of the affected firms to operate on the EU market, on those companies implementing certain due diligence requirements throughout their foreign supply chains, the end result is similar. If those supply chains do not meet EU standards, the goods produced cannot be sold on the EU market.

This re-orientation towards unilateralism is reflected in the Commission's trade policy documents. In 2020, the Commission announced an update to its trade policy under the banner of "Open Strategic Autonomy" (European Commission, 2020c). As explained by the Director-General of DG TRADE Sabine Weyand: "Open Strategic Autonomy is a mindset which means we act together with others, multilaterally, or bilaterally, wherever we *can*. And we act autonomously wherever we *must*" (Gehrke, 2022). The implications of this new strategic approach were fleshed out in the Trade Policy Review released in February 2021 (European Commission, 2021b). The Review frames Open Strategic Autonomy as resting on three pillars: resilience and competitiveness; sustainability and fairness; and assertiveness and rules-based cooperation. The third pillar is intended "to showcase the EU's preference for international cooperation and dialogue, but also its readiness to combat unfair practices and use autonomous tools to pursue its interests where needed" (European Commission, 2021b). This avowed readiness to use autonomous measures has thus not only come to colour the EU's trade policy, but also the many different types of environmental measures that have come to depend on EU trade policy for their enforcement.

Some of these dynamics have existed for some time. The EU has long been an international standard-setter *extraordinaire*. For instance, Bradford points to areas such as food, chemical and animal welfare as long-standing examples of EU regulatory action (Bradford, 2020, 170–231). Standards in these areas have had clear implications for trade in goods and some regulations, such as a ban on imports of hormone-treated beef, has led to trade disputes (World Trade Organisation, 2024a). However, in most areas EU standards have been less controversial and the Brussels Effect has allowed the EU to not only set standards and norms with respect to its own market but effectively influenced global standard-setting as well.

However, a unilateral – as opposed to multilateral – approach will always have trade-offs. The recently introduced laws mentioned above appear to be more controversial with the EU's trading partners, and the Brussels Effect may be weakening. Bradford emphasises that the Effect is only operative when five conditions are met: "market size, regulatory capacity, stringent standards, inelastic targets and non-divisibility" (Bradford, 2020, 25). While regulatory capacity, stringent standards and a focus on inelastic targets (in essence, consumer markets) are not in doubt with respect to the recently introduced laws, the relative size of the EU market is shrinking as other economies grow more rapidly. Moreover, supply chains may be "divisible", that is, foreign jurisdictions need not adopt the same stringent regulations wholesale or even at all, since some foreign companies in affected industries may be able to sell on the EU market, while others focus on non-EU markets. At the same time, the EU's foreign competitors – notably China – are focusing exports on other developing country markets (The Economist, 2024). By introducing additional regulatory requirements for EU businesses operating worldwide, these regulations may negatively impact EU companies' competitiveness in those markets. The new regulations may also upset trading partners which see them as contrary to their own regulatory preferences and overly burdensome (Beattie, 2024). By contrast, a multilateral approach could – in theory – have established harmonised rules across countries.

Therefore, while the EU's move towards greater assertiveness and unilateralism in the face of a dysfunctional multilateral system is understandable, and the integration of environmental objectives as a core part of the EU's trade agenda is likely necessary to achieve the goals of the green transition and the Paris Agreement, there is a potential trade-off with respect to EU competitiveness as well as overall economic efficiency.

As discussed in the next section, this same trade-off is also visible in the EU's bilateral and plurilateral trade agenda, which includes bilateral FTAs and different types of new forms of "partnerships".

## Environment and trade policy: complementary partnerships

With the largest network of trade agreements in the world, the EU has long negotiated formal trade agreements to further its economic and political goals. In recent years, however, the EU's trade agreements have also aimed at ensuring sustainable bilateral partnerships. Over time, trade and sustainable development (TSD) chapters have grown to become a key pillar of the EU's trade agreement negotiating mandate (see e.g., Council of the European Union, 2018). The EU has made it clear that trade with third countries must not only bring about economic and political value, it must also protect the EU's core values of democracy, the rule of law, Human Rights, basic freedoms and sustainability (European Commission, 2019). The first EU trade agreement that included a TSD chapter was reached with South Korea and ratified in 2015 (provisionally applied since 2011) (EU-Korea FTA, 2011, Chapter 13). This chapter was quite simple and straightforward in that the parties agreed to strengthen their trade relations and cooperation in ways that promote sustainable development, but did not seek to harmonise their labour or environment standards. The TSD chapter largely reaffirmed the EU's and South

Korea's commitments to effectively implementing the International Labour Organization (ILO) Conventions and multilateral environmental agreements to which they are a party and obliged them to uphold the levels of protection in their labour and environmental laws by not failing to enforce them or reducing them in a manner affecting bilateral trade or investment. The TSD chapter also established an institutional mechanism to discuss matters falling under the TSD chapter, including the resolution of disputes.

Since the ratification of the EU-South Korea FTA, all of the EU's subsequent trade agreements – a total of ten agreements in force and a further six waiting to be ratified – have contained more extensive TSD chapters. Each TSD chapter is tailored to the specific needs of the EU's trading partner. However, as a minimum, all agreements require the parties (i) to effectively implement ILO conventions and environmental agreements, including the Paris Agreement on Climate Change, (ii) not deviate from environmental or labour laws to encourage trade or investment, (iii) sustainably trade natural resources, such as timber or fish, (iv) combat illegal trade in threatened and endangered species of fauna and flora, (v) encourage trade that supports tackling climate change, (vi) cooperate for a shift to a circular and resource-efficient economy, and deforestation-free supply chains and (vii) promote responsible practices, such as corporate social responsibility.

Whilst negotiating high TSD standards in its FTAs is often heralded by the EU (EC, 2024a), the main challenge going forward will be to ensure that the commitments agreed on paper can actually be enforced in practice and are not merely symbolic. Recognising the importance of this, the Commission adopted a "green and just economic growth" strategy with respect to its trade agreements in 2022 (European Commission, 2022). That approach explains that the Commission will aim to adopt time-bound roadmaps in its trade agreements, step-up monitoring and implementation of TSD commitments, allow civil society to lodge complaints, extend dispute settlement compliance and as a last resort, adopt trade sanctions (i.e., a removal or trade concessions) for material breaches.

The application of this approach is likely going to make trade agreement negotiations even more challenging and hence time-consuming, particularly with (developing) countries with low sustainability and environmental standards. Indeed, the EU's TSD demands have caused considerable difficulties for the ratification of the EU-Mercosur trade deal (AP, 2024). While the EU has long dangled access to the "world's biggest market" in front of its negotiating counterparts to convince them to agree on TSD-related demands (Devuyst, 2014, 1; Carnegie Europe, 2023) the increasing regulatory burden that is being placed on third-country businesses wanting to enjoy EU market access (see Section "Environment and trade policy: complementary unilateralism") has made it ever more difficult for the Commission to insist on its green and just economic growth TSD standards with foreign governments. This has, for example, been a key issue in the EU-Mercosur negotiations (Reuters, 2023).

As a consequence, the more rigorous the EU's TSD standards become, the higher the risk the EU may not be able to conclude new and ambitious trade agreements. Only one trade agreement has so far been signed in accordance with the green and just economic growth strategy, namely the EU-New Zealand FTA (2024). That agreement is considered the new gold standard for EU trade agreements as it contains many "firsts", such as sanctionable commitments to the Paris Agreement and core ILO labour standards, a dedicated provision on trade and fossil fuel subsidies reform, zero tariffs on green goods from the get-go, and many novel provisions relating to the circular economy, deforestation, carbon pricing and the protection of the marine environment (EU-New-Zealand FTA, 2024, Chapter 19). While these provisions are indeed an achievement, going forward, the EU will need to apply caution during its trade negotiations with other countries that have lower TSD protection. Indeed, the EU may increasingly find itself having to weigh up whether forfeiting on certain aspects of the TSD chapter is worthwhile in the grander scheme of signing a trade agreement. In this regard, the trade negotiations with Australia were abandoned as the EU's "stance ended up being too rigid and protectionist" for the Australian government (Carnegie Europe, 2023).

Perhaps in part due to the complexity, rigidity and formality surrounding the negotiations of a trade agreement, which can take several years, up to a decade, or longer to negotiate and involve many components (such as market access, intellectual property rights, subsidies, energy, digital and financial services, and dispute settlement), the EU has recently turned to new avenues of engagement with third countries as part of its external trade policy. The EU has signed *ad hoc* mini partnerships, cooperation agreements and clubs that all seek to use, in varying ways, a combination of trade and investment measures to further the EU's green objectives as well as its net-zero industrial transition. They thus all centre around sustainable value chains, green energy and net-zero products and raw materials.

Starting in 2023, the EU has signed on an *ad hoc* basis, mini partnerships or cooperation agreements with third countries on topics of common strategic importance. For example, instead of re-opening formal trade negotiations, the EU signed a Green Partnership with South Korea to deepen cooperation on climate action, clean energy and environmental protection (European Commission, 2023a). Furthermore, shortly after the fall-out of formal trade negotiations with Australia, the EU and Australia signed a partnership agreement on sustainable critical and strategic minerals (European Commission, 2024b). Similar agreements have also been signed with Norway (European Commission, 2024c), Namibia (European Commission, 2023b), Kenya (European Commission, 2023c), Kazakhstan (European

Commission, 2023d) and Kyrgyz Republic (European Commission, 2024d). All of these agreements serve to ensure that the EU has the resources and partnerships to realise the green and digital transitions, whilst also promoting high environmental, social and governance standards and practices.

The EU has also been instrumental in establishing "clubs" with like-minded third countries on the topic of climate and critical raw materials. During the German G7 Presidency in December 2022, the G7 announced the creation of a "Climate Club", which aimed at helping the international community take joint steps towards achieving a climate-friendly transformation (European Parliament, 2023b) and to overcome limitations in the consensus-based multilateral climate policy arrangements. The Climate Club thus aims to make the decision-making process faster and to target non-compliance more easily. The club now has 36 members and operates on an informal basis with the OECD acting as interim secretariat. It is worth noting that many countries quickly signed up to the club under the impression that it could potentially exempt them from the EU's CBAM (European Parliament, 2023b; Euractiv, 2021). Yet, this has not materialised. The stated aims of the club are to promote green steel, industrial decarbonisation and international partnerships (European Parliament, 2023b; Climate Club, 2023).

The Commission has also launched the Coalition of Trade Ministers on Climate, which is the first Ministerial-level global forum dedicated to trade, climate and sustainable development issues (Coalition of Trade Ministers on Climate, 2023). It aims at fostering global action to promote trade policies that can help address climate change through both local and global initiatives. In 2024, the Ministers agreed on a first set of voluntary actions as part of its agenda (Coalition of Trade Ministers on Climate, 2023).

The Critical Raw Material Club was also established in the context of the EU's Green Deal Industrial Plan. It aims to bring together raw material consumers (i.e., resource-hungry countries like the EU) and resource-rich countries to ensure global security of supply through a competitive industrial base and of Clean Tech/Net-Zero Industrial Partnerships (European Commission, 2023e). While the Club is still in its infancy and has been folded into the US-led Mineral Securities Partnership Forum (European Commission, 2024e), the underlying objective of the EU is to ensure that resource-rich countries commit to providing the EU with raw materials needed for its green and digital transition (potentially on a priority-basis to the detriment of other countries) in exchange for EU funding in the resource-rich country's raw material value chains.

Finally, the EU has established Trade and Technology Councils (TTCs) with the United States (Office of the US Trade Representative, 2024) and India (European Commission, 2023f) to foster bilateral cooperation and coordinate approaches to key global trade, economic and technology issues. TTC meetings are convened regularly and have led to the adoption of standard development and alignment, funding projects and novel approaches to tackling environmental issues through trade (European Commission, 2024f). For example, the EU and the United States are negotiating a Global Arrangement on Sustainable Steel (and Aluminium) as part of the sustainable and resilient supply chain pillar of its TTC. The idea is to ensure that once the EU's and the US' respective safeguard measures on imports of steel products lapse, global overcapacity in steel and aluminium does not end up on the EU and US markets. Some of the ideas that are being considered include conditioning market access of steel and aluminium to Global Arrangement members, whereby membership criteria would be based on the member's average emissions intensity of steel/aluminium products, their share of near- or low-zero emission procurement, and a commitment to not contribute to non-market excess capacity (Gijs and Aarup, 2023). However, at the time of writing, EU-US agreement has remained elusive.

While these new models of engagement with third countries create many benefits, such as avoiding the lengthy negotiation and ratification processes of a fully fledged trade agreement with a sustainability chapter and targeting environmental issues with innovative solutions, they also bring many challenges. Indeed, the success of *ad hoc* mini partnerships and cooperation agreements to further the green transition depend on their timely implementation, and in this regard, the non-binding nature of the agreements mean that the EU is restricted from compelling its trade partners. Furthermore, while the Climate and Critical Raw Material Clubs may offer faster decision-making processes, without proper management and as these clubs grow bigger, the environmental measures agreed in the framework of these clubs might encounter similar implementation issues as those long experienced under multilateral, consensus-based treaties. Also, quite importantly, some of the environmentally driven measures being discussed in these clubs and in the TTCs could raise issues under international trade rules, particularly when they envisage providing trade advantages and incentives to countries within the club and/or negatively discriminating against non-member countries.

As explained in the next section, tensions between environmental and trade measures are also relevant in the context of the EU's recently developed industrial policy. Indeed, there are instances where the nexus between sustainability and trade objectives can become competitive rather than complementary.

## Environment and the EU's industrial policy: contradictory aims

Over the last couple of years, the EU has started pursuing an increasingly ambitious set of industrial policies. These have important spill-over effects and implications for the EU's environmental policy. The EU had previously refrained from developing an industrial policy given that such policies were perceived as bad economics, following the ascent of neoliberalism towards the end of the 20th century. The changing economic consensus and revival of such policies in the United States by the Biden administration as a response to China's dominance in net-zero product supply chains, led the EU to change its approach and adopt its own industrial policy so as to nurture green industries. Given the EU's limited competences, trade powers – both old and new – have provided important tools for pursuing such policies. However, as the EU uses its trade powers to protect its industries in sectors linked to the net-zero transition, this results in higher prices for such net-zero products and thus a more costly and potentially slower green transition.

Since the announcement of its Green Deal Industrial Plan (European Commission, 2023g), the EU, and especially the Commission, have embarked on a project to maintain and develop the EU's industrial base, in particular with regard to net-zero technologies. This is also visible in the European Economic Security Strategy, published by the Commission in June 2023 (European Commission, 2023h). However, due to the EU's lack of competence in fiscal and spending matters, the EU has not been able to pursue an industrial policy in the same way as most national governments have, namely by pouring money into projects wherever needed.

Instead, the EU has had to operate within the scope of the competences granted to it under the EU treaties. In order to support EU green industries, the EU has thus developed specific new legislation, in particular its Critical Raw Materials Act (European Parliament and Council, 2024a) and Net-Zero Industry Act (European Parliament and Council, 2024b). While these legislative acts aim to support and develop the EU's green industries mainly by developing new financing instruments and reducing red tape for investment, there are tensions between the EU's net-zero targets and the trade-related instruments envisaged in those legislative acts.

This is apparent from the provisions on public procurement and auctions for the deployment of renewable energy sources in the Net-Zero Industry Act. In the context of public procurement and auctions, contracting authorities must take into account the "resilience" of offers for certain net-zero technologies like solar panels, battery storage technologies, heat pumps, etc. "Resilience" means that when a specific net-zero technology (or its main component) from one third country supplies more than a certain threshold (50% for auctions and 40% for public procurements) of the EU's market, specific mechanisms kick in. For public procurement under EU law, the mechanism requires contracting authorities to condition that no more than 50% of the value of the specific net-zero technology (and its main component) originates from that third country. For auctions to deploy renewable energy, the mechanism requires authorities to apply a lower weighting of points to offers incorporating goods that originate from that specific third country. In other words, the EU wants to make it more difficult for net-zero technologies to be used in public projects when the net-zero technologies come from a country on whom the EU has an overreliance (otherwise called a "strategic dependency").

In addition, the Net Zero Industry Act requires Member States to take "resilience" considerations into account when developing consumer incentive support schemes for net-zero products, such as solar panel or electric vehicle rebates. As with other initiatives, these restrictions on access to public procurement procedures, renewable energy auctions and support schemes could lead to higher prices for net-zero technologies and renewable energy in the EU and thus a slower roll-out of the green transition. As further explained below, this would particularly be the case for solar panels, which are now mostly supplied from China.

The EU's new industrial legislative acts are also complemented by the EU's long-standing trade defence instruments. The EU has imposed an increasing number of trade remedies on imports of net-zero products or the inputs needed to manufacture such products. Traditionally, trade remedies measures have been used to protect industries mainly in commodities, such as steel and chemicals (World Trade Organisation, 2024b). Trade remedies consist of anti-dumping and anti-subsidy (or "countervailing") duties, as well as safeguard measures. They are expressly allowed under the rules of the WTO and specific regulations enable the Commission to use them under EU law when certain conditions are met. To impose these measures under EU law, the Commission must demonstrate that the imported products are being dumped (i.e., sold on the EU market at a cheaper price than on their domestic market) or subsidised, that they cause material injury to the Union industry of that product, and that the Union interest calls for intervention. Investigations leading to the imposition of trade measures usually start based on a request from the Union industry, but they can also be opened by the Commission on its own initiative.

In recent years, there has been a rise in trade defence measures on environmental products and their inputs. The Commission has progressively put a greater emphasis on the need to secure production of these products and materials in the EU, as opposed to reaching net-zero goals. One of the earliest investigations targeting green products was the imposition of both anti-dumping and anti-subsidy measures on imports of solar cells and modules from China in 2013 (Council of the European Union, 2013, 66). In this investigation, the Commission took into account the needs of the user industry and the volatility of demand for the product and ultimately limited the duration of the measures to two years instead of the usual five. The Commission also accepted price undertakings from exporting producers, which meant that the Chinese exporters could avoid paying duties if they sold above a minimum import price.

However, even though the duties were extended in 2017 for additional 18 months (again instead of the usual 5 years) (European Commission,

2017, 1, 131) they were insufficient to protect the EU solar industry from collapse. Today, China dominates essentially the entire solar PV value chain (US Department of Energy, 2022, 8). In that regard, the solar panels case has become somewhat of a cautionary tale both among EU officials and commentators (Boullenois et al., 2023).

The solar panels approach has therefore not been repeated in subsequent cases. The Commission has instead imposed trade remedies following complaints from the relevant Union industry on the entire wind turbine value chain – from the steel plates and the wind towers to the fibres needed to manufacture the blades. It did so without giving much importance to the impact these measures would have on the energy transition, despite complaints from downstream green industries and users. More recently, the Commission opened on its own motion (which is rare) an anti-subsidy investigation on imports of electric vehicles from China and on 30 October 2024 imposed definitive countervailing duties on imports into the EU (European Commission, 2023i). More of such cases on products necessary to reach net-zero are to be expected given that the EU has stressed that trade remedies will be used as one of the main tools to nurture domestic industries in those sectors (European Commission, 2023g). This focus on protecting industries needed for the green transition is thus likely to continue and even accelerate in the coming years. At the same time, the Commission has refrained from initiating another case on imports of solar cells and modules despite calls to do so (European Parliament, 2024), recognising the need for such green imports.

In addition to trade remedies, the Commission has also taken action to protect its net-zero industries under the new Foreign Subsidies Regulation (2022), in particular by hindering the participation of foreign manufacturers of net-zero products in public procurements in the EU. The Foreign Subsidies Regulation, which entered into force in 2023, requires companies that have received financial assistance from non-EU governments to notify the Commission of such assistance when taking part in a merger or acquisition as well as a public procurement procedure in the EU. When the Commission considers – as a result of these notifications – that the tenderer has received foreign subsidies which distort the EU internal market, it can exclude the tenderer from the bidding process.

The Commission has already investigated two bids made by Chinese solar panel producers for a project in Romania, which led them to withdraw their tenders from fear of reputational harm before the Commission could take any decision (European Commission, 2024g). It has also been investigating *ex officio* (that is in the absence of any notification) contracts already awarded to Chinese wind turbines for a wind farm in Greece (European Commission, 2024h).

This focus by the Commission on net-zero technologies under the public procurement instrument of the Foreign Subsidies Regulation is having ramifications outside of these specific instances. Foreign manufacturers, in particular from China, might be backing out of participating in public procurements due to the administrative red tape, potential remedies and related reputational harm. This could result in more expensive goods being procured.

These trends are clearly a symptom of the Commission's greater wariness of economic dependencies (European Commission, 2023h), influenced by recent occasions when trading partners have exploited EU reliance on both imports and exports (European Commission, 2023j; OECD, 2024). This has led the EU to put more emphasis on resilient green technologies supply chains, in particular by nurturing its own industries. As this section has shown, this turn towards greater industrial policy efforts results in a potential trade-off between encouraging indigenous Union industries, and the attainment of the EU's green transition objectives at a faster pace, and EU policy might continue in that way.

## Conclusion

This chapter has looked at the various ways in which EU trade and environmental policy interact with each other. The recent wave of EU trade-related sustainability laws is a vivid example of how trade instruments can complement environmental objectives and vice-versa. This has been an intentional decision of the EU. In the face of a weakened multilateral system, the EU has taken it upon itself to ensure that environmental goals are fully integrated into its trade agenda and that trade tools can be utilised to further the green transition. Working at a unilateral level in this regard is not without its trade-offs, however. Indeed, maintaining access to the EU's large market may encourage foreign manufacturers and governments to "green" their supply chains and regulatory environment, but overly burdensome compliance obligations might also lead companies to focus on non-EU markets, where environmental standards may be lower. This could lead to negative impacts on the EU's competitiveness in those markets, and the EU paying a higher price for the net-zero transition.

When examining the EU's bilateral and regional trade policy, it is clear that this has been used to complement the EU's environmental agenda. The EU's ever-growing FTAs network boasts ambitious sustainability chapters, which aim to ensure that trade between the EU and third countries protects the EU's core values, of which compliance with multilateral environmental and labour agreements, ensuring sustainable value chains and promoting corporate responsibility form part. Such ambitious sustainability provisions in FTAs and partnerships with third countries nonetheless create important challenges, such as greater difficulties in reaching agreement with developing country trading partners. As a result, new models of engagement with third countries, like the mini cooperation partnerships and clubs, have also come to complement the EU's trade partnership to deploy trade and investment tools to promote green goods and net-zero industries alongside sustainable value chains.

The EU is also trying to harmonise its new industrial policy with its green and net-zero agenda, by incorporating new and old instruments to support its net-zero industries. However, in doing so, the nexus between sustainability and industrial objectives can become competitive rather than complementary. The EU's industrial policy acts (such as the Net-Zero Industry Act and the Critical Raw Materials Act), recent trade remedies practice and the implementation of the Foreign Subsidies Regulation, illustrate the tension between boosting the EU's own net-zero industrial base on the one hand to avoid dependencies, and achieving a more rapid green transition, on the other hand.

Moving forward, EU trade and environmental policy is likely to continue to remain heavily integrated. Indeed, at the time of writing, the re-elected President of the European Commission Ursula von der Leyen has released the political guidelines for the new Commission, which touch upon many of the overlaps in the EU's trade and environment agenda discussed in this chapter (von der Leyen, 2024). For instance, it includes the idea of "Clean Trade and Investment Partnerships" (von der Leyen, 2024, 8). At the same time, the guidelines contain many references to the need to boost EU "competitiveness". This is being echoed by the EU industry, such as in their proposal for a new Industrial Deal to complement the Green Deal by empowering industry to make the necessary investments for decarbonisation, while boosting competitiveness (see e.g., Antwerp Declaration, 2024). Thus, while the EU will doubtless continue to seek to leverage the synergies between environment and trade policies to speed up the green transition, this also suggests a recognition of the multiple trade-offs related to competitiveness and efficiency discussed in this chapter. The EU will thus likely focus more on green industrial policies as discussed in the final section of this chapter and perhaps less on regulations and sustainability provisions in FTAs as discussed in the first two sections.

## Acknowledgement

* The views expressed are those of the authors and do not necessarily reflect those of their institutional affiliation. This chapter is a collaborative effort but Marcus Gustafsson was mainly responsible for the drafting of the first Section, Elyse Kneller the second Section, and Victor Crochet the last one.

## References

Antwerp Declaration. (2024). The Antwerp Declaration for a European Industrial Deal. Available at: https://antwerp-declaration.eu/ (accessed December 2024).

AP. (2024). France's Macron Tells Brazilian Execs that Prospective Mercosur-EU Deal Is 'Terrible' and Outdated. *AP News*. Available at: https://apnews.com/article/macron-brazil-mercosur-european-union-7b20e27d3c65b6509080d1dc66b52cc8 (accessed December 2024).

Beattie, A. (2024). The Colonialist Overtones of EU's Green Trade Crusade. *Financial Times*. Available at: https://www.ft.com/content/d6766577-7549-4d36-ab3d-c8cc6e09a407 (accessed December 2024).

Blenkinsop, P. (2023). At WTO, Growing Disregard for Trade Rules Shows World Is Fragmenting. *Reuters*. Available at: https://www.reuters.com/business/wto-growing-disregard-trade-rules-shows-world-is-fragmenting-2023-10-02/ (accessed December 2024).

Boullenois, C., Kratz, A., & Goujon, R. (2023). Opening Salvo: The EU's Electric Vehicle Probe and What Comes Next. *Rhodium Group*. Available at: https://rhg.com/research/opening-salvo-the-eus-electric-vehicle-probe-and-what-comes-next/ (accessed December 2024).

Bradford, A. (2020). *The Brussels Effect: How the European Union Rules the World* (Oxford University Press: New York, 2020)

Carnegie Europe. (2023). Europe's Trade Ambitions Hit Reality. Available at: https://carnegieendowment.org/europe/strategic-europe/2023/12/europes-trade-ambitions-hit-reality?lang=en (accessed December 2024).

Climate Club. (2023). 36 Members Launch Climate Club as a High-Ambition Forum to Turbocharge Decarbonisation of Key Industry Sectors. Available at: https://climate-club.org/wp-content/uploads/2023/12/Climate-Club-COP-28-Full-Launch-Press-Release-01.12.2023-1.pdf (accessed December 2024).

Coalition of Trade Minsiters on Climate. (2023) About us. Available at: https://www.tradeministersonclimate.org/ (accessed December 2024).

Council of the European Union. (2013). Council Implementing Regulation (EU) No 1239/2013 Imposing a Definitive Countervailing Duty on Imports of Crystalline Silicon Photovoltaic Modules and Key Components (i.e. Cells) Originating in or Consigned from the People's Republic of China O.J.L 325/66. Available at: https://eur-lex.europa.eu/eli/reg_impl/2013/1239/oj/eng

Council of the European Union. (2018). Negotiating Directives for a Free Trade Agreement with Australia. *7663/18 ADD 1 DCL 1*. Available at: https://www.consilium.europa.eu/media/35794/st07663-ad01dc01-en18.pdf (accessed December 2024).

Denamiel, T., Schleich, M., & Reinsch, W.A. (2024). Insight into the 13th WTO Ministerial Conference. *CSIS*. Available at: https://www.csis.org/analysis/insight-13th-wto-ministerial-conference (accessed December 2024).

Devuyst, Y. (2014). European Union Law and Practice in the Negotiation and Conclusion of International Trade Agreements. *Journal of International Business and Law, 12*(2), Article 13.

Euractiv. (2021). Germany's Scholz Proposes 'Climate Club' to Avoid Trade Friction. 27 May. Available at: https://www.euractiv.com/section/climate-environment/news/germanys-scholz-proposes-climate-club-to-avoid-trade-friction/ (accessed December 2024).

European Commission. (2017). Commission Implementing Regulation (EU) 2017/367 Imposing a Definitive Anti-Dumping Duty on Imports of Crystalline Silicon Photovoltaic Modules and Key Components (i.e. cells) Originating in or Consigned from the People's Republic of China. *Official Journal of the EU*, L 56, 131.

European Commission. (2019). Global Europe: The Value of Free and Fair Trade. Available at: https://www.consilium.europa.eu/en/eu-free-trade/#group-Principles-and-values-wkN9IQ72ES (accessed December 2024).

European Commission. (2020a). Commission Implementing Regulation (EU) 2020/492 Imposing Definitive Anti-Dumping Duties on Imports of Certain Woven and/or Stitched Glass Fibre Fabrics Originating in the People's Republic of China and Egypt. *Official Journal of the EU*, L 108, 1.

European Commission. (2020b). Commission Implementing Regulation (EU) 2020/776 Imposing Definitive Countervailing Duties on Imports of Certain Woven and/or Stitched Glass Fibre Fabrics Originating in the People's Republic of China and Egypt. *Official Journal of the EU*, L 189, 1.

European Commission. (2020c). Communication from the Commission to the European Parliament, the European Council, the Council, the European Economic and Social Committe and the Committee of the Regions. *Europe's Moment: Repair and Prepare for the Next Generation* (COM/2020/456 final). Available at: https://eur-lex.europa.eu/legal-content/EN/TXT/?qid=1590732521013&uri=COM:2020:456:FIN (accessed December 2024).

European Commission. (2021a). Commission Implementing Regulation (EU) 2021/2239 Imposing a Definitive Anti-Dumping Duty on Imports of Certain Utility Scale Steel Wind Towers Originating in the People's Republic of China. *Official Journal of the EU*, L 450, 59.

European Commission. (2021b). Communication from the Commission to the European Parliament, the Council, the European Economic and Social Committee and the Committee of the Regions. *Trade Policy Review - An Open, Sustainable and Assertive Trade Policy* (COM/2021/66 final). Available at: https://eur-lex.europa.eu/legal-content/EN/ALL/?uri=COM%3A2021%3A66%3AFIN (accessed December 2024).

European Commission. (2022). Commission Unveils New Approach to Trade Agreements to Promote Green and Just Growth. Available at: https://ec.europa.eu/commission/presscorner/detail/en/ip_22_3921 (accessed December 2024).

European Commission. (2023a). European Green Deal: EU and Republic of Korea Launch Green Partnership to Deepen Cooperation on Climate Action, Clean Energy and Environmental Protection. Available at: https://ec.europa.eu/commission/presscorner/detail/en/ip_23_2816 (accessed December 2024).

European Commission. (2023b). Global Gateway: EU and Namibia Agree on Next Steps of Strategic Partnership on Sustainable Raw Materials and Green Hydrogen. Available at: https://ec.europa.eu/commission/presscorner/detail/en/ip_23_5263 (accessed December 2024).

European Commission. (2023c). EU and Kenya Sign Ambitious Economic Partnership Agreement with Strong Sustainability Provisions. Available at: https://ec.europa.eu/commission/presscorner/detail/en/ip_23_6632 (accessed December 2024).

European Commission. (2023d). EU-Kazakhstan Strategic Partnership Becomes Operational. 19 May. Available at: https://ec.europa.eu/commission/presscorner/detail/en/ip_23_2815 (accessed December 2024).

European Commission. (2023e). The Green Deal Industrial Plan: Putting Europe's Net-Zero Industry in the Lead. Available at: https://ireland.representation.ec.europa.eu/

news-and-events/news/green-deal-industrial-plan-putting-europes-net-zero-industry-lead-2023-02-01_en (accessed December 2024).
European Commission. (2023f). EU-India: New Trade and Technology Council to Lead on Digital Transformation, Green Technologies and Trade. Available at: https://ec.europa.eu/commission/presscorner/detail/en/IP_23_596 (accessed December 2024).
European Commission. (2023g). Communication from the Commission to the European Parliament, the European Council, the Council, the European Economic and Social Committee and the Committee of the Regions. *A Green Deal Industrial Plan for the Net-Zero Age* (COM(2023) 62 final). Available at: https://eur-lex.europa.eu/legal-content/EN/TXT/?uri=CELEX%3A52023DC0062 (accessed December 2024).
European Commission. (2023h). Joint Communication to the European Parliament, The European Council and the Council On "European Economic Security Strategy". JOIN/2023/20 final. Available at: https://eur-lex.europa.eu/legal-content/EN/TXT/?uri=CELEX%3A52023JC0020&qid=1687525961309 (accessed December 2024).
European Commission. (2023i). Commission Implementing Regulation (EU) 2024/2754 of 29 October 2024 Imposing a Definitive Countervailing Duty on Imports of New Battery Electric Vehicles Designed for the Transport of Persons Originating in the People's Republic of China. *Official Journal of the EU*. L, 2024/2754, 29.10.2024/ Available at: https://eur-lex.europa.eu/eli/reg_impl/2024/2754/oj (accessed December 2024).
European Commission. (2023j). Speech by President von der Leyen on EU-China relations to the Mercator Institute for China Studies and the European Policy Centre. Available at: https://ec.europa.eu/commission/presscorner/detail/en/speech_23_206 (accessed December 2024).
European Commission. (2024a). EU-New Zealand Trade Agreement Enters into Force, Opening New Opportunities for EU Exporters. Available at: https://ec.europa.eu/commission/presscorner/detail/en/ip_24_2388 (accessed December 2024).
European Commission. (2024b). EU and Australia Sign Partnership on Sustainable Critical and Strategic Minerals. Available at: https://ec.europa.eu/commission/presscorner/detail/en/ip_24_2904 (accessed December 2024).
European Commission. (2024c). EU and Norway Sign Strategic Partnership on Sustainable Land-Based Raw Materials and Battery Value Chains. Available at: https://ec.europa.eu/commission/presscorner/detail/en/ip_24_1654 (accessed December 2024).
European Commission. (2024d). The European Union and the Kyrgyz Republic Sign an Enhanced Partnership and Cooperation Agreement (EPCA). Available at: https://ec.europa.eu/commission/presscorner/detail/en/ip_24_3462 (accessed December 2024).
European Commission. (2024e). EU and International Partners Agree to Expand Cooperation on Critical Raw Materials. Available at: https://ec.europa.eu/commission/presscorner/detail/en/IP_24_1807 (accessed December 2024).
European Commission. (2024f). EU-US Trade and Technology Council (2021–2024). Available at: https://digital-strategy.ec.europa.eu/en/factpages/eu-us-trade-and-technology-council-2021-2024 (accessed December 2024).

European Commission. (2024g). Statement by Commissioner Breton on Withdrawal of LONGi Solar Technologie GmbH and Shanghai Electric from Public Procurement Following the Commission's Opening of an Investigation under the Foreign Subsidies Regulation. Available at: https://ec.europa.eu/commission/presscorner/detail/en/STATEMENT_24_2570 (accessed December 2024).

European Commission. (2024h). Speech by Executive Vice President Vestager on Technology and Politics at the Institute for Advanced Study. Available at: https://ec.europa.eu/commission/presscorner/detail/en/speech_24_1927 (accessed December 2024).

European Parliament. (2023a). EU-Australia Trade Relations. Available at: https://www.europarl.europa.eu/thinktank/en/document/EPRS_BRI(2023)751418 (accessed December 2024).

European Parliament. (2023b). G7 Climate Club. Available at: https://www.europarl.europa.eu/RegData/etudes/ATAG/2023/739385/EPRS_ATA(2023)739385_EN.pdf (accessed December 2024).

European Parliament. (2024). Verbatim Report of Proceedings. 5 February 2024. Available at: https://www.europarl.europa.eu/doceo/document/CRE-9-2024-02-05-ITM-015_EN.html (accessed December 2024).

European Parliament and Council. (2022). Regulation (EU) 2022/2560 of the European Parliament and of the Council on Foreign Subsidies Distorting the Internal Market (Foreign Subsidies Regulation). O.J.L 330, 23. Available at: https://eur-lex.europa.eu/eli/reg/2022/2560/oj/eng (accessed December 2024).

European Parliament and Council. (2024a). Regulation (EU) 2024/1252 of the European Parliament and of the Council of 11 April 2024 Establishing a Framework for Ensuring a Secure and Sustainable Supply of Critical Raw Materials and Amending Regulations (EU) No 168/2013, (EU) 2018/858, (EU) 2018/1724 and (EU) 2019/1020. O.J.L 2024/1252. Available at: https://eur-lex.europa.eu/legal-content/EN/TXT/?uri=CELEX:32024R1252 (accessed December 2024).

European Parliament and Council. (2024b). Regulation (EU) 2024/1735 of the European Parliament and of the Council on Establishing a Framework of Measures for Strengthening Europe's Net-zero Technology Manufacturing Ecosystem and Amending Regulation (EU) 2018/1724 (Net Zero Industry Act) O.J.L 2024/1735. Available at: Regulation - EU - 2024/1735 - EN - EUR-Lex (accessed December 2024).

EU-Korea FTA. (2011). Free Trade Agreement between the European Union and Its Member States, of the One Part, and the Republic of Korea, of the Other Part. *Official Journal of the DoneEU*, L 127./6.

EU-New Zealand FTA. (2024). Free Trade Agreement between the European Union and New Zealand. *Official Journal of the EU*, 2024/866.

Gehrke, T. (2022). EU Open Strategic Autonomy and the Trappings of Geoeconomics. *European Foreign Affairs Review*, 27(Special Issue), 61–78.

Gijs, C., & Aarup, A.S. (2023). It's the EU and US against the Rest of the World in New Steel Club. *Politico*.Available at: https://www.politico.eu/article/steel-aluminum-tariffs-european-union-united-states-china-imports-trade/ (accessed December 2024).Gustafsson, M. & Crochet, V. (2021). At the crossroads of trade and environment *in EU Environmental Governance Current and Future Challenges* (A. Orsini & E. Kavvatha eds.) (Routledge: Abingdon, Oxon), 187-206.

Office of the United States Trade Representative. (2024). US-EU Trade and Technology Council (TTC). Available at: https://ustr.gov/useuttc (accessed December 2024).

Organisation for Economic Co-operation and Development (OECD). (2024). Trade Impacts of Economic Coercion. *OECD Trade Policy Papers*. Available at: https://www.oecd.org/en/publications/trade-impacts-of-economic-coercion_d4ab39b9-en.html (accessed December 2024).

Reuters. (2023). Brazil Says EU Deforestation Rules Hamper Mercosur Trade Deal Negotiation. *Reuters*. Available at: https://www.reuters.com/world/americas/brazil-says-eu-deforestation-rules-hamper-mercosur-trade-deal-negotiation-2023-11-08/ (accessed December 2024).

Schropp, S. (2024). International Trade Policy under Biden: The 'New' Washington Consensus and Its Discontents. *Mercatus Center, George Mason University*. Available at: https://www.mercatus.org/research/policy-briefs/international-trade-policy-under-biden-new-washington-consensus-and-its (accessed December 2024).

The Economist. (2024). Chinese Firms Are Growing Rapidly in the Global South. *The Economist*. Available at: https://www.economist.com/briefing/2024/08/01/chinese-firms-are-growing-rapidly-in-the-global-south (accessed December 2024).

United States Department of Energy. (2022). Solar Photovoltaics: Supply Chain Deep Dive Assessment. Available at: https://www.energy.gov/sites/default/files/2022-02/Solar%20Energy%20Supply%20Chain%20Report%20-%20Final.pdf (accessed December 2024).

von der Leyen, U. 2024. Europe's Choice: Political Guidelines for the next European Commission 2024–2029. Available at: https://commission.europa.eu/document/download/e6cd4328-673c-4e7a-8683-f63ffb2cf648_en?filename=Political%20Guidelines%202024-2029_EN.pdf (accessed December 2024).

World Trade Organisation. (2024a). DS26: European Communities — Measures Concerning Meat and Meat Products (Hormones). Available at: https://www.wto.org/english/tratop_e/dispu_e/cases_e/ds26_e.htm (accessed December 2024).

World Trade Organisation. (2024b). Trade Remedies Data Portal. Available at: https://trade-remedies.wto.org/en (accessed December 2024).

# 10

# CLEANING UP THE MESS

## Latest developments in the EU – an answer to environmental conflicts

*Angèle Minguet and Fabienne Haubold*

### Introduction

In the first edition of this edited volume, the chapter entitled "Cleaning up the mess" defined the notion of "environmental conflict" and presented the existing data on the subject, as well as the European instruments designed to address such conflicts both within and outside European territory—when these conflicts affect the European Union (EU) in one way or another (Minguet, 2021).

Since the first edition, the European Commission, the main body responsible for initiating European legislation, has proposed a series of legislative texts aimed at avoiding this type of conflict. In the first part of this chapter, we examine one example of an environmental conflict and its various aspects, provide an updated definition of the term, and offer updated figures on the phenomenon. The second part lists relevant legislative texts that have been discussed, adopted, or rejected by the EU since 2020. The third part offers reflections on recent case law examples related to environmental conflict.

### Environmental conflicts

#### *Definition and case studies*

Environmental conflicts are disasters that lead to organized, structured, and long-lasting public protests (Minguet, 2021). They may relate, among others, to the pollution of a site, to the restriction of certain natural resources and to air pollution. The term encompasses a wide range of activities that share

DOI: 10.4324/9781003479260-16

the common characteristic of causing actual, future, potential, or past environmental degradation. Most of the time, these are localized degradations. However, exceeding global planetary limits (Richardson et al., 2023) could also be included in this category (Box 10.1).

## BOX 10.1 TWO EMBLEMATIC ENVIRONMENTAL CONFLICTS

### Shell's Pollution on the Territory of the Ogoni People

Nigeria: Shell, half British and half Dutch, has been exploiting oil in Nigeria since 1956. Between 1976 and 1996, 4,647 spill incidents occurred, with a total of 2,369,470 barrels of oil lost in nature, only 23% of which were recovered. The situation worsened between 1997 and 2001, with an average of 419 oil spills occurring each year. Between 1993 and 1995, the government violently repressed Ogoni movements that were peacefully calling for environmental justice. Shell was accused of complicity. The United Nations (UN), the US Supreme Court, the African Commission on Human and People's Rights, Friends of the Earth International, Amnesty International, academics, and Nigerian poets have all contributed to the extensive debate on the environmental destruction and human rights abuses linked to Shell's activities. The company has been taken to court in Africa, the US, the UK, and The Netherlands. Some court decisions are still pending, while others have impacted judicial and political systems, not only in Africa but also in Western countries. This environmental conflict raises critical global issues that urgently need addressing, showing how the legacy of the Westphalian international order hampers the resolution of transnational conflicts.

### Chevron-Texaco's Contamination in the Amazonian Region of Ecuador, El Oriente

Between 1967 and 1990, the US company Texaco (now merged with Chevron) is believed to have spilled over 60 billion liters of toxic waste and approximately 650,000 barrels of oil in an area inhabited by Indigenous Peoples. This led to a dozen trials in Ecuador, the US, Canada, and Brazil. The plaintiffs' lawyers have been accused of corruption and fraud. While far from entirely successful, this case remains a source of hope for environmental justice defenders worldwide. Additionally, it contributed to deep political changes in Ecuadorian society and its political system.

Sources: Authors' own synthesis.

The use of the term is partly justified by the desire to map such types of degradation, collect data, and analyze them to identify structural solutions to address the root causes. Detailing environmental conflict cases involves specifying the type of conflict, the natural resources involved, the perpetrators, the type of reaction, the forms of resistance, and the actors involved. In the majority of cases, activists emerge to denounce the situation. They share a common objective to assign responsibility for the damages they witness, which they perceive as injustices.

For example, it was a sense of injustice that prompted a group of 18 media outlets from different European countries to launch in 2021 a large-scale investigative project called the "Forever Pollution Project" (see their official website: Forever Pollution, 2023). This project, initiated by the French newspaper *Le Monde*, aimed at visualizing the contamination by per- and polyfluoroalkyl substances, more commonly known as "PFAS," in Europe. The journalists feared such contamination would be widespread. In such a case (see Box 10.2), the term "environmental conflict" refers to several conflictual relations: (i) the journalists' long investigation that led to their denunciation; (ii) the lawsuit filed by Antwerp's civil society against the multinational 3M, producing PFAS; (iii) the ad hoc parliamentary commission created in Flanders for such a case; (iv) the appeal by 3M to the Belgian Council of State; (v) the disputes between political leaders regarding political responsibility, and possibility more, illustrating the polysemous nature of the term "environmental conflicts."

### BOX 10.2 THE PFAS CASE IN EUROPE

Nicknamed "forever chemicals," these toxic compounds are barely, if not at all, biodegradable. As the World Health Organization (2022) indicates, these human-made chemicals are used in a wide variety of applications due to their waterproof, grease-resistant, and stain-resistant properties. They are found in textiles, household products, firefighting, the automotive industry, food processing, construction, and electronics. Further epidemiological studies are needed to get a complete view of their effects on human and animal health. However, there are suspicions of harmful effects on the environment, human and animal health, including reduced immune responses, negative effects on ovaries (Ding et al., 2020), brain functions (Cao and Ng, 2021), and possibly cancer (Steenland and Winquits, 2021; Ding et al., 2020).

According to data processed in an investigation project, the "Forever Pollution Project" (The Guardian, 23 February 2023), nearly 23,000 sites across Europe are contaminated to a degree requiring public authority intervention, while contamination has reached "dangerous levels for health (over 100 nanograms per liter) at more than 2,300 locations." The project concludes that few places in Europe remain untouched by this contamination, with little public awareness of it.

The project also revealed that Belgium has the highest PFAS pollution levels in Europe, particularly around the 3M PFAS production site in Zwijndrecht, Flanders, active since the mid-1970s. The Guardian reported that

> people living within a 15 km radius of the site have been advised not to eat eggs laid in their gardens and to avoid home-grown vegetables. Additionally, 70,000 people living within a 5 km radius of the factory were offered blood tests to detect the presence of PFAS.

The case is contentious because:

- The company filed an appeal with the Belgian Council of State against the Flemish administration's decision to suspend the polluting production processes (VRT News, 8 July 2021). This appeal was rejected (RTBF, 5 November 2021).
- The Flemish Parliament set up an inquiry commission to determine political responsibility and the extent of 3M's liability. It concluded that the multinational was mainly responsible, not the relevant political bodies—though this last point was not unanimously accepted within the Flemish parliament (VRT, 28 March 2022).
- The citizens' collective Darkwater3M, representing more than 500 families living near the 3M factory in Zwijndrecht (Antwerp), filed a lawsuit against the US headquarters of the chemical company (Le Soir, 19 December 2023). The first hearings began in April 2024 (RTBF, 10 April 2024).

According to The Guardian,

> 3M claims it will restore the site and has 'signed an agreement with the Flemish region... with an investment of 571 million euros.' The company has also announced its intention to stop producing PFAS 'and work towards ending the use of PFAS in its entire product portfolio by the end of 2025.

Sources: Cao and Ng (2021); Ding et al. (2020); Le Soir (2023); RTBF (2024); Steenland and Winquist (2021); VRT News (2021).

### *A few numbers and major trends*

The members of the Environmental Justice Organizations, Liabilities, and Trade (EJOLT) project (Temper and Del Bene, 2016) offers an initial overview of environmental conflict cases worldwide. This project was initiated between 2010 and 2015 by a network of seven universities and activists from 15 non-governmental organizations (NGOs) from various parts of Europe, Africa, India, and Latin America, led by the Universitat Autònoma de Barcelona. As of September 2024, the EJOLT database has recorded 4,169 environmental disputes in 183 countries, all spontaneously reported by civil society.

According to such data, the most contested resources are, in descending order: land (1,574 cases, 37%), water (1,150 cases, 27%), electricity (861 cases, 20%), while the raw materials generating the most conflicts are: oil (451 cases, 10% of conflicts), gold (385 cases, 9%), and coal (420 cases, 10%). Additionally, there are 50 other contentious resources, ranging from diamonds to coffee. It is worth noting that these categories are not exclusive: a single conflict can involve multiple resources, with an average of nearly three resources per conflict. For example, an oil company may pose a problem to local inhabitants not only concerning oil but also regarding land and water that it monopolizes or pollutes for its activities (Table 10.1).

**TABLE 10.1** Percentages of Environmental Conflicts by Products or Natural Resources

| | % |
|---|---|
| Land | 37.8 |
| Water | 27.6 |
| Electricity | 20.7 |
| Crude oil | 10.8 |
| Coal | 10.1 |
| Gold | 9.2 |
| Industrial waste | 8.7 |
| Tourism services | 8.6 |
| Domestic municipal waste | 8.5 |
| Natural Gas | 7.9 |
| Chemical products | 7.3 |
| Biological resources | 6.3 |
| Copper | 6.2 |
| Ecosystem Services | 6.2 |
| Fish | 5.3 |
| Timber | 5.2 |
| Silver | 4.3 |
| Uranium | 4.2 |

*Source*: Ejatlas (2023).

The EJOLT data also identifies 120 companies, all of which are "repeat offenders" (involved in at least two conflicts). These companies are mostly private, although some are state-owned. Of the 120 companies, just under 40% come from the "Global North." Together, these 120 companies are involved in around 1,339 conflicts. On average, each of these companies is involved in 11 conflicts, but some far exceed this number. Among the multinationals involved in the largest number of environmental conflicts, Shell ranks first with 84 cases, followed by Chevron Corporation (61 cases), ExxonMobil Corporation (34 cases), Rio Tinto (33 cases), Monsanto Corporation (30 cases), BHP Billiton (27 cases), and Barrick Gold Corporation (15 cases). Together, these companies generate nearly 65% of the environmental conflicts listed by EJOLT. All these top-conflictive companies are from the "Global North," but the way the EJOLT database is structured does not allow for directly determining the proportion of conflicts generated by companies from the "Global North" in the "Global South." Furthermore, it cannot be concluded that these are Western companies polluting the Global South. For example, Shell is involved in conflicts both in the Global South (Nigeria, Peru, South Africa, Bolivia, Egypt, Thailand, Indonesia, etc.) and in the Global North (Australia, Canada, the USA, France, Sweden, The Netherlands, etc.).

Although the EJOLT database does not allow for conclusions about the role of multinationals, EJOLT academic members have published analyses on topics such as the specific situation of Indigenous Peoples and women in these conflicts. For example, the vulnerability of indigenous communities to extractivism and their marginalization due to their ethnic origin has been studied by EJOLT scientists, who also claim that their mobilization is part of a global environmental justice movement often allied with other actors—such as feminist movements, human rights movements, post-colonial movements, or civil rights movements (Martínez-Alier, 2023a; Temper and Martínez-Alier, 2013). Regarding women, the analyses of the EJOLT database conclude that women (many of whom come from indigenous or marginalized communities) are often at the forefront of fights against environmental degradation, advocating for the protection of land, water, forests, and indigenous rights. They are frequently the target of repression, violence, and even murder by state forces, corporations, and criminal groups seeking to exploit natural resources (Venes et al., 2023; Martínez-Alier, 2023b; Tran, 2021).

The EJOLT database also provides information about the social impacts of these conflicts. Generally, it appears that people or communities involved in these conflicts are forced to relocate (2,345 cases recorded in 2024, or 56%), confronted with corruption (1,709 cases, or 40%), and face violence (1,032 cases, or 25%). In some cases, the damage leads to the loss of livelihoods (2,922 cases, or 70%), increased violence in the area (1,032 cases, or 25%), increased alcoholism or prostitution (651 cases, or 15%), or the militarization of the area (1,186 cases, or 28%). Moreover, the NGO Global Witness reports nearly 2,000 land and environmental defenders killed between 2012 and 2022 for their efforts to protect the planet.

Finally, the database reveals substantial differences worldwide. 12% of EJOLT cases occurred in the EU. This figure is significant considering that the EU represents 3% of the world's land area and 7% of the world's population. According to EJOLT data, there are three times more environmental conflicts per capita in the EU than elsewhere; or at least they are three times more frequently reported.

These differences also concern the reasons behind the conflicts, as developed in Box 10.3.

**BOX 10.3 PERCENTAGE OF ENVIRONMENTAL CONFLICTS BY CATEGORY OF ACTIVITY OR CONFLICT**

| *Categories* | % |
|---|---|
| Fossil Fuels and Climate Justice/Energy | 19.6 |
| Mineral Ores and Building Materials Extraction | 18.9 |
| Biomass and Land Conflicts (Forests, Agriculture, Fisheries and Livestock Management) | 13.3 |
| Water Management | 12.4 |
| Infrastructure and Built Environment | 10.7 |
| Waste Management | 7.7 |
| Industrial and Utilities conflicts | 7.2 |
| Biodiversity conservation conflicts | 3.8 |
| Nuclear | 3.7 |
| Tourism Recreation | 2.2 |

*Source*: Ejatlas (2023).

**BOX 10.4 PLANNED RELOCATIONS AS AN ADAPTATION STRATEGY TO CLIMATE CHANGE**

Zoé Briard UCLouvain Saint-Louis – Bruxelles

In September 2024, Central Europe was hit by Storm Boris. This event made multiple casualties and affected several countries on the continent. Facing the increase of extreme weather-related events due to the consequences of climate change, one possibility is to relocate people to protect them from these events. For instance, following the extreme flood event of June 2016 at the Simbach torrent, in Germany, 11 households have been relocated. This relocation was implemented as a part of an adapted structural alleviation scheme and the homeowners of these households received financial compensation (Mayr et al. 2020).

Planned relocation refers to the planned process of moving a group of people from their place of origin to one or more new locations, with the intent for this move to be permanent. This process is distinct from the simultaneous and sudden decision of several people to move (Bower and Weerasinghe 2021; Cantor 2023). As for how they are referred to, planned relocations are described by various terms, such as strategic realignments, resettlements, or managed retreats. These labels typically consist of two words that combine an intention with a movement-related term (Ferris and Bower 2023, p. 2). In terms of timing, planned relocations in the context of climate change and natural disasters can either be preemptive or reactive. In practice, these processes fall somewhere along a continuum between these two extremes (Bower and Weerasinghe 2021, p. 9).

Processes of planned relocation take place in the context of development projects, armed conflicts, natural disasters, and the impacts of climate change (Cantor 2023). Since the Cancún Agreements, they have been officially recognized as a form of adaptation to climate change. The Cancún Agreements identified migration, displacement, and planned relocations as the three forms of climate mobility (2010). Since then, immobility, whether forced or voluntary, has commonly been considered the fourth form of climate mobility (van der Geest et al. 2023). Even if this might seem counterintuitive at first, adding immobility to forms of mobilities allows to address crucial questions to better understand mobilities in environmental contexts. For instance, immobility highlights the existence of 'trapped populations' meaning that some may not have the resources necessary to be mobile and therefore are trapped despite the conditions surrounding them (Zickgraf 2021). On the contrary, reasons like place attachment might encourage or lead to 'voluntary' immobility (Adams 2016). Despite growing attention to this issue, planned relocations remain the least studied form of climate mobility to date (Ferris and Bower 2023).

Planned relocations related to climate change are taking place everywhere in the world (Ferris and Bower 2023). While Europe is not the primary focus of migration and climate scholarship (Nash 2023), there are already instances of climate-related relocations in Europe. In Austria, 500 households have been relocated from the Danube floodplains over the past 50 years (Schindelegger et al. 2021). In France, the coastal area of Vias is facing severe erosion, prompting an experimental relocation scheme (Courtoy 2022; see also Simbach example above). These examples are likely to become more common in the years to come, as current projections on climate change suggest that planned relocations will increase (Ferris and Bower 2023).

In this context, the European institutions are increasingly paying attention to climate mobility (Hahn and Fessler 2023). However, the Commission principally develops policies and programmes that address climate mobilities through external action, as if such mobilities were not occurring within European territory (Hahn and Fessler 2023). As for planned relocations, there is no mention of them. An exception is the Climate-ADAPT website, a partnership between the European Commission and the European Environment Agency, which includes 'retreat from high-risk areas' as an adaptation option (Climate-ADAPT 2023).

Most of the time, these processes are internal, meaning they take place within the borders of a single state (Ferris and Bower 2023). This raises questions about the role of the EU in these processes, as states are free to develop their own relocation strategies. With few regulations adopted worldwide, obligations and regulations need to be clarified. Other EU regulations may apply. In the event of floods, Member States implement the Water Framework Directive (Directive 2000/60/EC) or the Floods Directive (Directive 2007/60/EC). These directives impose specific obligations on states, such as conducting Preliminary Flood Risk Assessments, identifying and mapping flood risks, and then providing measures to reduce these risks in a Flood Risk Management Plan. These assessments could be crucial for a relocation decision.

Additionally, planned relocations must comply with human rights law. Given the significant impact on those being relocated, these processes are generally considered as a last resort. Several human rights may be at risk during planned relocations, including the right to life, freedom of movement, and the right to property, to name a few. All these rights are enshrined in the EU Charter of Fundamental Rights.

These considerations highlight the importance of focusing on the obligations that must be respected in cases of planned relocations and underscore the need to better understand these processes to ensure that future relocations are in line with the EU project.

Cite this box as: Briard, Z. (2025). Planned relocations as an adaptation strategy to climate change. In A. Orsini & E. Kavvatha (Eds.), *EU environmental governance: Current and future challenges* (Box, pp. xx–xx). Routledge.

Sources: Adams (2016); Bower and Sanjula (2021); Cancùn Agreement (2010); Cantor (2023); Climate-ADAPT (2023); Courtoy (2022); Ferris and Bower (2023); Hahn and Fessler (2023); Mayr et al. (2020); Nash (2023); Schindelegger et al. (2021); van der Geest et al. (2023); Zickgraf (2021); Ferris and Weerasinghe (2021).

## Legislative responses from the EU

This section provides an overview of European legislative texts intended to address the issue of environmental conflicts or that prove useful for those who consider themselves victims of such conflicts. We distinguish two scenarios: environmental damage occurring within European borders and environmental damage occurring outside European borders but involving European companies.

### *Damage within the EU: the precautionary and polluter-pays principles*

The Treaty on the Functioning of the EU (2012) stipulates that the

> Union policy on the environment shall aim at a high level of protection, taking into account the diversity of situations in the various regions of the Union. It shall be based on the precautionary principle and the principle that preventive action should be taken, that environmental damage should as a priority be rectified at source and that the polluter should pay.
>
> *(Article 191)*

The precautionary principle is reflected in numerous EU directives, regulations, and decisions aimed at ensuring environmental protection in the Member States.

Moreover, the EU provides for environmental impact assessments of public and private projects (Directive 2011/92/EU), as well as the environmental impact assessment of plans and programs (Directive 2001/42/EC, also known as the SEA Directive). The EU also recognizes that prevention is better than cure. This is embodied in the precautionary principle (Article 191 of the TFEU) that "aims to ensure a higher level of environmental protection through preventative decision-making in the case of risk." To this end, large companies are required to disclose information regarding their internal social and environmental policies (Directives 2014/95/EU and 2013/34/EU). Finally, since the EU's transposition of the Aarhus Convention in 2006, European citizens are supposed to have access to environmental justice, as well as access to information and participation in decision-making processes when it comes to environmental issues (Regulation (EC) No. 1367/2006). The EU has also ensured that victims of (environmental) crimes benefit from a minimum level of rights, support, and protection (Directive 2012/29/EU).

### *Damage within the EU: transparency requirements for private actors*

Replacing the Non-Financial Reporting Directive, the Corporate Sustainability Reporting Directive (Directive 2022/2464/EU) of 14 December 2022 requires the Member States to introduce measures ensuring that sustainability information includes details on plans and strategies aimed at ensuring that a company's business model and strategy are compatible with the transition to a sustainable economy, the 1.5° target of the Paris Agreement, and the European objective of climate neutrality by 2050.

This directive allows investors and other stakeholders to access information necessary to assess companies' impacts on people and the environment.

While this directive does not directly address justice for those harmed by environmental damage, it is expected by civil society to act as a form of pressure on European companies to improve their consideration of environmental and societal issues. As this information becomes public, companies are subjected to the scrutiny of civil society, consumers, investors, and policymakers—their reputations being at stake.

### *Damage within the EU: protection for those seeking justice*

Directive 2024/1069/EU on the protection of individuals participating in public debate against manifestly unfounded or abusive legal proceedings ("strategic lawsuits against public participation") with a cross-border impact, initiated against natural or legal persons, was adopted especially for journalists and human rights' defenders, because of their participation in the public debate. In this directive, the environment and climate change are considered as matters of public interest. Under this directive, strategic lawsuits against public participation are identified as

> initiated by powerful entities, such as individuals, lobbying groups, businesses, and government bodies. They often involve a power imbalance between the parties, with the plaintiff in a financially or politically stronger position than the defendant. While not an essential component of such cases, a power imbalance, when it exists, significantly increases the harmful and chilling effects of legal proceedings altering public debate.
>
> *(Recital 10 of the Commission's directive proposal)*

This directive ensures that the complaint mechanisms provided by other legislation are effective.

### *Damage within the EU: dispute management*

Within the EU, the use of national courts is encouraged for resolving environmental conflicts. Article 19(1), second sentence of the Treaty on the EU imposes on Member States to "provide remedies sufficient to ensure effective legal protection in the fields covered by Union law." In environmental matters, this principle has been strengthened since the adoption in 1998 of the Aarhus Convention on access to information, public participation in decision-making, and access to justice in environmental matters. Environmental litigation is primarily conducted before national courts, even when EU law is involved (Brosset and Truilhé, 2018).

Several regulations also aim to resolve jurisdiction or law-related conflicts between Member States. Once a conflict extends beyond a Member State's borders, issues regarding competent jurisdiction and applicable law arise. Regulation (EU) No 1215/2012, also known as the Brussels I (recast) Regulation, determines the jurisdiction of the Member States' courts in cases of civil and commercial disputes with an international element, ensuring that any decision rendered in one Member State is recognized and enforced in the others without special procedures. This regulation stipulates, as a general rule (Article 4), that "persons domiciled in the territory of a Member State, whatever their nationality, shall be sued in the courts of that Member State." However, exceptions are possible and a person domiciled in a Member State can be sued in another Member State, particularly (Article 7.2) for tortious acts (delict or quasi-delict), before the court where the harmful event occurred or may occur.

Once the competent jurisdiction is identified, the applicable law must be determined. Regulation No 864/2007 of the European Parliament and of the Council of 11 July 2007 on the law applicable to non-contractual obligations (Rome II Regulation) provides (Article 7) that for environmental damage, the applicable law is the law of the country where the damage occurs or is likely to occur (lex loci delicti), unless the party seeking compensation bases their claim on the law of the country in which the event giving rise to the damage occurred. In this case, they must prove that the event causing the damage took place in the other country, which can be difficult. For environmental damage at the border between two countries, the origin can still be easily proven. This is less evident when it comes to proving that a decision was made in one or the other office of a multinational company whose documents are classified.

For the purposes of the Regulation, "damage" refers only to the direct consequences of the event causing it (not indirect consequences, such as financial losses due to the damage that may be located elsewhere). Currently, any legal entity under the jurisdiction of EU Member States must prevent or remedy environmental damage where applicable (Directive 2004/35/EC) or face charges (polluter-pays principle) or criminal prosecution under EU law (Directive (EU) 2024/1203 on the protection of the environment through criminal law, replacing Directives 2008/99/EC and 2009/123/EC).

### *Damages outside the territory of the EU: the initial legal context*

At the international level, large transnational companies and states often opt for amicable resolution of their conflicts, which allows for a certain level of discretion. This practice is legally enshrined on the international stage by Article 33 of the United Nations Charter. However, in recent years, an increasing number of environmental conflicts have been brought before courts. While we will focus here on cases brought before courts in the strict sense of the term, it is worth noting that environmental conflict litigation can be broadly interpreted to include not only permanent international courts (such as the International Court of Justice, international criminal tribunals, or international human rights courts) but also quasi-judicial international institutions such as the Compliance Committee of the Aarhus Convention or the "F4" Committee of the UN Compensation Commission (implementing

Iraq's reparation for environmental damage resulting from the invasion of Kuwait), as well as domestic courts ruling on matters of private international law or international criminal law (Sohnle, 2016).

In accordance with Regulation (EU) No. 1215/2012, EU Member States' courts have jurisdiction over entities established in their territory (Article 63). In principle, this is also true when the acts occur extraterritorially (Álvarez Rubio and Yannibas, 2017, p. 18), although not explicitly stated. When an environmental conflict involves a non-EU entity (most often a defendant not domiciled in an EU Member State), the application of the Brussels I (recast) Regulation is limited. Its Article 6 states: "If the defendant is not domiciled in a Member State, the jurisdiction of the courts in each Member State shall be determined by the law of that Member State." Therefore, the regulation applies, but it refers judges to their national rules of jurisdictional conflict (with a few very limited exceptions).

It is commonly accepted that

> unless the state concerned has concluded an agreement with the EU, the Brussels instruments do not provide any jurisdictional protection to persons domiciled in third countries. The jurisdiction of the courts over such persons is determined solely by the national law of the Member State in question.
>
> *(Hartley and Duval, 2018)*

In 2017, Daniel Augenstein and Nicola Jägers predicted that "in the foreseeable future, the rules of jurisdiction governing third-country defendants in cases of corporate human rights violations will be those of the EU Member States" (Augenstein and Jägers, 2017). This is the promise the Corporate Sustainability Due Diligence Directive (Directive 2024/1760 or CS3D) of June 13, 2023, tries to concretize.

### *Damages outside the territory of the EU: corporate due diligence*

The objective of the CS3D is to promote sustainable and responsible corporate behavior in corporations' activities and throughout their global value chains. The innovative aspect of the rationale of this directive was that it should enable people who consider themselves to be the "environmental" victims of a European company or of a supplier of European companies to seek redress before a court of justice of an EU Member State.

In its final version, adopted by the Council, the Parliament, and the Commission, the CS3D aims to establish three types of obligations for companies falling under its scope (Article 1):

- "Obligations regarding actual or potential negative impacts on human rights and the environment concerning their own activities, the activities of their subsidiaries, and operations carried out by their business partners in the companies' value chains."

- "Liability for failure to meet the obligations mentioned above."
- "The obligation for companies to adopt and implement a transition plan for climate change mitigation, aimed at ensuring, with all possible efforts, the compatibility of the business model and strategy with the transition to a sustainable economy and the limitation of global warming to 1.5°C, in accordance with the Paris Agreement."

Much debate centered on the definition of "negative environmental impacts" (Article 3, point b)) during the negotiations for the adoption of the directive. The adopted definition does not refer to other existing definitions in EU legislation, such as that of the Regulation on the establishment of a framework to facilitate sustainable investments (Regulation 2020/852/EU), but rather lists prohibitions and obligations in the annex, without referring to iconic international agreements such as the Paris Agreement. Referring to the Paris Agreement would have meant that the European corporations could have been responsible, to some extent, for the $CO_2$ emissions of their providers and along their value chain. This provision was not retained.

Article 14 details a complaint procedure for victims. Member States are expected to ensure that companies allow individuals (and certain entities) to file complaints "where these persons or entities have legitimate concerns regarding actual or potential negative impacts related to the companies' activities, the activities of their subsidiaries, or the activities of their business partners in the companies' value chains." Article 27 also requires Member States to

> determine the regime of sanctions, including pecuniary sanctions, applicable to violations of the national provisions adopted pursuant to this directive and take all measures necessary to ensure the implementation of these sanctions. These sanctions must be effective, proportionate, and dissuasive.

Moreover, Article 29 calls on Member States to ensure that a company can be held liable for damage caused to a natural or legal person under certain conditions and ensures that the cost of proceedings is not prohibitive for claimants seeking justice. It also provides for a right to full reparation. Finally, Article 30 is designed to protect whistleblowers.

Member States have two years from its entry into force (July 26, 2024) to transpose the CS3D, with implementation being progressive depending on the categories of companies concerned. It is therefore too early to fully assess the impact of this directive on the market. It is hoped that companies present in the EU will pressure their suppliers, and that civil society will take legal action when companies fail to meet the obligations set by the directive. In doing so, lawsuits based on this new legal framework are likely to highlight a series of previously unknown or upcoming conflicts.

### *Damages outside the territory of the EU: focus on deforestation and climate*

In line with efforts to minimize the negative effects of the European market outside its borders, the European Commission under Ursula von der Leyen (2019–2024) has also proposed two texts aimed at the external environmental and climate impact of the European market. Both have been adopted.

The first is Regulation (EU) 2023/1115 of the European Parliament and of the Council of 31 May 2023 on making available on the Union market and exporting from the Union certain commodities and products associated with deforestation and forest degradation. This regulation aims to (a) minimize the Union's contribution to deforestation and forest degradation worldwide and thereby help reduce global deforestation, and (b) reduce the Union's contribution to GHG emissions and biodiversity loss. It specifically targets the placing on the EU market and export from the EU of "commodities" suspected of causing deforestation such as cattle, cocoa, coffee, palm oil, rubber, soy, and timber.

The second is Regulation 2023/956 of the European Parliament and Council of May 10, 2023, establishing a carbon border adjustment mechanism (CBAM, see also Chapter 9). This legislation is one of the key proposals of the "Fit for 55" package aimed at achieving the EU's climate targets by 2030 on the path to climate neutrality. It seeks to avoid "the inherent GHG emissions" of certain goods (such as iron/steel, cement, fertilizers, aluminum, hydrogen, and electricity) "when imported into the customs territory of the Union to prevent carbon leakage, thereby reducing global carbon emissions and supporting the objectives of the Paris Agreement by introducing measures that encourage operators from third countries to reduce their emissions" (Article 1).

## Case law in the EU

Alongside these legislative advances in Europe, many lawsuits have taken place on this issue, serving as an incentive for legislative development. The most emblematic cases of the recent years are presented below by case categories.

### *Environmental case law: case law and national territory*

Among others, several lawsuits have begun before Belgian courts following confirmed cases of soil pollution in Flanders. Fluvia, one of the emergency zones in West Flanders, is the subject of several parallel legal proceedings before the Court of First Instance in Kortrijk, following decontamination requests from the Flemish Region. The pollution of several plots of land with firefighting foams containing PFAS, which were not prohibited at the time of use, is under scrutiny. The federal government has also been summoned to the case pending the appointment of one or more experts. Limited for now to a single emergency zone in Flanders, these cases are likely to multiply in the coming months and years, forcing different levels of government to collaborate in creating an efficient compensation mechanism to address PFAS pollution cases.

### *Environmental case law: national case law and extraterritoriality*

One of the most interesting judicial cases in recent years is the case "Oguru Efanga Vereniging Milieudefensie v. Shell Petroleum n.v. (zaak 1) and Royal Dutch Shell PLC (zaak 2)" in The Netherlands.

This case involves large-scale pollution, as presented in Box 10.2. Between 1993 and 1995, the government violently repressed the Ogoni movements, which were peacefully demanding environmental justice. Shell was accused of complicity. The UN, the US Supreme Court, the African Commission on Human and People's Rights, Friends of the Earth International, Amnesty International, scholars, and Nigerian poets all took part in the extensive debate sparked by the environmental destruction and human rights violations related to Shell's activities. This case resulted in trials in Africa, the US, the UK, and The Netherlands.

In The Netherlands, the court decision was particularly novel as the Hague Court held Shell Nigeria liable for the pollution caused by oil spills that occurred between 2004 and 2007. This ruling is significant for two reasons. Firstly, the court ruled the company guilty, although Shell argued that sabotage was the cause of the oil leaks. The court considered that Shell needed to prove it was sabotage, which it failed to do. Secondly, the company was found liable for negligence, not directly by itself, but by its subsidiary Shell Nigeria. Simultaneously, the British-Dutch parent company was ordered to install equipment to prevent future damage.

### *Climate case law: the decisions of the ECtHR (2024)*

Recently, several environmental conflicts have pitted citizens or civil society organizations against states, with the former expecting the latter to take all necessary steps to protect their citizens from climate change by addressing its causes. In this context, on April 9, 2024, the European Court of Human Rights (ECtHR) delivered a Grand Chamber ruling on three climate-related cases. In these three cases, the applicants, European citizens, complained about the negative effects of climate change on their living conditions, effects amplified by the states' inaction.

Two of the complaints were deemed inadmissible (Carême v. France and Duarte Agostinho et al. v. Portugal), primarily for procedural reasons (lack of victim status in one and failure to exhaust domestic remedies in the other), but the complaint by the Swiss senior women's association (Verein KlimaSeniorinnen et al. v. Switzerland) was admitted. The Court found that on the one hand, Article 8 of the Convention enshrines a right to effective protection by state authorities against the serious adverse effects of climate change on life, health, well-being, and quality of life. Therefore, Switzerland failed to meet its obligations by not quantifying the national limits applicable to GHG emissions and not meeting its past emission reduction targets. On the other hand, in violation of the right of access to a court (Article 6§1 ECtHR), Swiss courts did not examine the applicants' complaints regarding the effective implementation of mitigation measures provided for by Swiss law in force.

This ruling marks a significant jurisprudential development, as the Court departs from its traditional environmental case law and strengthens legal avenues for achieving climate justice through the ECtHR. By doing so, it sends a message to European policymakers that they must intensify their efforts to combat climate change. However, the Court's climate jurisprudence is still in its early stages. As the Court itself pointed out, failure to meet one of the criteria established in this judgment will not automatically result in a violation of the Convention in the future, as such violations will always require a comprehensive assessment of national climate measures.

### *Climate case law: the Urgenda case in The Netherlands (2018)*

In The Netherlands, the Supreme Court ruled on December 20, 2019, in a similar case initiated in 2015 by the Urgenda Foundation against the Dutch state. The foundation accused the Dutch government of setting insufficient GHG reduction targets that endangered citizens' fundamental rights. After winning in both the first instance and on appeal, the foundation saw its victory confirmed by the Dutch Supreme Court, which ruled based on Articles 2 and 8 of the ECtHR that the state is responsible for protecting its citizens from the causes and consequences of climate change, thereby recognizing a new human right.

While it is regrettable that the ruling does not provide sanctions for non-compliance, this unprecedented judgment also enshrines the militant use of the law, creating a new path for civil society to seek redress, making it an inspiring precedent for ongoing or future climate litigation across Europe. The consequences of this case have spread far beyond The Netherlands. Since then, other EU Member States have been sued on similar legal grounds, including France (Dalloz, 29 January 2020) and Belgium.

### *Climate case law: the Belgian Climate Case (2015-ongoing)*

Belgium is also experiencing a "climate case" at the national level. A citizen action was launched in 2015 by the NGO Klimaatzaak and aimed at forcing Belgian authorities to uphold their international climate promises. It led to a first ruling by the Brussels Court of First Instance on June 17, 2021. The court ruled that Belgium's climate policies were inadequate and condemned the federal state and regions. This ruling was upheld by the Brussels Court of Appeal on November 30, 2023, against the Belgian state, the Flemish Region, and the Brussels Region (Dengis, 2024). The court ruled that inaction constituted a violation of Articles 2 (right to life) and 8 (right to respect for private and family life) of the ECtHR and a fault under Articles 1382 and 1383 of the old Belgian Civil Code (extra-contractual liability). Only the Walloon Region escaped condemnation, largely due to its commitment to reduce GHG emissions, set by decree. This landmark ruling also imposes a binding and quantified reduction of GHG emissions, namely the European goal of a 55% reduction by 2030 compared to 1990. This is a first in Belgium, as the judges of first instance had not gone previously so far, fearing to violate the separation of powers' principle.

The court based its reasoning on the fact that it is acknowledged that judicial authorities have the power to prevent and repair illegal infringements by public authorities on individual rights by ordering the executive branch to take measures to stop infringement (without violating the separation of powers' principle). However, it recognizes that "the measures thus ordered cannot deprive the public authority of the choice of measures to implement to achieve the ordered result" (p. 142), which is not the case with the imposed injunction. The Flemish Region was the only one to file an appeal with the Court of Cassation against the ruling of the Court of Appeal, but the Brussels Region eventually joined the appeal. Their main argument is that the 55% reduction target by 2030 is unattainable and unaffordable, and that there is no support for an ambitious climate policy in Belgium. The case is still under way.

## Conclusion

Environmental conflicts, whether involving local communities or transnational corporations, highlight the ongoing struggle between economic development and environmental sustainability. Through landmark case studies, such as Shell's impact on Nigeria, we see the long-lasting harm that corporate negligence can cause. The EU, through legislative measures like the precautionary principle, the CS3D, the CBAM, and the regulation on deforestation, is progressively taking steps to address these conflicts and protect both human rights and the environment.

Future trends now depend on the extent to which these brand-new laws will be used by citizens and civil society organizations, and to which extent the culture within the private sector will keep pace with the move toward practices that are more respectful of the environment and people. Nevertheless, it remains clear that substantial challenges lie ahead, especially as global companies continue to evade responsibility through complex legal frameworks and transnational operations.

The rise in litigation and increasing pressure from civil society indicates that the global movement for environmental justice is gaining momentum. Cases like the PFAS pollution in Belgium and climate litigation across Europe serve as key examples of this shift. To mitigate future conflicts, greater accountability, transparency, and international cooperation are necessary. Only by addressing the root causes of environmental degradation and ensuring strong legal frameworks can we hope to resolve these environmental conflicts.

## Bibliography


Adams, H. (2016). Why Populations Persist: Mobility, Place Attachment and Climate Change. *Population and Environment*, 37(4), 429–448. doi: 10.1007/s11111-015-0246-3.

Augenstein, D., & Jägers, N. (2017). Judicial Remedies, The Issue of Jurisdiction. In Álvarez Rubio, J., & Yiannibas, K. (eds.), *Human Rights in Business. Removal of Barriers to Access to Justice in the EU* (pp. 7–38). London: Routledge.

Brosset, E., & Truilhé, E. (2018). L'accès au juge dans le domaine de l'environnement: le hiatus du droit de l'Union européenne. In *Revue des droits et libertés fondamentaux,* Chron. N°07. https://revuedlf.com/droit-ue/lacces-au-juge-dans-le-domaine-de-lenvironnement-le-hiatus-du-droit-de-lunion-europeenne/

Bower, E., & Weerasinghe, S. (2021). *Leaving Place, Restoring Home, Enhancing the Evidence Base on Planned Relocation Cases in the Context of Hazards, Disasters, and Climate Change*, Geneva, Platform on Disaster Displacement.

Cancun Agreement. (2010). Decision 1/CP.16, FCCC/CP/2010/7/add.1.

Cantor, D. J. (2023). Conceptualizing "Relocation" Across Displacement Contexts. *Journal of International Humanitarian Legal Studies*, *15*, 1–29. doi: 10.1163/18781527-bja10083.

Climate-ADAPT. (2023). Adaptation Option: Retreat from High-Risk Areas. https://climate-adapt.eea.europa.eu/en/metadata/adaptation-options/retreat-from-highrisk-areas.

Courtoy, M. (2022). "To Leave Is to Die": States' Use of Mobility in Anticipation of Land Uninhabitability. *German Law Journal*, *23*(7), 992–1011. doi: 10.1017/glj.2022.69.

Cao, Y., & Ng, C. (2021). Absorption, Distribution, and Toxicity of Per- and Polyfluoroalkyl Substances (PFAS) in the Brain: A Review. *Environmental Science. Processes & Impacts*, *23*(11), 1623–1640. doi:10.1039/d1em00228g

Carême v. France (Application nº 7189/21) ECtHR 9 April 2024 (Grand Chamber).

Coolsaet, B. (ed.). (2020). *Environmental Justice*. London: Routledge. doi:10.4324/9780429029585

Council of Europe. (1950). European Convention on Human Rights, Signed in Rome, 4 November 1950. Entered into Force on 3 September 1953. *Council of Europe Treaty Series* No. 5. https://www.echr.coe.int/documents/d/echr/convention_ENG

Dalloz Actualité. (2020). Le quotidien du droit. 29 January 2020. « Suite et fin de l'affaire Urgenda: une victoire pour le climat ». https://www.dalloz-actualite.fr/flash/suite-et-fin-de-l-affaire-urgenda-une-victoire-pour-climat

Dengis D, (2024). L'affaire Climat: quand la justice rappelle nos politiques à l'ordre. https://www.canopea.be/laffaire-climat-quand-la-justice-rappelle-nos-politiques-a-lordre/

Ding, N., Harlow, S. D., Randolph, J. F., Jr, Loch-Caruso, R., & Park, S. K. (2020). Perfluoroalkyl and Polyfluoroalkyl Substances (PFAS) and their Effects on the Ovary. *Human Reproduction Update*, *26*(5), 724–752. doi:10.1093/humupd/dmaa018

Duarte Agostinho and Others v. Portugal and 32 Other States (Application no 39371/20) ECtHR (Grand Chamber) 27 September 2023.

European Union. (2000). Directive 2000/60/EC: Directive of the European Parliament and of the Council of 23 October 2000 Establishing a Framework for Community Action in the Field of Water Policy. *Official Journal of the European Union* L 327, 22 December 2000.

European Union. (2007). Directive 2007/60/EC: Directive of the European Parliament and of the Council of 23 October 2007 on the Assessment and Management of Flood Risks. *Official Journal of the European Union* L 288, 6 November 2007.

Ejatlas. (2023). PFAS (PFOS) Pollution by 3M, Belgium. https://ejatlas.org/conflict/pfas-pollution-by-3m-belgium

EJOLT. (2015). http://www.ejolt.org

European Union (EU). (2001). Directive 2001/42/CE: Directive of the European Parliament and of the Council of 27 June 2001 on the Assessment of the Effects of Certain Plans and Programmes on the Environment (Strategic Environmental Assessment Directive). *Official Journal of the European Union* L 197, 21 July 2001.

European Union. (2004). Directive 2004/35/CE: Directive of the European Parliament and of the Council of 21 April 2004 on Environmental Liability with Regard to the Prevention and Remedying of Environmental Damage. *Official Journal of the European Union* L 143, 30 April 2004.

European Union. (2006). Regulation No 1367/2006/EC: Regulation of the European Parliament and of the Council of 6 September 2006 on the Application of the Provisions of the Aarhus Convention on Access to Information, Public Participation in Decision-making and Access to Justice in Environmental Matters to EU Institutions and Bodies. *Official Journal of the European Union* L 264, 25 September 2006.

European Union. (2007). Regulation No 864/2007/EC: Regulation of the European Parliament and of the Council of 11 July 2007 on the Law Applicable to Non-Contractual Obligations (Rome II). *Official Journal of the European Union* L 199, 31 July 2007.

European Union. (2008). Directive 2008/99/CE: Directive of the European Parliament and of the Council of 19 November 2008 on the Protection of the Environment through Criminal Law. *Official Journal of the European Union* L 328, 6 December 2008.

European Union. (2009). Directive 2009/123/CE: Directive of the European Parliament and of the Council of 21 October 2009 Amending Directive 2005/35/EC on Ship-Source Pollution and on the Introduction of Penalties for Infringements. *Official Journal of the European Union* L 280, 27 October 2009.

European Union. (2011). Directive 2011/92/EU: Directive of the European Parliament and of the Council of 13 December 2011 on the Assessment of the Effects of Certain Public and Private Projects on the Environment (codified version). *Official Journal of the European Union* L 26, 28 January 2012.

European Union. (2012). Consolidated Version of the Treaty on the Functioning of the European Union. *Official Journal of the European Union* C 326, 26 October 2012, pp. 47–390.

European Union. (2012). Directive 2012/29/EU: Directive of the European Parliament and of the Council of 25 October 2012 Establishing Minimum Standards on the Rights, Support, and Protection of Victims of Crime. *Official Journal of the European Union* L 315, 14 November 2012.

European Union. (2012). Regulation No 1215/2012/EU: Regulation of the European Parliament and of the Council of 12 December 2012 on Jurisdiction and the Recognition and Enforcement of Judgments in Civil and Commercial Matters (Brussels I Recast). *Official Journal of the European Union* L 351, 20 December 2012.

European Union). (2013). Directive 2013/34/EU: Directive of the European Parliament and of the Council of 26 June 2013 on the Annual Financial Statements, Consolidated Financial Statements, and Related Reports of Certain Types of Undertakings. *Official Journal of the European Union* L 182, 29 June 2013.

European Union. (2014). Directive 2014/95/EU: Directive of the European Parliament and of the Council of 22 October 2014 Amending Directive 2013/34/EU as Regards Disclosure of Non-financial and Diversity Information by Certain Large Undertakings and Groups. *Official Journal of the European Union* L 330, 15 November 2014.

European Union. (2020). Regulation No 2020/852/EU: Regulation of the European Parliament and of the Council of 18 June 2020 on the Establishment of a Framework to Facilitate Sustainable Investment, and Amending Regulation (EU) 2019/2088 (EU Taxonomy Regulation). *Official Journal of the European Union* L 198, 22 June 2020.

European Union. (2022). Directive 2022/2464/EU: Directive of the European Parliament and of the Council of 14 December 2022 Amending Regulation (EU) No 537/2014, Directive 2004/109/EC, Directive 2006/43/EC, and Directive 2013/34/EU, as Regards Corporate Sustainability Reporting (CSRD). *Official Journal of the European Union* L 322, 16 December 2022.

European Union. (2023). Regulation No 2023/956/EU: Regulation of the European Parliament and of the Council of 10 May 2023 Establishing a Carbon Border Adjustment Mechanism (CBAM). *Official Journal of the European Union* L 150, 16 June 2023.

European Union. (2024). Directive 2024/1069/EU: Directive of the European Parliament and of the Council of 11 April 2024 on Protecting Persons Who Engage in Public Participation from Manifestly Unfounded Claims or Abusive Court Proceedings (Strategic Lawsuits against Public Participation). *Official Journal of the European Union* L series, 16 April 2024.

European Union. (2024). Directive 2024/1203/EU: Directive of the European Parliament and of the Council of 11 April 2024 on the Protection of the Environment through Criminal Law and Replacing Directives 2008/99/EC and 2009/123/EC. *Official Journal of the European Union* L series, 30 April 2024.

European Union. (2024). Directive 2024/1760/EU: Directive of the European Parliament and of the Council of 13 June 2024 on Corporate Sustainability Due Diligence. *Official Journal of the European Union* L series, 5 July 2024.

Ferris, E., & Bower, E. (2023). *Planned Relocations: What We Know, Don't Know, and Need to Learn, Researching Internal Displacement.* https://researching internaldis placement.org/short_pieces/planned-relocations-what-we-know-dont-know-and-need-to-learn/.

Ferris, E., & Weerasinghe, S. (2020). *Promoting Human Security: Planned Relocation as a Protection Tool in a Time of Climate Change. Journal on Migration and Human Security*, *8*(2), 134–149. Doi:HYPERLINK "https://doi.org/10.1177/233 1502420909305"10.1177/2331502420909305.

Forever Pollution. (2023). https://foreverpollution.eu/

Global Witness. (2023). https://www.globalwitness.org/fr/press-releases-fr/almost-2000-land-and-environmental-defenders-killed-between-2012-and-2022-protecting-planet-fr/

Hahn, H., & Fessler, M. (2023). *The EU's Approach to Climate Mobility: Which Way Forward*? Brussels, European Policy Center.

Hartley, T. C., & Duval, C. (2018). La Convention de Bruxelles mise en perspective internationale. *Revue Critique de Droit International Privé, N° 3*(3), 485–493. doi:10.3917/rcdip.183.0485

Klimaatzaak v. Belgian State (Cour d'appel de Bruxelles, 30 November 2023, 2e ch. F – RG n° 2021/AR/1589–2022/AR/737–2022/AR/891).

Le Soir. (2023). Pollution aux Pfas: le siège américain de 3M poursuivi en justice par un collectif citoyen. 19 December 2023. https://www.lesoir.be/556426/article/2023-12-19/pollution-aux-pfas-le-siege-americain-de-3m-poursuivi-en-justice-par-un

Mayr, B., Thaler, T., & Hubl, J. (2020). Successful Small-Scale Household Relocation after a Millennial Flood Event in Simbach, Germany 2016. *Water, 12*(1), 1–19. doi: 10.3390/w12010156.

Martinez-Alier, J. (2014). Between Activism and Science: Grassroots Concepts for Sustainability Coined by Environmental Justice Organizations. *Journal of Political Ecology*, *21*, 19–60.

Martínez-Alier, J. (2023a). Indigenous Revival and Resistance around the World. In *Land, Water, Air and Freedom*. Cheltenham, UK: Edward Elgar Publishing (pp. 560–582). doi:10.4337/9781035312771.00032

Martínez-Alier, J. (2023b). Women Environmental Defenders Killed around the World. In *Land, Water, Air and Freedom*. Cheltenham, UK: Edward Elgar Publishing (pp. 68–91). doi:10.4337/9781035312771.00011

Martínez-Alier, J. (2023c). *Land, Water, Air and Freedom*. Cheltenham, UK: Edward Elgar Publishing. doi:10.4337/9781035312771

Minguet, A. (2021). Cleaning up the Mess? The EU's Answer to Environmental Conflicts Within and Beyond Its Borders. In Orsini, A., & Kavvatha, E. (eds.), *EU Environmental Governance*. London, UK: Routledge.

Nash, S. L. (2023). The View from the Fortress: European Governance Perspective on Climate Change and Migration. In Nicholson, C., & Mayer, B. (eds.), *Climate Migration. Critical Perspectives for Law, Policy and Research*. Oxford: Hart, 130–150.

Perez, A. C., Grafton, B., Mohai, P., Hardin, R., Hintzen, K., & Orvis, S. (2015). Evolution of the Environmental Justice Movement: Activism, Formalization and Differentiation. *Environmental Research Letters*, *10*(10), 105002. doi:10.1088/1748-9326/10/10/105002

Petel, M., & Vander Putten, N. (2023). The Belgian Climate Case. doi:10.59704/f9b1ef7c677ed70c

Richardson, K., Steffen, W., Lucht, W., Bendtsen, J., Cornell, S. E., Donges, J. F., … Rockström, J. (2023). Earth beyond Six of Nine Planetary Boundaries. *Science Advances*, *9*(37), eadh2458. doi:10.1126/sciadv.adh2458

RTBF. (2021). Pollution à Zwijndrecht: le Conseil d'Etat rejette le recours de 3M contre l'arrêt de sa production. 5 November 2021. https://www.rtbf.be/article/pollution-a-zwijndrecht-le-conseil-d-etat-rejette-le-recours-de-3m-contre-l-arret-de-sa-production-10873779

RTBF. (2024). Pollution aux PFAS à Zwijndrecht: première audience contre la société américaine 3M. 10 April 2024. https://www.rtbf.be/article/pollution-aux-pfas-a-zwijndrecht-premiere-audience-contre-la-societe-americaine-3m-11356928

Rubio, A., & Yannibas, K. (2017), Human Rights in Business. Removal of Barriers to Access to Justice in the European Union. Routledge, London. https://doi.org/10.4324/9781315269467

Schindelegger, A., Seebauer, S., & Thaler, T. (2021). *Planned Relocation from Danube Floodplains in Austria, Geneva*, IDMC (Internal Displacement Monitoring Center).

Sohnle, J. (2016). Le règlement international des différends environnementaux: argumentaire raisonné, raisonnable et avec raisons sur un accès amplifié au juge. In J. Bétaille (éd.), Le droit d'accès à la justice en matière d'environnement (1-). Presses de l'Université Toulouse Capitole. doi:10.4000/books.putc.1040

Steenland, K., & Winquist, A. (2021). PFAS and Cancer, a Scoping Review of the Epidemiologic Evidence. *Environmental Research*, *194*, 110690. doi:10.1016/j.envres.2020.110690

Temper, L., Avila, S., Bene, D. D., Gobby, J., Kosoy, N., Billon, P. L., ... Walter, M. (2020). Movements Shaping Climate Futures: A Systematic Mapping of Protests against Fossil Fuel and Low-Carbon Energy Projects. *Environmental Research Letters*, *15*(12), 123004. doi:10.1088/1748-9326/abc197

Temper, L., & Del Bene, D. (2016). Transforming Knowledge Creation for Environmental and Epistemic Justice. *Current Opinion in Environmental Sustainability*, *20*, 41–49. doi:10.1016/j.cosust.2016.05.004

Temper, L., & Martinez-Alier, J. (2013). The God of the Mountain and Godavarman: Net Present Value, Indigenous Territorial Rights and Sacredness in a Bauxite Mining Conflict in India. *Ecological Economics: The Journal of the International Society for Ecological Economics*, *96*, 79–87. doi:10.1016/j.ecolecon.2013.09.011

The Guardian. (2023). Revealed: Scale of 'Forever Chemical' Pollution across UK and Europe. 23 February 2023. https://www.theguardian.com/environment/2023/feb/23/revealed-scale-of-forever-chemical-pollution-across-uk-and-europe

Tran, D. (2021). A Comparative Study of Women Environmental Defenders' Antiviolent Success Strategies. *Geoforum; Journal of Physical, Human, and Regional Geosciences*, *126*, 126–138. doi:10.1016/j.geoforum.2021.07.024

van der Geest, K., de Sherbinin, A., Gemenne, F., & Warner, K. (2023). Editorial: Climate Migration Research and Policy Connections: Progress since the Foresight Report. *Frontiers in Climate, 5*. doi: 10.3389/fclim.2023.1231679.

Venes, F., Barca, S., & Navas, G. (2023). Not victims, but Fighters: A Global Overview on Women's Leadership in Anti-Mining Struggles. *Journal of Political Ecology*, *30*(1). doi:10.2458/jpe.3054

Verein KlimaSeniorinnen and Others v. Switzerland (Application no 53600/20) ECtHR (Grand Chamber) 24 April 2024.

VRT News. (2021). 3M loosde te grote hoeveelheden PFOS in de Schelde en de overheid wist dat, maar reageerde jarenlang amper. 8 July 2021. https://www.vrt.be/vrtnws/nl/2021/07/05/3m-loosde-grote-hoeveelheden-pfos-in-de-schelde-en-de-overheid/

VRT News. (2022, 28 March). La commission d'enquête flamande pointe 3M comme le principal responsable de la pollution au PFOS. https://www.vrt.be/vrtnws/fr/2022/03/28/la-commission-d_enquete-flamande-pointe-3m-comme-le-principal-re/

WHO (World Health Organization). (2022). *Guidelines for Drinking-Water Quality: Fourth Edition Incorporating the First and Second Addenda*. Genève, Switzerland: WHO.

Zickgraf, C. (2021). Theorizing (im)mobility in the Face of Environmental Change. *Regional Environmental Change, 21*(4). DOI: 10.1007/s10113-021-01839-2.

# PART 6

# Upcoming challenges for the European Union

# 11

# ARCTIC GOVERNANCE – AN UPDATED FIELD OF EU ACTION

*Eleni Kavvatha*

## Introduction

Over the last decade, the Arctic region has become an increasingly critical area of focus in international governance due to its unique environmental, geopolitical, and social characteristics. The Arctic region does not have a single, universally agreed-upon definition, instead various exist and are all used extensively. Figures 11.1 and 11.2 illustrate the complexity of defining the Arctic region. While Figure 11.1 presents a political map of the Arctic, scientific definitions often emphasize climatic conditions or dominant vegetation types. In contrast, political bodies such as the Arctic Council (AC) apply different criteria to set southern boundaries for the purposes of their reporting as shown in Figure 11.2. Other terms commonly used to describe the Arctic include the northern polar region, the circumpolar north, or simply the north, and these are often used interchangeably.[1] The boundaries of the Arctic can be defined in various ways, including using the 10°C July isotherm, vegetation zones like tundra, or political boundaries such as those set by the Conservation of Arctic Flora and Fauna.

The term "European Arctic"[2] is also not universally defined, but it is commonly understood to encompass the northernmost regions of Europe, including parts of Norway, Sweden, Finland, Russia, and occasionally Greenland (Raspotnik & Stępień, 2020). Spanning across eight countries in the European continent, including three European Union (EU) Member States (Denmark, Finland, and Sweden), the European Arctic is home to

DOI: 10.4324/9781003479260-18

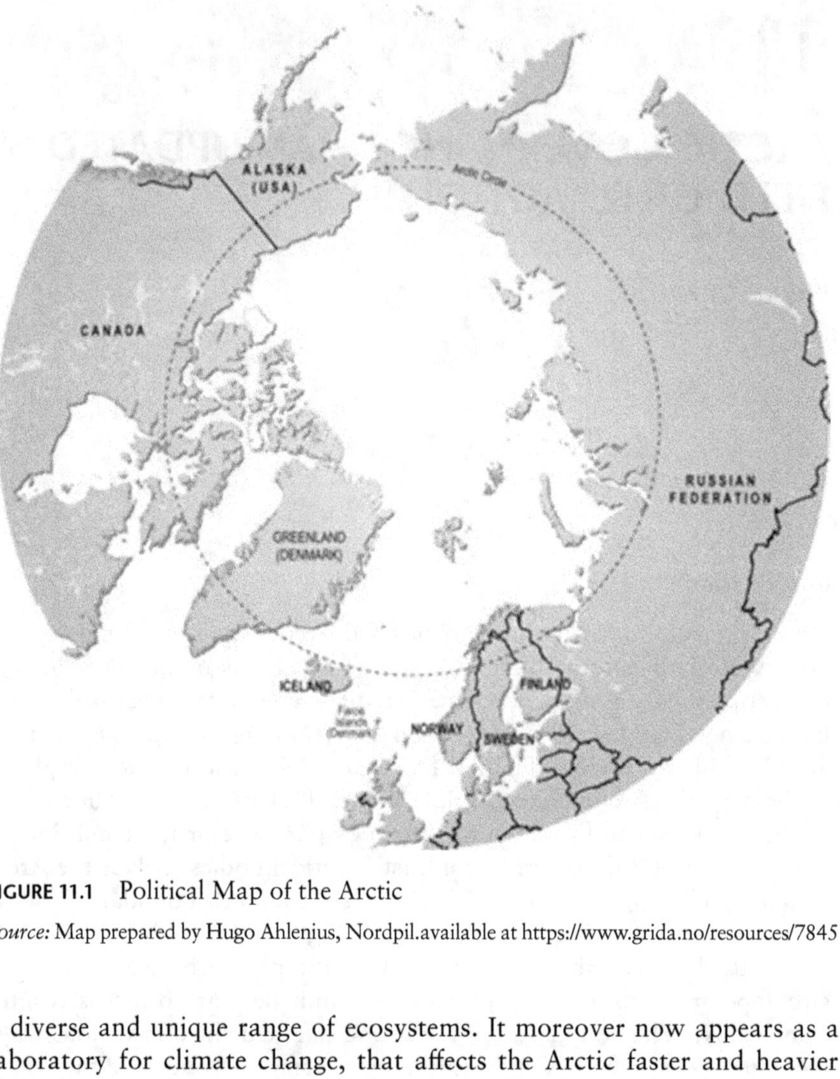


**FIGURE 11.1** Political Map of the Arctic

*Source:* Map prepared by Hugo Ahlenius, Nordpil.available at https://www.grida.no/resources/7845

a diverse and unique range of ecosystems. It moreover now appears as a laboratory for climate change, that affects the Arctic faster and heavier than the rest of the world (Landrum & Holland, 2020) with change in the Arctic affecting the whole planet (Crépin et al., 2017). The Arctic also holds immense geopolitical and economic significance, particularly in light of the vast mineral resources it contains, and of its strategic location as global warming opens new shipping routes (Alvarez et al., 2020). Finally, regarding social unique characteristics, a large part of the Arctic region is historically populated by indigenous communities that suffer from these evolutions, requiring new forms of governance suitable to address these challenges and strengthen the position of those living in the region (Østerud & Hønneland, 2014) (Box 11.1).

**FIGURE 11.2** Map of the Arctic by various definitions

*Source:* Nilsson, A. E., & Christensen, M. (2019). Arctic geopolitics, media and power. doi:10.4324/9780429199646 Map prepared by Hugo Ahlenius, Nordpil.

### BOX 11.1 INDIGENOUS POPULATION IN THE ARCTIC

Approximately 1 million people, or about 9% of the total Arctic population, are indigenous, representing more than 40 different ethnic groups. However, there is no universal definition of an indigenous person across the Arctic, meaning the population statistics are based on the varying national definitions of indigenous groups. As a result, official records may not consistently distinguish indigenous populations. The highest concentration of indigenous people is found in the Canadian Arctic and Greenland, where they make up over 75% of the population. On the other hand, regions such as the Yukon, southern Northern Quebec, and parts of Labrador and Newfoundland (Canada) have lower indigenous populations. In most other parts of the Arctic, Indigenous Peoples account for less than half of the total population, with exceptions in areas like Sakha (Russia) and the northern and southwestern regions of Alaska (USA), where indigenous groups represent 50%–75% of the population. Iceland, the Faroe Islands, and Svalbard do not have indigenous populations.

Source: Nordregio, https://nordregio.org/maps/indigenous-population-in-the-arctic/

The European Arctic in particular is of strategic importance for the EU, not only because of the Union's Global Strategy, which, including the Arctic as part of its broader foreign policy and security concerns, requires the area to remain a region of peace, security, and cooperation, but also because the EU itself is in the Arctic and not outside or nearby, with three Arctic states being EU members and two more being part of the European Economic Area (Boedeltje & van Houtum, 2011). As a result, the EU has become increasingly active in the field of Arctic Governance, developing comprehensive policies aimed at balancing environmental protection, sustainable development, and the safeguarding of Indigenous Rights (Hossain, 2015).

The EU's engagement in Arctic governance is motivated by a combination of environmental concerns and strategic interests. The Arctic is often described as the "barometer" of global climate change due to the disproportionate impact that global warming has on the region. The region is warming at more than twice the rate of the global average, leading to significant environmental changes that have both regional and global implications (IPCC, 2023). These changes include the rapid melting of sea ice, which contributes to rising sea levels and alters global weather patterns, and the thawing of permafrost, which releases greenhouse gases that exacerbate climate change. Given these dynamics, the EU has identified the Arctic as a key area in its broader environmental and climate policies, particularly under the framework of the European Green Deal (European Commission, 2021a,b). Moreover, the EU, through its project Arctic Climate Change, Economy and Society (ACCESS), carried out a transdisciplinary assessment of the physical impacts of climate change on the Arctic Ocean and the resulting socioeconomic impacts within the Arctic region. The project focused on key economic activities: shipping, tourism, sea food production, and natural resource extraction up to 2050 (NERC, 2015; Crépin et al., 2017a; Gascard et al., 2017).

Beyond environmental considerations, it is also the Arctic's geopolitical significance that has driven EU interest in the region. The Arctic is a region of increasing geopolitical competition, particularly raising the interests of major powers such as Russia, the United States, and China, who are all vying for influence and access to the region's resources. The EU, with its Member States and associated countries within the Arctic Circle, has a vested interest in ensuring that the region remains stable, peaceful, and cooperative (Boedeltje & van Houtum, 2011). The EU's Arctic policy reflects this strategic interest by advocating for international cooperation, multilateral governance, and the rule of law in the region, while also promoting the sustainable management of Arctic resources (Conde Pérez & Yaneva, 2016).

In addition to environmental and geopolitical concerns, the protection of Indigenous Peoples' rights and livelihoods is a dimension that is becoming more and more apparent in the EU's Arctic policy. Indigenous Peoples such as the Sámi and the Inuit, have long been stewards of the region's unique ecosystems and possess invaluable traditional knowledge regarding sustainable resource management. As climate change accelerates and economic activities, such as resource extraction and shipping, intensify, indigenous communities are facing increasing threats to their land, culture, and ways of life (Hanaček et al. 2022). The EU's Arctic policy seeks to strike a balance between environmental protection, strategic interests, and the protection of Indigenous Rights, showcasing a willingness to uphold the principles of free, prior, and informed consent in projects that may affect Indigenous territories.

But how did the EU get this far? Though the Arctic region is yet to become a number one area of priority for the EU, the Union's Arctic policy has indeed evolved significantly over the past two decades. Initially focused on environmental concerns, particularly in relation to pollution and climate change, the EU's approach has gradually expanded to encompass economic, social, and geopolitical dimensions, especially in the European Arctic (Conde Pérez & Yaneva, 2016). This evolution is evident in key policy documents such as the 2016 Joint Communication on "An Integrated European Union Policy for the Arctic," which emphasized climate change, sustainable development, and international cooperation as the three pillars of EU Arctic policy (Stang, 2016). In 2021, the EU updated its Arctic policy to reflect the rapidly changing Arctic environment and geopolitical landscape. The updated policy focuses on making the Arctic more sustainable, peaceful, and prosperous, and emphasizes the EU's role in promoting a sustainable Arctic through initiatives such as the European Green Deal, while also supporting the rights of Indigenous Peoples (European Commission, 2021a,b). The Critical Raw Materials Act, though not specifically mentioning the Arctic region, reflects a shift in the importance the Arctic holds for the EU for its energy efficiency, as an important share of raw materials are located in the European Arctic (European Commission, 2020).

This chapter aims to provide a comprehensive analysis of the EU's role in Arctic governance, examining the strategic importance of the region for the EU, the evolution of its Arctic policy, and the environmental, socioeconomic, and governance challenges it faces. This chapter also considers the impact of EU policies on Arctic communities, particularly Indigenous Peoples like the Sámi, and how these policies contribute to global environmental governance. By exploring these themes, this chapter seeks to highlight the complexities and opportunities associated with the EU's Arctic strategy, as well as its implications for the future of the region. It shows that the Arctic region's strategic importance for the EU has increasingly come into focus during the EU' policy formation, as the region's significance is multifaceted, encompassing vital aspects such as geopolitical stability, environmental sustainability, resource availability, and the well-being of indigenous communities (European Commission, 2021a,b).

## The Evolution of EU Arctic policy

The EU's Arctic policy has undergone significant evolution over the past two decades, driven by the increasing recognition of the Arctic's environmental, economic, and geopolitical significance. Initially, the EU's engagement was primarily environmental, but it has since expanded to encompass broader socioeconomic and geopolitical dimensions, reflecting the complex and rapidly changing nature of the Arctic region (Conde Pérez & Yaneva, 2016).

### *Phase 1: Early recognition and initial research (Pre-2000s)*

The environmental importance of the Arctic began to gain global attention in the latter half of the 20th century. By the 1980s and 1990s, researchers and environmentalists started to understand the Arctic as a sensitive barometer for the planet's broader environmental health. Early studies identified the Arctic's ecosystems as being particularly vulnerable to climate change, due to its high latitudes and extreme weather conditions. The region's role in regulating global climate, primarily through the reflection of sunlight via ice and snow, began to be emphasized. Initial scientific investigations during this period largely focused on understanding the dynamics of the Arctic's physical systems, including ice cover, atmospheric conditions, and ocean currents. These studies laid the groundwork for what would become an increasingly urgent global dialogue on climate change in the coming decades. Though the concept of Arctic warming was known, its implications were not fully understood.

The EU's initial involvement in Arctic affairs was predominantly through its environmental policies, which aimed to address the impacts of pollution and climate change in the region. The Habitats Directive (Council Directive 92/43/EEC) and the Birds Directive (Council Directive 79/409/EEC) were critical in this regard, as they provided a framework for the conservation of natural habitats and species across Europe, including those in the Arctic. These directives were instrumental in protecting Arctic biodiversity, particularly in the Nordic countries, which are home to numerous species that are highly sensitive to environmental changes (Stang, 2016). A significant example of early EU environmental action in the Arctic is also the implementation of the Natura 2000 network, which includes numerous sites in the Arctic region, particularly in Finland and Sweden. Natura 2000 is the largest coordinated network of protected areas in the world, established under the Habitats and Birds Directives. The inclusion of Arctic sites in this network highlights the EU's early recognition of the need to protect the fragile Arctic environment from the adverse effects of human activities and climate change (Koivurova et al., 2021).

Despite these early efforts, the EU's Arctic engagement was somewhat fragmented and lacked a coherent strategy. This began to change in the early 2000s, as the EU recognized the need for a more integrated approach to Arctic governance, driven by the increasing global attention on the region's environmental and geopolitical significance.

### *Phase 2: Recognition of arctic amplification and intensified research—EU early engagement (2000–2010)*

By the early 2000s, the phenomenon of Arctic amplification became widely recognized. This is the process by which the Arctic warms at a rate more than twice the global average, which has profound implications for both the region and the planet as a whole (Serreze & Barry, 2011). During this period, scientists observed the melting of sea ice, the thawing of permafrost, and major alterations in local ecosystems. These environmental changes have direct consequences for the global climate system, including rising sea levels and disrupted weather patterns across the globe (Serreze & Barry, 2014). As the effects of climate change became more evident, the need for coordinated international action to address Arctic environmental degradation grew more urgent.

In response, the EU began to shape more formalized policies aimed at reducing the region's vulnerability to climate change and supporting sustainable development in the Arctic. At this time, the EU's policy priorities began to align more explicitly with the growing body of scientific research showing that the Arctic was a key player in the global climate system (Conde Pérez & Yaneva, 2016). The Horizon 2020 program, launched in the early 2010s, marked a significant expansion of the EU's support to Arctic research. Through Horizon 2020, the EU funded numerous projects focused on understanding Arctic environmental changes and developing strategies to mitigate their impacts. Notable among these was the EU-PolarNet initiative, which sought to enhance coordination across European polar research efforts, consolidating efforts to study the Arctic and Antarctic regions, and generating critical data for evidence-based policymaking.

### *Phase 3: Strategic policy development and EU commitment to Arctic sustainability (2010–2020)*

By the 2010s, the EU had solidified its commitment to the Arctic environment through more formalized, high-level policies. The EU recognized the Arctic's critical role for global climate regulation and broader international environmental governance. As a result, a series of strategic positions and policies were developed that explicitly aimed to reduce the region's exposure to climate change, promote environmental sustainability, and support research.

The EU's observer status at the AC, which has held since 2013, is a critical aspect of its increasing commitment to Arctic governance. As an observer, the EU participates in the work of the Council's working groups and contributes to discussions on issues such as climate change, pollution, and sustainable development. However, the EU's quest for full membership has been complicated by geopolitical tensions and concerns over the EU's regulatory impact on Arctic affairs. For instance, the EU's ban on seal products, introduced in 2009, was met with strong opposition from Canada and Norway, who argued that it undermined the livelihoods of Indigenous Peoples in the Arctic (Boedeltje & van Houtum, 2011).

During that phase, a major milestone in the evolution of the EU's Arctic policy was the publication of the 2016 Joint Communication on the Arctic titled "An Integrated European Union Policy for the Arctic" (Stang, 2016). It marked the first comprehensive attempt by the EU to articulate its interests and policy objectives in the region. The communication identified three priority areas: climate change and environmental protection, sustainable development, and international cooperation. The document called for enhanced efforts to reduce black carbon emissions, a significant contributor to Arctic warming. In the area of sustainable development, the EU emphasized the need to balance economic growth with environmental sustainability. It advocated for the responsible exploitation of Arctic resources, including fisheries, minerals, and energy, while ensuring that these activities do not compromise the region's environmental integrity. It recognized the importance of supporting the economic development of Arctic communities, particularly

Indigenous Peoples, through initiatives that respect their rights and cultural heritage (Stang, 2016). Recognizing that the Arctic is a region of shared interest, the Joint Communication highlighted the importance of international cooperation in addressing Arctic challenges. The EU committed to working closely with Arctic states, Indigenous Peoples' organizations, and other stakeholders to promote a stable and secure Arctic. The document also reiterated the EU's desire to become a full member of the AC, although this has been met with resistance from some Arctic states, particularly Russia and Canada (Hossain, 2015).

In 2019, the European Green Deal was introduced as a comprehensive framework for addressing climate change across Europe, with significant implications for the Arctic (Nanni et al., 2024) (see also Chapters 2 and 5). While the Green Deal focused on the entire European continent, it included provisions for the protection of the Arctic, acknowledging the region's vulnerability and importance. This period also saw the formalization of the EU's Arctic policy, which emphasized environmental protection, the sustainable use of resources, and the promotion of Indigenous Rights (Lebel & Nilsson, 2024).

### *Phase 4: Updated Arctic policy (2021)*

In response to the rapidly changing Arctic environment and geopolitical landscape, the EU updated its Arctic policy with the 2021 Joint Communication titled "A Stronger EU Engagement for a Peaceful, Sustainable, and Prosperous Arctic" (European Commission, 2021a,b). This updated policy reflects the EU's recognition of the need to intensify its efforts to address the growing challenges and opportunities in the Arctic. A key feature of the 2021 policy is its emphasis on sustainability. The EU reaffirmed its commitment to making the Arctic more sustainable by calling for enhanced efforts to combat climate change in the Arctic, including reducing greenhouse gas emissions and supporting the transition to renewable energy sources in Arctic communities (Orsini & Kavvatha, 2020). The 2021 policy also places a stronger emphasis on supporting the rights and well-being of Indigenous Peoples in the Arctic. The EU committed to increasing its engagement with indigenous communities and organizations, particularly in areas related to cultural preservation, land rights, and sustainable development. The policy recognizes the vital role that Indigenous Peoples play in Arctic governance and aims to ensure that their voices are heard in decision-making processes at all levels. Geopolitical considerations are also more prominent in the 2021 policy, reflecting the increasing strategic importance of the Arctic in global affairs. The EU expressed its concern over the growing militarization of the Arctic and the potential for conflict in the region. The policy calls for the Arctic to remain a zone of peace and stability and advocates for increased dialogue and cooperation among Arctic and non-Arctic states to prevent escalation of tensions as they happened in the past (European Commission, 2021a,b).

Further international commitments were also made during this period. The EU's active support for the Paris Agreement reinforced its stance on climate action, aligning its internal policies with the global effort to limit global warming to below 2°C. The establishment of marine protected areas (MPAs) in the Arctic ocean during this period was another notable achievement, aiming to protect marine biodiversity and prevent unsustainable exploitation of Arctic resources.

The evolution of the EU's Arctic policy reflects a growing awareness of the region's strategic importance and the need for a more integrated and comprehensive approach to Arctic governance. From its early focus on environmental protection to its current emphasis on sustainability, Indigenous Rights, and geopolitical stability, the EU's Arctic policy has become a key component of its broader environmental and geopolitical objectives. As the Arctic continues to undergo rapid change, the EU's role in shaping the future of the region will be crucial in ensuring that it remains a zone of peace, stability, and sustainable development.

## The quadruple objective of the EU Arctic policy

The EU's policy objectives in the Arctic, though mostly focused on the European Arctic region, are guided by its broader commitments to global environmental, economic, geopolitical, and cultural considerations.

### *Environmental objectives: climate change mitigation and biodiversity conservation*

Arctic amplification has profound implications not only for the Arctic but also for global natural systems, at it creates rising sea levels and altered weather patterns. The EU has positioned itself as a leader in climate diplomacy, pushing for global reductions in greenhouse gas emissions through frameworks like the Paris Agreement, while also committing to its own ambitious goals for climate neutrality by 2050, as outlined in the European Green Deal (European Commission, 2021a,b). In addition, the EU emphasizes climate change mitigation efforts as part of its Arctic policy, recognizing the region's vulnerability and its pivotal role in the global climate system.

In line with its climate goals, the EU-PolarNet research project (see above) developed strategies for mitigating the impacts of climate change on Arctic ecosystems. This initiative is a part of the EU's broader efforts to promote adaptive strategies that protect Arctic communities and biodiversity from the severe effects of environmental change (European Commission, 2021a,b). A central aspect of the Arctic environmental agenda is the EU's additional focus on biodiversity conservation, which is crucial given the Arctic's unique and vulnerable species and ecosystems. The EU has made biodiversity conservation a cornerstone of its Arctic policy, recognizing that protecting these ecosystems is vital for maintaining global biodiversity, thus supporting the establishment of MPAs in Arctic waters as part of its Biodiversity Strategy for 2030. These efforts aim to protect at least 30% of European seas, including Arctic waters, from degradation caused by human activities such as overfishing and pollution, through legally binding conservation measures (European Commission, 2021a,b). For example, the EU has advocated for the protection of the Central Arctic Ocean, which is increasingly accessible due to melting ice. The EU's push for MPAs in this area aims to prevent overfishing and protect the fragile marine ecosystem from the impacts of increased human activity (European Commission, 2021a,b). Another example is the EU's involvement in the Circumpolar Biodiversity Monitoring Programme, which monitors the status and trends of Arctic biodiversity. This program provides critical data that informs conservation policies and helps to mitigate the impacts of climate change on Arctic species and habitats (Koivurova et al., 2021). Through continued support for research, international collaboration, and adaptive policy development, the EU is poised to play a central role in safeguarding the Arctic's environmental health, ensuring that the region's ecological integrity is preserved for future generations (Lebel & Nilsson, 2024).

### *Economic objectives: sustainable resource management, tourism, shipping, and fisheries*

The Arctic's vast natural resources present both opportunities and challenges for the EU. In addition to unvaluable natural ecosystems, the region is rich in resources such as oil, gas, minerals, and fisheries, which are increasingly accessible due to the melting of Arctic ice. However, the extraction and use of these resources raise significant environmental and social concerns.

In 2009, the Arctic was estimated to hold around 13% of the world's undiscovered oil and 30% of its undiscovered natural gas reserves (Gautier et al., 2009). These resources are increasingly accessible as Arctic ice retreats, leading to growing interest from global powers in exploiting them. Additionally, the Arctic's mineral resources, including rare earth elements, are critical for the EU's technological and industrial needs, particularly in the context of the EU's green and digital transitions (European Commission, 2020).

The EU Critical Raw Materials Act, adopted in 2023, represents a pivotal step in the EU's efforts to secure a sustainable and strategic supply of critical raw materials. These materials, which include rare earth elements, lithium, and cobalt, are crucial for the manufacturing of renewable energy technologies, electric vehicles, and digital devices ((EU) 2024/1252). The Act emphasizes the need for sustainable extraction practices, calling for stringent environmental standards to mitigate the negative impacts of mining. While these provisions are a positive step, they remain insufficient if not accompanied by concrete actions to protect the rights of Indigenous Peoples (Bogojević, 2024; Tröster, Papatheophilou, & Küblböck, 2024).

The EU's commitment to sustainable development in the Arctic, as outlined in its Green Deal, also intersects with the objectives of the Critical Raw Materials Act. The EU has set ambitious goals to achieve climate neutrality by 2050, which will require the integration of green technologies that depend on critical raw materials. While this presents an opportunity for the EU to promote sustainable mining practices, it also underscores the importance of ensuring that the extraction of these materials does not come at the expense of the Sámi people's rights and the Arctic environment (Tröster, Papatheophilou, & Küblböck, 2024). As a result, the EU's approach to resource management in the Arctic emphasizes sustainability, responsibility, and the involvement of local communities, particularly Indigenous Peoples.

Fisheries are another crucial resource in the Arctic, providing food, livelihoods, and economic opportunities for local communities. However, overfishing and the impacts of climate change pose significant threats to Arctic fish stocks. The EU has implemented several policies to promote sustainable fisheries management in the Arctic, aligning with its Common Fisheries Policy and international agreements. The EU has been a strong proponent of the Agreement to Prevent Unregulated High Seas Fisheries in the Central Arctic Ocean, which was signed in 2018 by Arctic and non-Arctic states, including the EU. This agreement is a precautionary measure to protect fish stocks in the central Arctic Ocean, which are becoming more accessible as ice cover recedes (European Commission, 2021a,b). In addition to international agreements, the EU has also supported research and monitoring initiatives aimed at understanding the impacts of climate change on Arctic fisheries. For example, the EU's Horizon 2020 program has funded projects such as Blue-Action, which studies the effects of Arctic warming on global weather patterns and marine ecosystems. The findings from such research are crucial for developing adaptive management strategies that can sustain Arctic fisheries in the face of environmental changes (Koivurova et al., 2021).

The EU also plays a role in managing the environmental impacts of increased shipping activity in the Arctic, which has risen due to melting sea ice. It promotes sustainable shipping practices through its support of the International Maritime Organization regulations, such as the Polar Code, which sets out safety and environmental standards for ships operating in polar waters. The EU has advocated for stricter regulations on the use of heavy fuel oil in Arctic waters, due to its high risk of causing severe environmental damage in the event of a spill (Stang, 2016).

Arctic tourism is another area where the EU seeks to balance economic opportunities with environmental protection. As tourism in the Arctic grows, there is increasing pressure on fragile ecosystems and local communities. The EU's policies promote sustainable tourism practices that minimize environmental impacts and support the socioeconomic development of indigenous communities. For instance, the EU has supported projects that develop guidelines for sustainable tourism in the Arctic, ensuring that tourism activities do not harm local ecosystems or disrupt the traditional ways of life of Indigenous Peoples. These efforts are part of the broader EU Sustainable Development Goals framework, which aims to promote sustainable economic growth while protecting the environment (European Commission, 2021a,b).

### *Geopolitical objectives: Arctic governance for all?*

The Arctic's geopolitical importance has also grown considerably over the past decade, driven by the region's strategic location, resource potential, and the changing global power dynamics. The Arctic is a region where the interests of major global powers converge, leading to increasing geopolitical competition (Heininen et al., 2019). For the EU, which includes three Arctic states (Denmark, Finland, and Sweden) and maintains close relations with other Arctic states (Norway and Iceland), stability and cooperation in the European Arctic in particular are of paramount importance. The presence of EU Member States in the Arctic necessitates a coordinated and coherent approach to regional governance, balancing national interests with broader EU objectives (European Parliament, 2020). For instance, Denmark's sovereignty over Greenland, an autonomous Arctic territory, gives the EU a significant stake in Arctic affairs, particularly regarding natural resource management and environmental protection (Lebel & Nilsson, 2024).

The Arctic itself has become a focal point of geopolitical competition, particularly among Russia, the United States, and China. Russia, which controls a large portion of the Arctic coastline, has invested heavily in its Arctic military capabilities and infrastructure, viewing the region as a critical area for its national security and economic development (Østhagen, 2021). In contrast, China's growing interest in the Arctic, exemplified by its "Polar Silk Road" initiative, has raised concerns among Arctic states about increased competition over resources and influence in the region (Lanteigne, Koivurova, & Nojonen, 2020). The EU, while not a direct Arctic power, has advocated for a rules-based international order in the Arctic, emphasizing the need for cooperation and dialogue to prevent conflict and ensure sustainable development.

The EU's response to Russia's assertive Arctic policies illustrates the geopolitical significance of the region for the Union. Following Russia's annexation of Crimea in 2014 and its subsequent military buildup in the

Arctic, leading to the war between Russia and Ukraine, the EU imposed sanctions on Russia, affecting bilateral cooperation in the Arctic (Conley & Rohloff, 2015; Barron, 2022). These restrictive measures were applied by the EU in multiple rounds and packages and gradually became the cornerstone of the EU's policy toward Russia (Pertiwi, 2024). The current situation between Russia and Ukraine creates extreme friction and halt a series of interaction efforts between Russia and the EU, though over time some communication regarding Arctic issues seems to have resumed, through multilateral forums such as the AC, in an effort to maintain stability and cooperation in the region. Collaboration comes mostly in areas that are considered of "less significance" and is still hindered by sanctions imposed on Russia by the EU on several levels, but also dictated by the EU's dependence on fossil fuels and gas from Russia (Berlin, 2022; Heininen, 2019; Pertiwi, 2024).

The geopolitical significance of the Arctic has grown as melting ice opens new shipping routes and exposes untapped natural resources. The EU's geopolitical interests in the Arctic are intertwined with its environmental and economic goals, particularly in terms of governance and regional cooperation (Nanni et al., 2024). The EU has been a vocal advocate for the strengthening of international governance frameworks to ensure that the Arctic is managed in a way that benefits all stakeholders, including indigenous communities, while preventing conflicts over resource claims. In this context, the EU supports the AC, the primary intergovernmental forum for Arctic cooperation, and works to enhance collaboration between Arctic and non-Arctic states. Furthermore, the EU is active in international regulatory bodies, such as the International Maritime Organization, where it advocates for the adoption of stringent environmental standards for shipping in the region (Olaniyi et al., 2024).

### *Cultural objectives: protecting Indigenous Rights*

The Arctic is home to Indigenous Peoples who have lived in the region for thousands of years. They encompass around 4 million people, including Sámi, Inuit, and Nenets (Einarsson et al. 2004). The Sámi people, who inhabit regions of Norway, Sweden, Finland, and Russia, a territory under the name Sápmi (Figure 11.3), are the only indigenous group recognized within the EU.[3] The Sámi make up a large part of the overall Indigenous population in the Arctic, with an estimated population between 50,000 and 100,000 (UN, 2024). The EU has supported various initiatives to protect Sámi culture and rights, including funding for language preservation programs and the promotion of Sámi traditional livelihoods such as reindeer herding (Toivanen, 2001; Scarpa, 2014). The EU recognizes the importance of protecting the rights and cultures of these communities, particularly in the face of rapid environmental and economic changes.

**FIGURE 11.3** Sápmi – the land of the Sámi. Grey shading indicates the area inhabited by the Sámi People today

*Source:* Drawing: I.-M. Mulk & Ellinor M. Hoff In: Kleppe, E. J. & Mulk, I. M. (2014). Emic and etic perspectives on the history of collecting Sami material culture, p.4. Available at : https://www.researchgate.net/publication/242262122_Emic_and_etic_perspectives_on_the_history_of_collecting_Sami_material_culture

These communities have unique cultures, languages, and ways of life that are closely tied to the Arctic environment. However, climate change, resource extraction, and other external pressures pose significant challenges to the survival of these cultures (Keil, 2014). The EU's Arctic policy aims to also emphasize the importance of protecting the rights of Indigenous Peoples, including their land rights, cultural heritage, and traditional knowledge (European Commission, 2021a,b). The EU supports international frameworks such as the United Nations Declaration on the Rights of Indigenous Peoples, advocating for the inclusion of Indigenous Peoples in decision-making processes related to resource extraction and environmental protection and engages with indigenous organizations through international bodies such as the AC. However, this support is not always easy to implement and prioritize.

For example, cultural considerations are integral to the EU's Arctic policy, particularly in relation to the rights of Indigenous Peoples, notably the Sámi communities, who depend on the region's natural resources for their traditional ways of life (Herrmann & Heinämäki, 2017). However, in reality, the increasing extraction of resources and growing tourism industry in the European Arctic takes place more and more in regions within Sápmi, posing risks to both the environment and the cultural heritage of the Sámi who inhabit them. In Sweden for example, mineral resources exploited in the **Kallak mining project** in **Jokkmokk**, located in Swedish Lapland, are directly impacting the life of the Sámi people. This area is traditionally used by the Sámi for reindeer herding, and the proposed extraction of iron ore has led to significant concerns regarding the environmental and cultural impact on the Sámi communities. The project has faced strong opposition from the Sámi, who argue that mining activities threaten their way of life and the ecological balance of the region (Tarras-Wahlberg & Southalan, 2022).

The EU has also engaged in attempts to address the growing issue of Arctic tourism, which, while providing economic opportunities, can also place significant pressure on local communities culture, as several tour operators use indigenous clothing, artifacts and images for profit in a non-appropriate way and without permission, creating false stereotypes on how the indigenous way of life and culture is (Kavvatha, 2023). As a result, in Finland, the EU supports the development of sustainable tourism practices that prioritize the well-being of indigenous communities and their right to self-determinate. Through funding for projects aimed at developing sustainable tourism guidelines, the EU aims to ensure that tourism in the Arctic does not disrupt the traditional lifestyles of Indigenous Peoples (European Commission, 2021a,b).

## EU challenges in Arctic governance and the way forward

### *Challenges in Arctic governance*

The EU faces several challenges in its efforts to govern and shape the Arctic region, stemming from geopolitical tensions, the complexity of multilateral cooperation, and policy coherence issues. These challenges are compounded by the rapidly changing environmental conditions in the Arctic, which demand immediate and coordinated responses.

The Arctic is a region of significant geopolitical tensions, with multiple actors vying for influence (Heininen, 2014). The EU, despite being a key player in global governance, faces challenges in asserting its influence in the

Arctic, particularly in light of competing interests from major powers such as Russia, the United States, and China. Russia's militarization of the Arctic and its assertive stance on resource claims create a tense environment where the EU's emphasis on environmental protection and sustainable development can be overshadowed by security concerns (Konyshev & Sergunin, 2012). Moreover, the EU's status as an observer in the AC, limits its direct influence on decision-making processes and its relationship with Arctic states. While the EU advocates for peaceful cooperation and sustainable development, it must navigate the differing priorities of Arctic states, which may prioritize economic exploitation and sovereignty over environmental concerns (Filimonova, Obydenkova, & Rodrigues Vieira, 2023).

The complexity of cooperation adds to the difficulty the EU faces in imposing its agenda on the region. Governance in the Arctic is characterized by a complex web of multilateral agreements and organizations, including the AC, the United Nations Convention on the Law of the Sea, and various bilateral agreements among Arctic states. The EU's role in these frameworks is often indirect, which can lead to challenges in policy implementation and coordination (Radaelli, 2002). The fragmentation of such governance structures also poses a significant challenge. The AC, though influential, does not have the authority to enforce binding agreements, and its decisions are made by consensus among the eight Arctic states, which can lead to policy gridlock (Filimonova, Obydenkova, & Rodrigues Vieira, 2023). This fragmentation is further exacerbated by the diverging priorities of Arctic states compared to the EU, particularly regarding resource exploitation, environmental protection, and Indigenous Rights.

Ensuring policy coherence across different levels of governance—EU, national, regional, and local—is another significant challenge. The EU's Arctic policies must align with the environmental and socioeconomic strategies of its Member States, particularly those within the Arctic Circle, such as Denmark (via Greenland), Finland, and Sweden. However, inconsistencies can arise when national interests, such as economic development or resource extraction, conflict with EU-wide environmental objectives (Maggetti & Trein, 2019). For example, Denmark's economic interests in exploiting Greenland's mineral resources can sometimes clash with the EU's broader environmental goals, creating challenges in policy coherence and implementation. Similarly, the EU's ambitious climate goals, outlined in the European Green Deal, include activities such as mining, and oil extraction for economic stability and as mentioned before in the **Kallak mining project in Jokkmokk** may face resistance from regional stakeholders the most prominent ones being the Indigenous People in the Arctic who rely on traditional industries like fishing and reindeer. Box 11.2 illustrates the need to ensure coherency between economic development and indigenous practices.

### BOX 11.2 CHALLENGES OF RESOURCE EXTRACTION—IMPACT OF EU ECONOMIC DEVELOPMENT POLICIES ON INDIGENOUS COMMUNITIES

Economic development in the Arctic is a double-edged sword for indigenous communities. On the one hand, it offers opportunities for employment, infrastructure development, and increased economic activity. On the other hand, if not managed responsibly, it can lead to the erosion of Indigenous Rights, cultural degradation, and environmental harm (Gladun et al., 2021).

While the EU supports economic development in the Arctic, there is growing concern over the impact of resource extraction on indigenous communities (Bogojević, 2024). These activities often occur on lands that are crucial for indigenous practices such as reindeer herding, fishing, and hunting, leading to tensions between economic interests and the preservation of indigenous ways of life (Kavvatha, 2023). The EU Critical Raw Materials Act and the expansion of mining activities in the European Arctic with fast-track procedures that it brings, has been particularly contentious. While the Act seeks to reduce the EU's dependency on non-EU sources, it also raises important questions about the environmental, social, and cultural implications of raw material extraction, particularly in regions such as the Arctic, where the Sámi people live (Bogojević, 2024).

The Sámi have consistently expressed concerns that their traditional land rights and cultural heritage are at risk from large-scale resource extraction projects. For instance, in Sweden and Finland, Sámi communities have been engaged in legal battles to prevent mining activities on their ancestral lands, arguing that such projects violate their rights under international frameworks such as the United Nations Declaration on the Rights of Indigenous Peoples. In response to the Sámi concerns, the EU has included provisions in its Arctic policy such as encouraging prior information, consultation and consent processes as well as impact assessments, aimed at ensuring the participation of Indigenous Peoples in decision-making processes regarding the development of natural resources. Though the EU aims to thoroughly acknowledge the importance of respecting the rights of Indigenous Peoples, the application of these principles to resource extraction in the Arctic remains a contentious issue (Alvarez, Yumashev, & Whiteman, 2020).

Sources: Alvarez, Yumashev, & Whiteman (2020); Bogojević (2024); Gladun, Nysten-Haarala, & Tulaeva (2021); Kavvatha (2023).

## The way forward: strengthening EU Arctic governance

To address these challenges, the EU would need to adopt a more proactive and coherent approach to Arctic governance, emphasizing multilateral cooperation, policy integration, and the protection of Indigenous Rights and environmental sustainability.

The EU could strengthen its influence in the Arctic by enhancing its role in multilateral organizations and forging strategic alliances with Arctic states. Active participation in the AC, despite its observer status, can be augmented by collaborating with like-minded states on initiatives that align with EU objectives, such as promoting MPAs and advancing climate research (Koivurova et al., 2021). Moreover, the EU could work to strengthen the international legal frameworks that govern the Arctic, by advocating for more stringent environmental protection and supporting the development of binding agreements on issues such as Arctic shipping and resource extraction. Collaborating with non-Arctic states such as China, that share environmental concerns can also help the EU build a coalition that promotes sustainable development in the region (Biedermann, 2021).

To ensure policy coherence, the EU would need to better integrate its Arctic policies with broader EU strategies, such as the European Green Deal and the EU Biodiversity Strategy for 2030. This integration requires close coordination with Member States, regional governments, and local communities, particularly in Arctic regions like Greenland, where the intersection of economic development and environmental protection is most pronounced (European Commission, 2021a,b). The EU could also develop tailored policies that address the specific needs and challenges of Arctic regions, recognizing the unique environmental, social, and economic conditions of these areas. For instance, promoting sustainable tourism and responsible fisheries management can help balance economic development with environmental protection, ensuring that local communities benefit from EU policies while preserving the fragile Arctic ecosystem (Orsini & Kavvatha, 2020).

*Finally,* the EU could prioritize the rights and well-being of Indigenous Peoples in the Arctic, particularly the Sámi, whose traditional livelihoods and cultural heritage are threatened by environmental changes and economic development. This can be achieved by ensuring that indigenous voices are included in decision-making processes at all levels of governance, from local to international (European Commission, 2021a,b). It could also support initiatives that preserve indigenous cultures and knowledge systems, recognizing their critical role in sustainable resource management and environmental stewardship. Programs that promote language preservation, education, and economic opportunities for indigenous communities can help mitigate the adverse impacts of development and climate change, fostering resilience and cultural continuity (Kavvatha, 2023). One good example of such a program already in place, is the 'Filling the EU-Sápmi knowledge gaps' project, which operates in the cross-border region of North-East Finland, Upper Norrland in Sweden and parts of Norway. The project lets Sámi people undertake education courses and participate in traineeships.

## Conclusion

This chapter has explored the EU's strategic involvement in Arctic governance, addressing the intersections of environmental, economic, geopolitical and cultural dimensions. The rapid transformations occurring in the Arctic, driven by both environmental and geopolitical shifts, highlight the region's critical role in global affairs. The EU's Arctic policy seeks to balance these complex factors, though it faces several challenges in integrating environmental, economic, and social objectives within a coherent governance framework.

From an environmental perspective, the EU has made considerable strides in advocating for sustainable resource management and climate change mitigation in the Arctic. The EU's leadership in global climate agreements, such as the Paris Agreement, and its focus on Arctic-specific initiatives, such as the EU-PolarNet research program and MAPs, exemplify its commitment to preserving Arctic ecosystems (European Commission, 2021a,b). However, the EU's policy has not been without challenges. The increasing accessibility of Arctic resources due to climate change presents the EU with a difficult dilemma: how to protect the region's fragile environment while accommodating growing demand for energy and minerals? Furthermore, the region's rapid warming exacerbates the environmental challenges, requiring more adaptive and anticipatory governance approaches. In this context, the EU's capacity to adapt its policies to the shifting environmental realities of the Arctic will be pivotal for ensuring long-term sustainability.

In the economic domain, the EU has made efforts to support sustainable development in the Arctic, particularly regarding resource extraction. However, economic growth in the Arctic, driven by resource extraction and shipping, brings with it risks of environmental degradation and disruption of local livelihoods, especially for Indigenous Peoples. The EU's policies must address these risks while promoting economic opportunities in the region, a challenge that requires careful attention to the socioeconomic needs of both indigenous communities and external stakeholders. Additionally, the EU's support for sustainable shipping practices and its advocacy for environmental standards demonstrate a proactive approach to mitigating the environmental impacts of increased maritime activity in the region (Stang, 2016). Despite these efforts, the EU faces difficulties in aligning the diverse economic interests of Member States and non-EU Arctic stakeholders, particularly in areas like resource extraction and transportation.

Geopolitically, the Arctic has become a contested space, with major global powers vying for influence as the region's importance grows due to its potential resource wealth and strategic shipping routes. The EU's geopolitical role in Arctic governance is multifaceted, involving diplomatic engagement with Arctic and non-Arctic states through forums like the AC. While the EU has positioned itself as a promoter of peace and multilateral cooperation in the region, the geopolitical tensions arising from the ambitions of Russia, the United States, and China present significant challenges for the EU's Arctic strategy. Moreover, the EU's ability to navigate these tensions while advancing its environmental and economic goals requires careful diplomacy and robust multilateral cooperation. The challenge lies in maintaining a constructive role in the face of competing national interests and the increasing militarization of the Arctic by other powers. The EU's success in this dimension will depend on its ability to leverage its soft power and align its Arctic governance strategy with broader international frameworks for peace and security.

The cultural dimension, particularly the rights and interests of Indigenous Peoples, especially the Sámi, remains a central pillar of EU Arctic policy. The EU has expressed a strong commitment to respecting the rights of indigenous communities and supporting their participation in decision-making processes related to Arctic governance. This is evident in the EU's support for initiatives that promote the protection of indigenous cultures and traditional livelihoods, as well as its efforts to include indigenous perspectives in policy development (European Commission, 2021a,b). However, the increasing industrialization of the Arctic, coupled with environmental degradation, poses risks to the cultural heritage and traditional lifestyles of Indigenous Peoples. Resource extraction, infrastructure development, and the expansion of tourism all have the potential to disrupt these communities. Thus, while the EU has made progress in acknowledging Indigenous Rights, ensuring that these communities have a meaningful voice in policy decisions remains a complex challenge. The EU must continue to prioritize the participation of Indigenous Peoples in governance processes and address the sociocultural impacts of its policies.

The overall success of the EU's Arctic governance strategy will depend on its ability to navigate these challenges, particularly the tensions between environmental protection, economic development, and cultural preservation. The EU's role in Arctic governance, while influential, faces significant hurdles. These include the fragmentation of governance structures across different levels, from EU institutions to national governments and regional actors, as well as the difficulties in harmonizing policy priorities among diverse Arctic stakeholders. Moreover, the increasing presence of external powers in the Arctic region adds a layer of complexity to EU efforts to maintain peace and cooperation. As the Arctic continues to evolve, the EU must work to enhance multilateral cooperation, integrate its Arctic policies with broader

EU strategies, and remain responsive to the evolving environmental and geopolitical realities of the region.

Looking to the future, the EU should aim to adopt a long-term, adaptive approach to Arctic governance, prioritizing climate adaptation and mitigation strategies that address the region's changing environmental conditions. This will involve strengthening international partnerships, especially with indigenous communities, to ensure that governance frameworks are equitable and effective in safeguarding both the Arctic ecosystem and the rights of its inhabitants. Technological innovations to reduce the carbon footprint of activities in the Arctic, alongside the integration of sustainable practices in economic development, will be critical to maintaining the resilience of the Arctic in the face of accelerating climate change. The EU's commitment to these goals will shape the future of Arctic governance and its contribution to global environmental protection.

## Notes

1 However, when referring to the region specifically, the term "Arctic" is capitalized as a proper noun. Some countries also have their own names for the northernmost areas within their borders. For instance, Norway refers to its northern areas as the "High North," while Russia calls it the "Far North," and Canada designates the region north of the Arctic Circle as the "Far North" as well.

2 The European Arctic is defined from Greenland in the West to the Ural Mountains in Russia to the East. In this part of the Arctic the Sami peoples live in Northern Norway, Sweden, Finland and Russia; the Inuit peoples live in Greenland, Canada and the US/Alaska; and the Nenets peoples live in Nenets Autonomous Okrug and the Yamalo-Nenets Autonomous Okrug in northwest Russia. A small percentage of the Komi peoples can also be found in this region.

3 In Norway, Sweden, and Finland, the Sámi are recognized as indigenous peoples under national laws. Norway has the Sami Act, which acknowledges the Sami as an indigenous people with distinct cultural rights and Sweden and Finland recognize the Sámi in their constitutions, with Finland specifically designating the Sami as "the indigenous people of Finland" in its Constitution (Section 17).

## References

Alvarez, J., Yumashev, D., & Whiteman, G. (2020). A Framework for Assessing the Economic Impacts of Arctic Change. *Ambio*, *49*(2), 407–418. doi:10.1007/s13280-019-01211-z

Barron, K. (2022). The Annexation of Crimea and EU Sanctions: An Ineffective Response. *The Arbutus Review*, *13*(1). doi:10.18357/tar131202220760

Berlin, M. P. (2022). *The Effects of Sanctions*. Free Network Policy Bries Series, May Edition.

Biedermann, R. (2021). China's Impact on the European Union's Arctic Policy: Critical Junctures, Crossovers, and Geographic Shifts. *Asia Europe Journal*, *19*(4), 467–487. doi:10.1007/s10308-021-00605-7

Boedeltje, F., & van Houtum, H. (2011). Brussels is Speaking: The Adverse Speech Geo-Politics of the European Union Towards Its Neighbours. *Geopolitics*, *16*(1), 130–145. doi:10.1080/14650045.2010.493791

Bogojević, S. (2024). The European Green Deal, the Rush for Critical Raw Materials, and Colonialism. *Transnational Legal Theory*, *15*(4), 600–615. doi:10.1080/20414005.2024.2399408

Conde Pérez, E., & Yaneva, Z. V. (2016). The European Arctic Policy in Progress. *Polar Science*, *10*(3), 441–449. doi:10.1016/j.polar.2016.06.008

Conley, H. A., & Rohloff, C. (2015). *The New Ice Curtain*. Center for Strategic & International Studies.

Crépin, A.-S., Karcher, M., & Gascard, J.-C. (2017). Arctic Climate Change, Economy and Society (ACCESS): Integrated Perspectives. *Ambio*, *46*(S3), 341–354. doi:10.1007/s13280-017-0953-3

Einarsson, N., Larsen, J. N., Young, O., & Nilsson, A. E. (Eds.). (2004). *Arctic Human Development Report 2004*. Stefansson Arctic Institute. https://oaarchive.arctic-council.org/items/f6c63158-401c-4a14-a8d2-5bc9f10710bfsvs.is+3

European Commission. (2020 a). *Study on the EU's List of Critical Raw Materials: Critical Raw Materials Factsheets*. https://op.europa.eu/en/publication-detail/-/publication/c0d5292a-ee54-11ea-991b-01aa75ed71a1/language-en

European Commission. (2021a). *Overview of EU Actions in the Arctic and their Impact: Final Report*. EPRD Office for Economic Policy and Regional Development. https://eprd.pl/wp-content/uploads/2021/06/EU-Policy-Arctic-Impact-Overview-Final-Report.pdf

European Commission. (2021b). The European Union and the Arctic: Policy Objectives and Engagement with Indigenous Peoples and Geopolitical Dynamics. https://eur-lex.europa.eu/legal-content/EN/TXT/?uri=JOIN:2021:27:FIN

European Parliament & Council of the European Union. (2024). Regulation (EU) 2024/1252 of the European Parliament and of the Council Establishing a Framework for Ensuring a Secure and Sustainable Supply of Critical Raw Materials and Amending Regulations (EU) No 168/2013, (EU) 2018/858, (EU) 2018/1724 and (EU) 2019/1020. https://eur-lex.europa.eu/legal-content/EN/TXT/?uri=CELEX%3A32024R1252

Filimonova, N., Obydenkova, A., & Rodrigues Vieira, V. G. (2023). Geopolitical and Economic Interests in Environmental Governance: Explaining Observer State Status in the Arctic Council. *Climatic Change*, *176*, 50. doi:10.1007/s10584-023-03490-8

Gascard, J.-C., Crépin, A.-S., Karcher, M., & Young, O. R. (2017). Facets of Arctic Change. *Ambio*, *46*(S3), 339–340. doi:10.1007/s13280-017-0952-4

Gautier, D. L., Bird, K. J., Charpentier, R. R., Grantz, A., Houseknecht, D. W., Klett, T. R., ... Wandrey, C. J. (2009). Assessment of Undiscovered Oil and Gas in the Arctic. *Science*, *324*(5931), 1175–1179. doi:10.1126/science.1169467

Gladun, E., Nysten-Haarala, S., & Tulaeva, S. (2021). Indigenous Economies in the Arctic. *Elementa*, *9*(1), 00088. doi:10.1525/elementa.2019.00088

Hanaček, K., Kröger, M., Scheidel, A., Rojas, F., & Martinez-Alier, J. (2022). On Thin Ice – The Arctic Commodity Extraction Frontier and Environmental Conflicts. *Ecological Economics: The Journal of the International Society for Ecological Economics*, *191*(107247), 107247. doi:10.1016/j.ecolecon.2021.107247

Heininen, L. (2014). Northern geopolitics: Actors, interests and processes in the circumpolar Arctic. In R. C. Powell & K. Dodds (Eds.), *Polar geopolitics? Knowledges, resources and legal regimes* (pp. 241–258). Edward Elgar Publishing. https://doi.org/10.4337/9781781009413.00023Heininen, L., Everett, K., Padrtova, B., & Reissell, A. (2019). *Artcic Policies and Strategies-Analysis, Synthesis, and Trends.* International Institute for Applied Systems Analysis. doi:10.22022/AFI/11-2019.16175

Herrmann, T. M., & Heinämäki, L. (2017). Experiencing and Safeguarding the Sacred in the Arctic: Sacred Natural Sites, Cultural Landscapes and Indigenous Peoples' Rights. In Heinämäki, L., & Herrmann, T. (eds.), *Experiencing and Protecting Sacred Natural Sites of Sámi and other Indigenous Peoples* (pp. 1–8). Springer Polar Sciences. doi:10.1007/978-3-319-48069-5_1

Hossain, K. (2015). EU engagement in the Arctic: Do the policy responses from the Arctic states recognise the EU as a legitimate stakeholder? *Arctic Review on Law and Politics*, *6*(2), 89–110. https://doi.org/10.17585/arctic.v6.97IPCC Summary for Policymakers. (2023). *Climate Change 2022 – Impacts, Adaptation and Vulnerability* (pp. 3–34). doi:10.1017/9781009325844.001

Kavvatha, E. (2023). *Balancing on Ice: Policy Related to Indigenous Tourism in the European Arctic – The Case of Sápmi.* UiT the Arctic University Open Research Archive. https://munin.uit.no/handle/10037/29435

Keil, K. (2014). The Arctic: A New Region of Conflict? The Case of Oil and Gas. *Cooperation and Conflict*, *49*(2), 162–190. doi:10.1177/0010836713482555

Koivurova, T., Hoel, A. H., Humpert, M., Kirchner, S., Raspotnik, A., Śmieszek, M., & Stępień, A. (2021). *Overview of EU actions in the Arctic and their impact.* European Union. https://eprd.pl/wp-content/uploads/2021/06/EU-Policy-Arctic-Impact-Overview-Final-Report.pdf

Konyshev, V., & Sergunin, A. (2012). The Arctic at the Crossroads of Geopolitical Interests. *Russian Politics and Law*, *50*(2), 34–54. doi:10.2753/rup1061-1940500202

Landrum, L., & Holland, M. M. (2020). Extremes Become Routine in an Emerging New Arctic. *Nature Climate Change*, *10*(12), 1108–1115. doi:10.1038/s41558-020-0892-z

Lanteigne, M., Koivurova, T., & Nojonen, M. (2020). China's Rise in a Changing World. In Koivurova, T. & Kopra, S. (eds.), *Chinese Policy and Presence in the Arctic* (pp. 5–24). Brill Nijhoff. doi:10.1163/9789004408425_003

Lebel, J., & Nilsson, A. E. (2024). EU Engagement in the Arctic: Challenges to Achieving Ambitions in an Area Outside Its Jurisdiction. *Arctic Review on Law and Politics*, *15*, 47–71. doi:10.23865/arctic.v15.6271

Maggetti, M., & Trein, P. (2019). Multilevel Governance and Problem-Solving: Towards a Dynamic Theory of Multilevel Policy-Making? *Public Administration*, *97*(2), 355–369. doi:10.1111/padm.12573

Nanni, U., DeRepentigny, P., Lundén, A., Popovaitė, V., Shen, Y., Basaran, I. K., … Wołoszyn, A. (2024). Redefining Arctic Boundaries in a Changing Climate: Interdisciplinary Perspectives on Governance Strategies. *Polar Geography*, *47*(2), 127–155. doi:10.1080/1088937X.2024.2359926

NERC. (2015). Report on Cross-Sectoral Synthesis of Economic, Policy and Governance Options for Sustainable Development. Deliverable D5.91 of the European Union FP7 Project Nr. 265863 ACCESS. https://www.access-eu.org/en/deliverables2/wp5.html

Olaniyi, E. O., Solarte-Vasquez, M. C., & Inkinen, T. (2024). Smart Regulations in Maritime Governance: Efficacy, Gaps, and Stakeholder Perspectives. *Marine Pollution Bulletin*, *202*(116341), 116341. doi:10.1016/j.marpolbul.2024.116341

Orsini, A., & Kavvatha, E. (Eds.). (2020). *EU environmental governance*. London, England: Routledge.

Østerud, Ø., & Hønneland, G. (2014). Geopolitics and International Governance in the Arctic. *Arctic Review on Law and Politics*, *5*(1), 1–22. doi:10.23865/arctic.v5.1044

Østhagen, A. (2021). The Arctic Security Region: Misconceptions and Contradictions. *Polar Geography*, *44*(1), 55–74. doi:10.1080/1088937x.2021.1881645

Pertiwi, L. A. (2024). The EU's Approach to Sanctions on Russia: A Critical Analysis of the Existing Literature. *Central European Journal of International & Security Studies*, *18*(3), 61–86. doi:10.51870/noex4475

Radaelli, C. (2002). The Domestic Impact of European Union Public Policy: Notes on Concepts, Methods, and the Challenge of Empirical Research. *Politique européenne*, *5*(1), 105–136. doi:10.3917/poeu.005.0105

Scarpa, F. (2014). The EU, the Arctic, and Arctic Indigenous Peoples. *The Yearbook of Polar Law Online*, *6*, 427–465.

Serreze, M. C., & Barry, R. G. (2011). Processes and impacts of Arctic amplification: A research synthesis. *Global and Planetary Change*, *77*(1–2), 85–96. doi:10.1016/j.gloplacha.2011.03.004

Serreze, M. C., & Barry, R. G. (2014). *Cambridge Atmospheric and Space Science Series: The Arctic Climate System* (2nd ed.). Cambridge University Press. doi:10.1017/cbo9781139583817

Stang, G. (2016). EU Arctic policy in regional context: Study.

Tarras-Wahlberg, H., & Southalan, J. (2022). Mining and Indigenous Rights in Sweden: What Is at Stake and the Role for Legislation. *Mineral Economics*, *35*(2), 239–252. doi:10.1007/s13563-021-00280-5

Toivanen, R. (2001). Saami in the European Union. *International Journal on Minority and Group Rights*, *8*, 303–323.

Tröster, B., Papatheophilou, S., & Küblböck, K. (2024). *In Search of Critical Raw Materials: What Will the EU Critical Raw Materials Act Achieve? An Analysis of Legal and Factual Implications of the CRMA*. Research Reports 18/2024, Austrian Foundation for Development Research (ÖFSE). doi:10.60637/2024-RR18

United Nations. (2024). *The Arctic*. United Nations Climate Change. https://lcipp.unfccc.int/about-lcipp/un-indigenous-sociocultural-regions/arctic

# 12

# YOUTH AND EU ENVIRONMENTAL POLICY

*Amandine Orsini*

## Introduction

European Union (EU) action on the environment is known to be ambitious (Rosamond and Dupont, 2021; Oberthür and Dupont, 2021), covering many environmental topics and asking for clear objectives to be met. Within this ambitious agenda, youth are a complicated and fascinating object of study. Complicated because, among others, no definition of youth exists worldwide, and the European level is no exception. When dealing with statistics, the EU defines "young people" as individuals aged between 15 and 29 years old. In 2020, they represented 16.5% of the EU population and their percentage is expected to decrease slightly in the future (15.1% in 2080, Eurostat, 2022). However, politically, several definitions of youth coexist in the EU. A 2016 report found six different ways to define the youth age group across the European continent (Perovic, 2016). These six ways are present across all the EU Member States (see Table 12.1) with considerable variability, especially regarding the lower age limit for youth with some countries including children and others not.

This complicates the understanding of youth[1] and youth policy. Despite such differences regarding the age range, data about youth and the environment in the EU converge on two important dimensions. First, data coming from surveys of youth in the EU show that youth have specific and strong preferences for environmental policy (see below), stronger than the ones of other age groups. This makes youth a particularly pro-environmental actor. Secondly, youth have been recently very vocal on environmental issues, especially on climate change, through the Fridays for Future (FFF) marches and protests (Wahlström et al., 2019). This tends to indicate that youth are not satisfied with the current situation.

DOI: 10.4324/9781003479260-19

**TABLE 12.1** Age Range Defining Youth in Different EU Member States

| *Youth Age Range (Lower – Upper Limit)* | *EU Member State* |
|---|---|
| 14/15/16 – 29/30 | Bulgaria, Croatia, Czech Republic, Denmark, Italy, Lithuania, Hungary, Germany, Poland, Slovenia, and Spain |
| 13/15/16 – 24/25 | Ireland, Latvia, and Sweden |
| 12/13 – 30 | Luxembourg, Malta, and Portugal |
| 12/14/15/16 – 32/35 | Greece, Cyprus, and Romania |
| 3/6/7 – 25/26/30 | France and Estonia |
| 0 – 25/29/30 | Austria, Belgium, Slovak Republic, Finland, and The Netherlands |

*Source*: based on Perovic 2016.

The puzzle this chapter aims at exploring is therefore the following: EU environmental policy is one of the most advanced in the world; and yet, European youth express discontent with regard to how European governments deal with environmental crises. How can we explain such a puzzle? The research hypothesis is that youth do not feel fully represented in the EU environmental policy process, which explains why they use protest strategies to make their voices heard. In order to explore such hypothesis, this chapter assesses the current state of the interactions between the EU and youth on environmental issues, with a more precise focus on how youth participate in environmental policymaking.

This chapter relies on a literature review and, most of all, as such literature is limited, on a selection of 20 anonymized interviews[2] conducted with European youth included in Member State delegations (see below) at the United Nations Framework Convention on Climate Change (UNFCCC) Conferences of the Parties (COPs) 26 (November 2022) and 27 (November 2023), at COP15 of the Convention on biological diversity in December 2022 and at the Stockholm + 50 and at the 2023 high-level political forum (HLPF) on sustainable development for the negotiations of the sustainable development goals (SDGs) in June 2022 and July 2023, for a total of about 11 hours of interviews.[3] More precisely, nine youth delegates were interviewed at the biodiversity COP; four at climate COPs; five at SDGs negotiations and one was interviewed both at a climate and a biodiversity COP and one both at a climate and an SDG meeting. Their age range varied from 20 to 28 with a medium age of 24; 10 were female and 10 male. 6 were from Belgium, 4 from Denmark, 2 from The Netherlands and Sweden, and 1 from France, Spain, Ireland, Estonia, Finland, and Germany.

The interviews were recorded, transcribed, and uploaded in Nvivo for qualitative analysis and coding. Questions were rather open and meant to create space for youth-based insights to be used as a unique and major source of information for mapping and understanding the political spaces of youth presence in EU environmental policy. Interview quotes were selected to illustrate the most important claims. In addition to conducting interviews, fieldwork missions at the different COPs and international meetings also enabled to observe European youth delegates during the negotiations.

This chapter is organized as follows: Section "The potential of youth for environmental policy in the EU" explains why youth are specific and important actors for EU environmental policy. Section "European youth and the environment: current policies and participatory mechanisms" presents the different policy instruments and representation mechanisms that link youth to environmental policymaking in the EU. Section "European youth and environmental policy : challenges" details the challenges of such instruments and mechanisms. The conclusion summarizes the findings and opens new research paths.

## The potential of youth for environmental policy in the EU

### *A greater environmental consciousness*

The 2022 youth flash barometer indicates that, for youth, "protecting the environment and fighting climate change (34%) must be the top priority of the European Year of Youth,[4] together with physical health and wellbeing (34%)" (European Commission, 2022).

Another recent survey, pan-European, conducted in 2021, indicates that: "Close to half (46%) of young Europeans[5] consider climate change to be among the most serious problems facing the world, which puts it in first place among problems listed, even in the midst of the COVID-19 pandemic.[6] In second place comes environmental degradation, including for example air pollution, deforestation and the extinction of animals (44%)" (Ipsos, 2021: 5). The survey details that

> more than eight out of ten young Europeans (84%) are fairly, very or extremely worried about climate change. Close to half (46%) are very or extremely worried. Few (16%) are not very worried or not worried at all about climate change.
>
> *(Ipsos, 2021: 6)*

What is of particular interest is that youth appear more concerned about the environment than older age groups, or former generations, across this survey. Among others, youth are less inclined to climate denial: "Less than one out of ten young Europeans (8%) deny climate change," while the proportion is of one in ten when including other age groups (Ipsos, 2021: 7).

Overall, research has shown that young Europeans have a more frequent perception of the climate change problem compared to older individuals (Skeirytė et al., 2022). Confirming the specificity of the young generation, Hurrelman and Albrecht (2021) explain that the Gen Z (individuals born in 1997-2012) generation has been found to be extremely vocal on the climate crisis. They, moreover, tend to suggest quite strong measures to correct it, in the streets, but also when involved in the policy process. As explained by one Swedish youth delegate:

> Yeah. So again, what I often hear from young leadership society is that we need to totally phase out fossil fuels and that it's quite urgent. So, I would say that is one of the demands that always comes up almost.
>
> *(interview 11)*

A youth delegate from Denmark, also representative of the Danish Youth Council explains the tendency is reflected at the national level as well: "with all these surveys, I have seen them, at least for Denmark, young people are more ambitious and more engaged than older generations" (interview 7).

### *A broader political repertoire*

Youth are not only worried about the environment; they also have developed a very varied political repertoire to engage in politics. Studies on environmental non-governmental organizations (NGOs) indicate that two main modes of political action[7] exist for civil society groups: insider strategies and outsider strategies (Parks et al., 2023). Youth have been using both, while also adapting them, and have added additional ones to make their voices heard.

Outsider strategies, also known as activism, have been the most visible recent strategies of youth. Outsider strategies take the form of petitions, actions, trials, and protests. With the FFF movement, initiated in Europe and rallied around worldwide, youth have been able to put pressure on governments: "Intense protests, such as the Fridays for Future movements, demonstrated public support across Europe (and beyond) for climate action, and increased bottom-up pressure on policymakers" (Dupont et al., 2020: 1105). One specificity of youth outsider strategies has been to rely highly on social media for publicizing and organizing protests. Among others, Tyson et al. 2021 show that the Gen Z generation is very active on social media for climate change.

Insider strategies for youth are also visible in EU environmental policymaking, even if they have been the object of fewer studies, a gap this chapter answers to. They are also known as advocacy. As developed below, specific youth policy and mandates have been created at the level of the Member States for youth to contribute to policymaking. Moreover, European youth have organized into NGOs such as youth-led organizations with the European Youth Forum being one of the most established (created in 1996) and active. These youth participate to consultations or political meetings, including local, national and international consultations such as the COP meetings mentioned above. Just as for outsider strategies, the position of youth as insiders has been noticed for being strong on environmental ambition. Even when involved in formal arenas, youth tend to keep their freedom of speech, as they are most of the time not linked to vested interests or work constraints. As explained by one youth delegate also involved in a National Youth Council:

> young people are, youth have quite a different perspective from older generations (…) we, young people, are not always as fixed by previous habits and by previous, you know, experience in our careers and so on so we are often quite flexible in what kind of system we can work within. And I think that kind of flexibility is important to have in the negotiations as well.
>
> *(interview 10)*

Both outsider and insider strategies are often used at the same time by political actors. Actually, "cyclical shifts from one mode to the other (from activism to advocacy) are evident, in distinct phases" (Parks et al., 2023: 99). During interviews, several youths engaged in official youth representation mandates confirmed being members of FFFs and joining the biodiversity or the climate marches when they were happening.

In addition to outsider and insider strategies, youth are known to have developed a third strategy to engage in environmental policy: practicing their environmental engagement by themselves. This strategy, known as "Do It Ourselves" politics (Pickard, 2022), is somehow close to the "leading by example" strategy used by the EU to convince other players of its environmental objectives, as identified by European studies scholars (Wurzel et al., 2019). It consists in adopting life choices in line with environmental commitment to show that they are feasible and beneficial. Zamponi et al. (2022), studying climate movements in 15 European cities, show a clear generational effect: younger cohorts have life choices, such as avoiding flying or limiting fast fashion, that are much more in line with environmental objectives. During our interviews, many youths confirmed being vegetarian and complained that COP venues were not vegetarian venues. Others complained that air conditioning was used at too strong levels.

All three strategies are compatible (Zamponi et al., 2022) and, combined with the use of social media, present youth as potentially equipped to be strong agents of change. Such change is oriented toward environmental objectives and beyond.

### *A specific attention to the importance of broader perspectives to protect the environment*

The literature on youth outsider strategies has soon demonstrated that youth have embraced objectives that go beyond environmental priorities. Talking about the claims expressed by FFF activists, scholars explain:

> Deriving from the concept of "environmental justice", they invented the so-called "climate justice" referring to the fact that climate change is not only a scientific challenge but also has social dimensions. For this reason, social issues such as gender equality, LGBTQ+ rights, the rights of people of color and indigenous peoples are always included during demonstrations.
> *(Magosi and Hammond, 2023: 124)*

Other studies show that fighting the environmental crisis is only the entry point for youth, especially for the Gen Z generation, to then question broader choices about the economy, Human Rights, and nature rights (Hurrelman and Albrecht, 2021).

This is reflected as well when one questions the expectations of youth during international environmental negotiations. Studying biodiversity negotiations, research has recently shown that youth support a triple extension of the sovereignty concept: a temporal one, whereby States would need to take rapid measures to protect current and future generations; a representative one, whereby States would protect all humans, minorities included; and an ecocentric one, whereby States would protect humans and nature (Orsini and Duque, 2024). As explained by one European youth delegate:

> A second demand that comes up is redefining well-being from being about economic growth and the GDP of a country to revolve more around the health of the people in the country, both physical and mental, and also the integrity of natural systems with a value in itself. So not only that ecosystems continue to provide resources to humans, but also that they are also healthy by themselves.

He details even further:

> when I speak to young people, they often raise that it is an issue of the humans affected rather than just some numbers on a sheet of paper. So that's the first reason to bring in that kind of value.
> *(interview 10)*

Environmental objectives are coupled with a much wider transformative agenda.

For Do-It-Ourselves strategies our interviews and fieldwork at international negotiations also confirm that youth are aware of the importance of recognizing gender, racial and cultural diversity as global politics are also intersectional.

While the consciousness, the repertoires, and the claims of youth appear to be wide, the next section questions whether and how the EU has developed political spaces to welcome such ambition for environmental policy. It particularly details insider strategies' spaces to understand to which extent youth have been integrated as proper agents in the policy process.

## European youth and the environment: current policies and participatory mechanisms

### *Current policies and participatory mechanisms at the EU level*

Youth policy is a national competence for the Member States. Therefore, the EU level of policymaking on youth might not appear as the right level of analysis to look at. And yet, youth policy in the EU is far from being inexistant. Rather, it is known to be "a complex but increasingly coherent journey" (Williamson, 2007).

Two fundamental treaties of the EU mention youth. Articles 165 and 166 of the 1957 Treaty on the Functioning of the EU are the basis for EU action in the youth field. Article 165 provides for EU action in order to encourage the development of youth exchanges and exchanges between youth workers, and to encourage the participation of young people in democratic life in Europe. Article 166 tasks the EU with facilitating access to vocational training and encouraging mobility of instructors and trainees, particularly young people. In addition, the 2000 Charter of Fundamental Rights of the EU only includes one article on youth[8]: article 32 providing for the protection of young people in the workplace.

These provisions are limited in number and, most of all, only cover very specific topics. Most actions refer to employment, mobility, and training; with only one on participation in democracy. This means that the overall framing of youth is a framing centered around its economic potential:

> Nevertheless, researchers also prevailingly observe that EU youth policies before the 2000s focused on economic governance and the provision of a skilled workforce. This view is well summarized by Geyer (2000: 202): 'In essence, the EU wants the young to be educated, skilled and then employed. Beyond that, it has paid no attention'.
>
> *(Pušnik and Banjac, 2023: 1186)*

Youth are perceived as recipients of policies, and not as political agents in their own rights.

Partially going beyond this limited understanding of youth, the EU has adopted additional soft-law, non-binding instruments in addition to the fundamental treaties. More precisely, the importance of involving further youth as a specific category of actors for EU policy was recognized and youth dialogues were already organized in the 1970s, after the major historical event of the May 1968 protests (Pušnik and Banjac, 2023). After their ad-hoc sporadic and irregular organization in the 2000s, since 2010, Structured Youth Dialogues have been organized in the framework of the first official EU Youth Strategy (2010–2018). Within its framework, the EU Youth Dialogues are organized in 18-month work cycles with the results discussed at the EU Youth Conferences that happen twice a year. They might lead to the adoption of new youth objectives.

Interestingly, these EU-level soft-law initiatives were strengthened after the FFF protests:

> After Greta Thunberg spoke in the European Parliament in 2019, the European Commission initiated several youth events attended by EU leaders (Tenti, 2019). FFF activists met with the European Commission President Ursula von der Leyen three times between 2018–2022 to discuss climate issues (Jack, 2022). The European Economic and Social Council (EESC) hosted Youth Climate and Sustainability Round Tables twice in 2021 in cooperation with the European Youth Forum and Generation Climate Europe (a youth-led organization). Regular dialogues with young people have also been organized since 2020 as part of the European Climate Pact (Gorman, 2021), 2022 was labelled the European Year of Youth, and 305 European Climate Pact Youth Ambassadors were nominated.
>
> *(Orsini and Kang, 2023: 92)*

In 2020, the EESC adopted an opinion to propose the inclusion of a youth delegate in the official EU delegation to UNFCCC COP meetings (see below). In addition, the EESC proposed including a youth delegate as an additional member of the EESC delegation which would hold observer status at such events (European Economic and Social Committee, 2020).

The new EU Youth Strategy adopted over the period 2019–2027 included additional youth mandates. In June 2021, as part of this strategy, the Commission appointed Biliana Sirakova as the first EU Youth Coordinator. Her role is to strengthen cross-sectoral cooperation within the Commission. Finally, in 2022, the EU launched the "Voice your Vision" platform to collect testimonies of all youth on the EU and its policy development, however these are mostly from children.

Two general comments can be drawn from this brief review of EU policies involving youth as political actors. First, several initiatives have been created or reinforced following youth protests. This confirms that external strategies have an important role to play in dynamizing policies. Second, none of these initiatives are specifically dedicated to environmental policy, at least not prior to 2019. Indeed, it is only recently that the topic has gained importance in EU youth policy, culminating in 2022 with a Joint Communication to the European Parliament and the Council, for a Youth Action Plan (YAP) in EU external action 2022–2027. This plan promotes meaningful youth participation and empowerment in EU external action for sustainable development, equality, and peace. The plan mentions the United Nations (UN) youth delegate program (see below) as an important program on these issues. In 2022, the EU has developed a one-year pilot project consisting of two EU-UN Youth Delegates who were, however, entitled mostly to follow the UN General Assembly meetings (Interview 6). The UN Youth Delegates Program otherwise mostly relies on its adoption and implementation by the Member States.

### *Current policies and participatory mechanisms at the Member State level*

Youth official representation in political positions, within the EU Member States, is a problem: "the current composition of European legislatures and executives does not even begin to approximate an appropriate generational representation" (Buchholz, 2021: 8). This confirms a worldwide tendency (Stockemer and Sundström, 2022), whereby decision-makers are much older than the people they are expected to represent. This is one reason why most EU Member States have created specific youth structures and mandates, to improve the representation of youth, in parallel to voting and to EU-level youth policy and mechanisms. The two most institutionalized structures and mandates mentioned during our interviews have been: The National Youth Councils and the UN National Youth Delegates Program.[9]

#### *National Youth Councils*

National Youth Councils are organizations created for youth to be better represented in national policies: "National Youth Councils exist in many forms and in many countries around the world, serving as coordinators of youth organizations on the national level, contacts with similar organizations in other countries, and the nexus of youth-government relations" (TakingITGlobal, 2006: 6). They are not specific to the EU, but the first National Youth Councils were created in Europe: "National Youth Councils in Europe (...) have the longest history and the most experience to learn from" (TakingITGlobal, 2006: 6) and are therefore taken as examples for other countries to create their own youth structures.

The mandates and functioning of the National Youth Councils depend on the Member States. They usually organize consultations with youth, ensure the representation of youth when solicited by governments (among others, for the EU Youth Dialogues (see above)), and all topics that might concern youth. Therefore, National Youth Councils, like EU-level youth policy, have only progressively embraced environmental topics within their mandates. In 2006, among the 11 National Youth Councils of EU members that existed at the time, only the National Youth Council from Portugal specifically had the environment as a priority topic. There exists no recent study of National Youth Councils and their current mandates, but the environment seems to now be a priority for all of them according to our interviews, with the National Youth Councils often in charge of selecting some youth delegates to international environmental negotiations.

### *UN National Youth Delegates Program*

Since 1995, the UN has been promoting the UN Youth Delegates program, in the framework of a broader UN Youth Strategy. Such a program invites governments to integrate youth delegates into their delegation to important international meetings. Initially, the targeted meetings were the UN General Assembly, but progressively, youth delegates have also been sent to climate and biodiversity international negotiations:

> In Ireland we have generalists, we just have UN (General Assembly) youth delegates. Other countries have youth delegates for HLPF (…) or else like UN delegates for sustainable development, youth delegates for gender equality. Denmark is a good one to look at that: I think they have 8 delegates and four thematic areas.
>
> *(Interview 20)*

The EU is seen as a good player within the overall program: "European countries are particularly well represented in this report,[10] which largely reflects the high level of activity of youth delegate programs in Europe, as well as capacities to establish such programs" (Lo Curlo et al., 2023: 5).

No comprehensive data exists on the current UN Youth Delegates within EU Member States' delegations to international environmental negotiations. A recent study on UNFCCC COPs indicates that:

> Among the 119 parties that sent youth delegates to climate COPs, 81 sent them to more than two meetings, and 15 of them are EU member states (Austria, Belgium, Denmark, Finland, the Netherlands, Sweden, France, Hungary, Latvia, Czech Republic, Slovenia, Poland, Germany, and Luxembourg).
>
> *(Orsini and Kang, 2023: 89)*

During fieldwork, a Finland youth delegate indicated that there were in total about 26 UN Youth Delegates at the biodiversity COP15, including the following for EU Member States' delegations: "Finland 2, Sweden 1, Norway 1, Denmark 2, Netherlands 1, France 1, Belgium, 1." While UN Youth Delegates for climate have been created several decades ago, the phenomenon is recent within biodiversity COPs, as Belgium is found to be the only EU country to have included youth delegates for two CBD COPs. All the UN Youth Delegates for biodiversity mandates were created only in 2022, for COP15 (this is the case for Finland, interview 8). Regarding the SDGs' negotiations, Belgium, Germany, The Netherlands, and Sweden are the only EU delegations to have included UN Youth Delegates for more than two Commission on Sustainable Development (CSD)/HLPF meetings. Youth Delegates from these countries, together with their Canadian equivalent, presented a joint statement at CSD16 (in 2008), urging other countries to include youth delegates in their delegations (Walter et al., 2008).

Such data show discrepancies in the way the UN Youth Delegates program is implemented in the different Member States. First, many EU countries did not have UN Youth Delegates in 2022, as was the case of Cyprus, Spain, France, and Greece (Lo Curlo et al., 2023). Second, some EU countries, like Belgium, The Netherlands, and Sweden, have a more comprehensive approach regarding youth participation in environmental issues, with UN Youth Delegates dedicated both to climate change and biodiversity; or biodiversity and the SDGs' negotiations, while other countries only have UN Youth Delegates for one topic.

### *Interactions between mechanisms*

There are correspondences between the National Youth Councils and the UN Youth Delegates mandates' selection:

> In some programs, particularly in Northern Europe, candidates must be nominated by a youth organization to be eligible to apply (Denmark, Finland, Iceland, Norway, Sweden). Candidates are not only considered as individual young persons, but as representatives of their nominating organizations. The rationale for nominations is based on a goal of ensuring balanced representation of young people and youth organizations involved in the youth delegate program.
>
> *(Lo Curlo et al., 2023: 16)*

In Ireland and Finland, it is actually the National Youth Council that proceeds to the UN Youth Delegate selection. Sweden follows a similar process: "In Sweden, we have a National Youth Council called LSU which elects one youth delegate for climate each year" (interview 10). This enables youth to select candidates among themselves: "There is a process where at least youth select the person that they think is most eligible" (interview 10).

This last quotation, and especially the "at least" expression, opens the door to questioning whether the detailed policies and representation mechanisms designed for youth within the EU really improve their representation for environmental policy. The next section details three challenges to the inclusion of youth in such policy, highlighted by the interviewees.

## European youth and environmental policy: challenges

### *Lack of continuity of youth participatory mechanisms*

One key issue is the lack of continuity of the above-detailed policies and participatory instruments first and foremost because of the lack of funding dedicated to them. The UN Youth Delegates program, when implemented by the Member States, most of the time leads to the creation of volunteering positions: "There are disparities in funding even in Youth Delegates programs in Europe. Some are getting hundreds of thousands (of euros) and others are getting six thousands. It really just depends on the government. Some are getting none at all" (interview 20). Actually, only 2 of the 20 UN Youth Delegates from EU Member States that we interviewed were paid for their mandate. Moreover, funding is always for short-term commitments and can be ended at any time. This has been the case for the Swedish UN Youth Delegates program, with the UN Children's Fund providing money at the very last minute for the youth delegate to be able to attend the planned HLPF meeting (interview 10).

In addition to funding, the way the program is institutionalized can be very different from one Member State to the other, and often lacunary. Already the decision to join the program is often ad-hoc and bottom-up: for example, it is when a UNFCCC COP got scheduled in Madrid that Spain decided to have a Youth Delegate. As explained by one Belgian Youth Delegate: "It also depends on how much energy people have in creating the youth delegates. So, for instance, the biodiversity position (for Belgium) was created by someone who was very motivated to work on biodiversity issues" (interview 18).

Moreover, many youth complain about the differences of institutionalization and seriousness of the different national programs:

> if you read the Secretary General policy brief about youth participation and youth inclusion, it says that 90 UN Member States[11] participate in the UN Youth Delegates program, and I can tell you categorically that since the inception of the UN Youth Delegates program, there has never, at any one stage or over the course of time that it has been implemented, there has never been 90 Member States engaged with the program.

> It has just not been the case. There are also Member States who are seen to engage with the program, they only take UN Youth Delegates here for three weeks and that is it. And to me that is not meaningful engagement. It needs to be prolonged; it needs to be sustainable; it needs to be maintained. It's not enough to say that if you have a UN Youth Delegate, you engage with the program. Just because you took them to one conference.
>
> *(interview 20)*

Another challenge expressed has been the lack of continuity in youth recruitment for the mandates, giving the impression that decision-makers "cherry pick" youth when they need one youth representative, instead of engaging with youth already established in youth representation dynamics, such as youth from the National Youth Councils for instance. As explained by one interviewee:

> Also, there was like a statement on behalf of the youth representative. But he wasn't a youth representative. He was a young person representing himself. He was some TikTok creator but he, but he doesn't have a mandate representing anyone. So like how is that youth representation rather than just a young person speaking? And so that is what we are very focused on that that once talking for youth actually having a mandate to talk for a broader youth.
>
> *(interview 18)*

One interviewee denounced clientelism in the choice of youth delegates, but the practice seems to be rather limited (interview 3). On the one hand, the 'control' of youth recruitment by established youth organizations could limit the possibility to include unorganized youth, especially in the EU Youth Dialogues (Pušnik and Banjac, 2023: 1183). On the other hand, such control ensures that youth are not represented by third party and can more directly reflect the positions of young people.

Finally, the lack of involvement throughout the whole process leading to international environmental discussions has been mentioned as a true limit. As explained by one interviewee:

> Yes what she says is right (referring to a youth colleague): if you want to bring youth to the COP, don't bring them only to the COP. Bring them to the delegation meetings the whole year, because otherwise you are not prepared. You won't know what is going on. And it doesn't really make sense. So youth participation is only working if youth is just put from the beginning of the process.
>
> *(interview 19)*

Interestingly, these challenges of continuity have generated adaptation strategies by the concerned youth. European youth are trying to push for all Member States to have UN Youth Delegates, to reinforce the importance of the program:

> Spain will be the one being the Presidency to come for Europe. So that will mean that the Presidency has the possibility to advocate for other things. So we are really having a compromise on making Spain making mandatory for all of the European countries' delegation to have youth delegates.
> *(interview 13)*

Interestingly, to create continuity, learning, and institutional memory, some EU countries like Belgium, France, The Netherlands, and Sweden have decided to nominate two youth delegates for climate or two delegates for biodiversity for two years, with half of them renewed each year. As explained by a Swedish youth:

> The mandate is for two years, and since one person is chosen each year, there will always be one of the youth delegates who is the senior and has one extra year of experience which I think is very good since it really helps to bring in the other youth delegates and introduce them to the role.[12]
> *(interview 20)*

On its hand Denmark is having a unique youth delegate on "climate and the environment" attending both biodiversity and climate COPs for better continuity in both processes.

Coordination among the different youth delegates from the Member States exists, but only informally, in the format of common chat platforms or emails. For Nordic countries, some youth delegates are integrated in the Nordic Council which is more institutionalized and enable them to more regularly prepare collectively for the negotiations.

### *Limited inclusion in decision-making*

While the above-mentioned policy mechanisms are great steps in including youth views in decision-making related to environmental objectives, these policies and mechanisms are still not fully inclusive. For instance, the UN Youth Delegates from EU Member States were initially not allowed to participate in the coordination meetings that the EU is organizing every morning and every evening during international environmental negotiations:

> Because the EU negotiates as one you know, so first of all is that there is the silent agreement that the youth delegates aren't allowed to be in the

> EU coordination meetings before COP and during the COP, but now it starts to like kind of break because there have been some youth delegates present at times, but it has been a very long time that no youth delegates are allowed to be there. So that's like one big problem because you know, like in the coordination meetings there's a lot of power used when they create their opinion.
>
> *(interview 8)*

Sometimes, youth delegates are not even included in the national delegation and do not get party badges; or are included but cannot participate in the coordination meetings of the delegation.

Moreover, it is rare that youth delegates be invited to take the floor to express their views, even in plenary sessions[13]:

> And then like in the actual negotiations, it's never, it's never youth delegates. And then in the plenary, in some other COPs like climate COPs and the SDG Forum in New York, some countries have started to do it: youth delegates can have half of the country's statement. So here for example the Netherlands did it. But I think here in CBD, compared to youth environment in general it is like a little bit less, so it was only the Netherlands who did it. But for example, at UNEA,[14] there were many more country delegates.
>
> *(interview 8)*

Again, the situation can vary from one country to the other. According to one interviewee, about four to five European youth delegates are really speaking during the negotiations. On this, Sweden is a particular case in which youth feel they have their say in the delegation and negotiations, showing that meaningful participation is feasible. But overall, interviews confirm previous studies that indicate: "While the EU's organizational support for youth participation is growing, youth delegates' activities are often limited to subsidiary roles such as observers, learners, or communicators without negotiation mandates" (Orsini and Kang, 2023: 92). The presence of youth is growing but their meaningful inclusion is not warranted: "Young people might be tokenized by the authorities to create an illusion of openness and progress" (Lelwic-Ojeda and Akintola, 2023: 5). One Belgian youth delegate explains that it can happen that governments "invite one young person to give a 5-minute speech to close an official event" just because it is fashionable (interview 18). Another youth delegate confirms, "Oh yeah, they always want to take selfies with youth" and also explains that youth is now developing strategies to avoid such situations that they have become aware of "you know, we are seeing through like the game it is" (interview 19).

A recent study on national youth policies in ten countries around the world indicates

> Leveraging the potential of the youth as agents of change, entrepreneurs, and innovators has become a central topic discussed by policymakers (...). Despite these prospects, young people are often viewed as inexperienced, disruptive, or lacking the required skills to meaningfully participate in decision-making processes.
>
> *(Lelwic-Ojeda and Akintola, 2023)*

While the study includes only one European country, it echoes many of the interviews conducted for this study.

### *The persistence of one-way communication*

Participation is hampered by a lack of symmetry in the way the information is transmitted between youth and decision-makers, with, most of the time, decision-makers trying to teach or train youth. For instance, during interactions, youth are often confined to a role of communication back home, and information diffusion (interview 12), especially on social media – with governments asking them to relay their activities on social media (interviews 5, 7, and 15, each from different countries) and not the other way around – that is, on the general evolution of the negotiations according to the governmental agenda, not according to what would be important for youth. This is related to the capture of youth identities into certain specific roles, that do not reflect the diversity of their interests and knowledge. Youth are used as communicators on social media:

> But in (my country), yeah, I participated in encounters with the politicians, which was clearly the point was to tweet that ...yeah, 'this minister has met young people'. Then there's your Twitter account that is fed and. And that's it. So yeah, it happens.
>
> *(interview 15)*

A key change would be for youth to be truly listened to as explained by one Belgium youth delegate:

> (Being there is) better than nothing, I would say. Because it's complicated. You see that it's the main issue here is that everyone is talking about youth. Like of course, like just now the speech of our Prime Minister saying that the youth is super important and we have to take them into account and please be engaged. But, like, we are trying to do it. So please listen to us. So

> that's why we are also asking to listen to the youth, but also to make them participate. Because there is often youthwashing here, like saying that they are hearing us, but they are not doing anything so yeah (...). But it is still a lot of things to do to be sure that we are listened.
>
> *(interview 14)*

One-way communication and the confinement of youth to certain roles are somehow related to a culture of senior-centric institutions:

> Current modes of youth participation are often determined from an adult[15]-centric perspective, where elders frequently minimize the value of youth contributions while expecting them to navigate systems designed solely by and for adults, and only when adults deem it indispensable or beneficial.
>
> *(Lelwic-Ojeda and Akintola, 2023: 5)*

## Conclusion

This chapter has investigated the interactions between youth and the EU on environmental policy. The initial hypothesis, that youth might resort to protest to express views they are not able to express by insider strategies is to some extent validated. While youth have an important political interest for environmental issues, that they perceive in a broad sense, the existing mechanisms and policies at the EU and national level are underdeveloped. While the available scientific literature on these instruments for youth to engage directly in politics is limited, the existing testimonies gathered for this research confirm that such mechanisms and instruments present some limits. EU initiatives toward youth, especially the National Youth Councils and the UN Youth Delegates Program are seen as positive, and often taken as examples. As explained by one interviewee "a lot of countries (outside the EU) don't even have youth delegates as far as I have understood" (interview 19). But this contribution also identifies ways for improving the EU involvement toward youth. Another key result is to highlight the importance of outsider strategies, such as protests, in fostering a policy response engaged toward more youth involvement.

To be sure, perfect participation is an ideal that is rarely met. But in the case of youth, their lack of resources (they are most of the times students, with no revenue and professional position), their willingness to push for more environmental ambition (and even more), and the very fact that they will be implementing policies in the future, would require more attention from policymakers willing to adopt fair, ambitious and feasible environmental objectives. What is required is a systemic change that is likely to be about changing or at least adjusting "The old is gold" syndrome that was already identified by research on age in national parliaments worldwide (Inter-Parliamentary Union, 2021: 41), and according to which individuals who have not reached a certain age have no value.

To some extent, the results resonate with the following assessment of Mariya Gabriel, Commissioner for Innovation, Research, Culture, Education and Youth, on the Year of Youth in 2022: "As we can see today, young people are increasingly involved and engaged in the future of societies. They care about peace, international cooperation, wellbeing, the environment. But they want to be listened to and heard more." If they will be able to do so through insider strategies, and not protests, depends on the capacities of current EU Member States to deepen their recent positive development of mechanisms to include youth as important agents of politics.

While putting light on the involvement of the EU in environmental policymaking, this chapter also opens several possible complementary research paths. First, with the perception of youth now partly analyzed, further research could work on gathering the experience of other international actors, and especially decision-makers, with regards to youth. Moreover, comparison with the current situation in other policy fields than the environmental one could be important to compare challenges as well as best practices.

## Notes

1 In this chapter, youth are defined according to the 16–35 age range. Sixteen is the usual lower age limit for youth vote across Europe and 35 is the upper limit for youth involved in the youth constituency (YOUNGO) of the UNFCCC.
2 A random number has been assigned to each interview throughout this chapter to preserve the anonymity of the interviewees.
3 This chapter does therefore not study all European youth but the specific proportion of them who are engaged for the environment at the international level and who occupy insider youth representation roles within the EU.
4 Which took place in 2022.
5 Age range 15–35 years old.
6 Spread of infectious diseases only arrives third (36%).
7 Voting remains the main insider strategy for any actor to engage in the political process. In this chapter we focus on additional strategies to understand their specificity. This is also due to the fact that EU environmental policy is also determined by international dynamics where vote is not a tool used for decision-making. In any case, recent research has measured that "younger individuals vote more consistently with their environmental concern than older voters" (Mannoni, 2024: 1).
8 As defined in this chapter, otherwise it has two articles on children.
9 Another program, the "Youth Delegates project" gives young people from the 46 Council of Europe Member States the possibility to follow the Council discussions. This format was not mentioned during our interviews though and is rather a format of observation, not of participation to policymaking.
10 A report on 44 National Youth Delegates Program out of which 24 are European.
11 In Ireland, they have a 18 month mandate with 6 months overlap.

12 On the one hand, plenary sessions are important because all States are represented; on the other hand, most of the important and hard discussions happen in the working groups. Youth seem to be more easily invited to provide statements at SDGs meetings.
13 The General Assembly of the UN Environment Programme.
14 Because most youth are adults, we prefer to refer to a senior-centric culture here.

## References

Buchholz, T. (2021). *Fixing the relationship between the European Union and its youth: The case for the creation of a European Youth Parliament* (Master's thesis). University College Maastricht. Retrieved from https://thilobuchholz.de/wp-content/uploads/2021/09/Capstone-BuchholzThilo-Public.pdf

Dupont, C., Oberthür, S., & von Homeyer, I. (2020). The Covid-19 crisis: A critical juncture for EU climate policy development? *Journal of European Integration*, *42*(8), 1095–1110. https://doi.org/10.1080/07036337.2020.1853117

European Commission. (2022, May 6). *Eurobarometer on the European Year of Youth: Young Europeans are increasingly engaged* [Press release]. Brussels. Retrieved from https://ec.europa.eu/commission/presscorner/detail/en/ip_22_2774

European Economic and Social Committee. (2020). *Towards structured youth engagement on climate and sustainability in the EU decision-making process* [Own-initiative opinion]. Retrieved from https://www.eesc.europa.eu/en/our-work/opinions-information-reports/opinions/towards-structured-youth-engagement-climate-and-sustainability-eu-decision-making-process-own-initiative-opinion

Eurostat. (2022). *Young people in Europe: A statistical summary* [Report]. Retrieved from https://ec.europa.eu/eurostat/web/products-catalogues/-/ks-06-22-076

Hurrelmann, K., & Albrecht, E. (2021). *Gen Z: Between climate crisis and coronavirus pandemic*. Routledge. https://doi.org/10.4324/9781003128700

Inter-Parliamentary Union. (2021). *Youth participation in national parliaments*. Report. Retrieved from https://www.eods.eu/library/IPU_2021_YouthParticipationParliament_EN.pdf

Ipsos. (2021). *Pan-European survey: Main multi-country report*. Study commissioned to Ipsos by the #ClimateOfChange project in the framework of its pan-European "End Climate Change, Start Climate of Change" #ClimateOfChange campaign. Retrieved from https://eeb.org/wp-content/uploads/2021/04/IPSOS-Multi-Country-Report-complete.FINAL_.pdf

Lelwic-Ojeda, A., & Akintola, L. (2023). *A comparative analysis on National Youth Policies*. WYDE Civic engagement project. Retrieved from https://youthdemocracycohort.com/wp-content/uploads/2024/03/A-comparative-analysis-on-National-Youth-Policies-Full-Layout.pdf

Lo Curlo, J., Lamberg, L., & Luts, L. (2023). *Global youth voices: A mapping report of youth delegate programs to the UN*. Future Agents Now research project. Retrieved from https://duf.dk/internationalt-arbejde/global-youth-voices

Magosi, O.-K., & Hammond, P. (2023). Climate crisis and youth vocal representatives of the climate discourse. In P. Pillók & L. Székely (Eds.), *Hard times create strong youth: The impact of the era of crisis on future generations* (pp. 117–136). National Youth Council of Hungary. ISBN: 9786150188645. https://doi.org/10.1007/s40647-020-00297-0

Mannoni, E. (2024). Pro-environmental voting: What it is, how to measure it, and its determinants among contemporary European voters. *Parliamentary Affairs*, *78*(1), 77–96. https://doi.org/10.1093/pa/gsae017

Oberthür, S., & Dupont, C. (2021). The European Union's international climate leadership: towards a grand climate strategy? *Journal of European Public Policy*, *28*(7), 1095–1114. https://doi.org/10.1080/13501763.2021.1918218

Orsini, A., & Duque, J. F. (2024). *Questionner les frontières de la souveraineté. La jeunesse et la crise environnementale mondiale*. Raisons Politiques. *96*(4), 15–29.

Orsini, A., & Kang, Y. H. (2023). European leadership and European youth in the climate change regime complex. *Politics and Governance*, *11*(2), 84–96. https://doi.org/10.17645/pag.v11i2.6500

Parks, L., della Porta, D., & Portos, M. (2023). Environmental and climate activism and advocacy in the EU. In T. Rayner, K. Szulecki, A. J. Jordan, & S. Oberthür (Eds.), *Handbook on European Union climate change policy and politics* (pp. 98–112). Edward Elgar.

Perovic, B. (2016). *Defining youth in contemporary national legal and policy frameworks across Europe*. Report. Retrieved from https://pjp-eu.coe.int/documents/42128013/47261653/Analytical+paper+Youth+Age+Bojana+Perovic+4.4.16.pdf/eb59c5e2-45d8-4e70-b672-f8de0a5ca08c

Pickard, S. (2022). Young environmental activists and Do-It-Ourselves (DIO) politics: Collective engagement, generational agency, efficacy, belonging and hope. *Journal of Youth Studies*, *25*(6), 730–750. https://doi.org/10.1080/13676261.2022.2046258

Pušnik, T., & Banjac, M. (2023). Dialogue with youth 'is not a dialogue among "elites"': Problematization of dialogue with unorganized youth in the EU. *Journal of Youth Studies*, *26*(9), 1182–1199. https://doi.org/10.1080/13676261.2022.2080537

Rosamond, J., & Dupont, C. (2021). The European Council, the Council, and the European Green Deal. *Politics and Governance*, *9*(3), 348–359. https://doi.org/10.17645/pag.v9i3.4326

Skeirytė, A., Krikštolaitis, R., & Liobikienė, G. (2022). The differences of climate change perception, responsibility and climate-friendly behavior among generations and the main determinants of youth's climate-friendly actions in the EU. *Journal of Environmental Management*, *323*, Article 116277. https://doi.org/10.1016/j.jenvman.2022.116277

Stockemer, D., & Sundström, A. (2022). *Youth without representation: The absence of young adults in parliaments, cabinets, and candidacies*. University of Michigan Press. https://doi.org/10.3998/mpub.11459940

TakingITGlobal. (2006). National Youth Councils: Their creation, evolution, purpose, and governance. Retrieved from https://www.ndi.org/node/23775

Tyson, A., Kennedy, B., & Funk, C. (2021, May 26). Gen Z, millennials stand out for climate change activism, social media engagement with issue. *Pew Research Center*. Retrieved from https://www.pewresearch.org/science/2021/05/26/gen-z-millennials-stand-out-for-climate-change-activismsocial-media-engagement-with-issue/

Wahlström, M., Kocyba, P., De Vydt, M., & de Moor, J. (2019, July 29). Protest for a future: Composition, mobilization and motives of the participants in Fridays for Future climate protests on 15 March 2019 in 13 European cities. *Protest Institute*. Retrieved from https://gup.ub.gu.se/publication/283193

Walter, K., Hassel, M., Dalunde, J., Develtere, L., Niessen, D., Bocquet, A., Dafoe, J., & Damen, M. (2008). Count the youth delegates, youth delegates count! *United Nations Department of Economic and Social Affairs Social Inclusion*. Retrieved from https://www.un.org/esa/socdev/unyin/documents/ydcsd08_article.pdf

Williamson, H. (2007). A complex but increasingly coherent journey? The emergence of 'youth policy' in Europe. *Youth & Policy*, *95*(Spring), 57–72. Retrieved from https://www.youthandpolicy.org/wp-content/uploads/2017/06/youthandpolicy95-1.pdf

Wurzel, R. K., Liefferink, D., & Torney, D. (2019). Pioneers, leaders and followers in multilevel and polycentric climate governance. *Environmental Politics*, *28*(1), 1–21. https://doi.org/10.1080/09644016.2019.1522033

Zamponi, L., Baukloh, A. C., Bertuzzi, N., Chironi, D., della Porta, D., & Portos, M. (2022). (Water) bottles and (street) barricades: The politicisation of lifestyle-centred action in youth climate strike participation. *Journal of Youth Studies*, *25*(6), 854–875. https://doi.org/10.1080/13676261.2022.2060730

# EU ENVIRONMENTAL GOVERNANCE REVISITED

## Conclusion

*Amandine Orsini and Eleni Kavvatha*

### Introduction

The second edition of this book on the European Union (EU)'s environmental governance emerges from a pressing need to reflect on the rapid changes and significant developments that have shaped the EU's environmental agenda in recent years. Surprisingly, considering the major shocks that appeared in the post-2020 era, such an agenda has still been rather prolific. Despite the pandemic, the energy crisis, and the escalating environmental crisis, the EU continues to play a pivotal role as a global leader in environmental governance by proposing innovative legal and policy frameworks. For sure, the picture is not completely positive, and the book chapters also identify points of attention. In this short conclusion, we precisely provide a transversal analysis of the EU's evolving environmental governance framework as detailed in the different chapters, highlighting both the successes and ongoing challenges the Union faces in addressing pressing environmental issues.

### Successes of EU environmental governance

#### *New important topics are regularly included on the EU's agenda*

A comparison of the two editions of the book reveals that new topics have appeared on the EU environmental agenda through time. Such agenda not only (i) adapts to ongoing dynamics, but also (ii) tackles new emerging challenges. On the first dimension, among others, the 'one health' approach (Chapter 8) has been reinforced after the COVID-19 pandemic, in order to improve coherence along policies that were so far fragmented, a fragmentation that

DOI: 10.4324/9781003479260-20

facilitated the sanitary crisis. In a similar dynamic, European instruments to include youth in the formulation of environmental policies (Chapter 12) have somehow been an attempt to respond to the large youth protests in the format of the climate marches initiated by Greta Thunberg in December 2018. Finally, initiatives to recognize rights for nature (Chapters 1 and 10) also respond to some extent to mobilizations by NGOs on this issue.

On the second dimension, marine biodiversity beyond national jurisdictions is an important new topic (Chapter 6) that appeared following technological improvements that have enabled new seabed mining activities. The EU has decided to embrace this new issue by being proactive for the adoption of a new international agreement on the topic. The EU's leadership in the International Ocean Governance framework further strengthens its commitment to global marine conservation and sustainable management. Also, the EU is increasingly strengthening its policies in the Arctic (Chapter 11), conscious of the emerging challenges shaking the region as a consequence of increased climate change and energy needs. Finally, while nuclear waste management is not a completely new topic, the EU Member States have recently put efforts to develop long-term solutions through a more transparent decision-making process (Chapter 3). Taken together, these chapters show that the EU environmental agenda is continuously evolving, flexible, and extendible enough to tackle new challenges.

### *Environmental governance beyond EU borders has been strengthened*

Several chapters indicate the growing interest of the EU for (i) environmental topics that are global per nature; (ii) initiatives that can reinforce the EU environmental agenda outside its borders. On the first dimension, marine biodiversity beyond national jurisdiction (Chapter 6) and climate change (Chapter 2) are environmental topics important for the EU territory, but they mostly go beyond it. And yet, the EU is investing considerable efforts to ensure that solid international commitments are taken on both topics.

On the second dimension, Chapter 9 explains how recent EU directives are meant to cover the activities of EU firms beyond the European territory. These instruments are essential to mitigate environmental conflicts and bring a greater environmental justice framework worldwide. Another key EU strategy has been to include environmental provisions in trade agreements. The EU has made efforts to incorporate sustainability criteria into trade agreements and foster green technology innovations (Chapter 10). Both chapters show that the ongoing tension between economic growth and environmental sustainability remains one of the EU's greatest challenges. The promotion of free trade often conflicts with environmental protection, and reconciling these competing imperatives requires innovative solutions and political will. This balance between economic imperatives and environmental protection remains a notable exception and success of EU environmental governance.

### *Inclusion is a priority*

Attentive to growing protests following the yellow vests movement, climate marches, and pandemic resistance, the EU is increasingly aware of the necessity to keep all civil society actors on board of its governance choices. On climate policy, the EU has positioned itself as a global leader in climate action, with a clear commitment to the Paris Agreement. Chapter 2 outlines the core principles of EU climate policy, which aim for significant greenhouse gas reductions while balancing the needs of affected populations and industries. The Union has been proactive in integrating climate relocations into its policy frameworks, demonstrating its commitment to a green transition.

Chapter 4 on renewable energy communities, Chapter 12 on youth, and Chapter 11 on the Arctic and the importance of Indigenous Peoples show the real concern of the EU to include a wide variety of non-state actors in its environmental governance efforts. However, achieving just transitions and managing the social and economic impacts remains an important priority and even a challenge.

## Challenges facing EU environmental governance

### *Convincing other players*

In many respects, this book gives a very specific, and special, view of environmental governance. A view from a particularly active international player, the EU. However, the EU will not be able to sort out worldwide environmental problems on its own. This is one reason why it is increasingly orienting its actions beyond its borders (see above). But this is unlikely to be enough. Arctic governance (Chapter 11) requires collaboration with other global powers to protect the region's fragile ecosystems. Environmental rights (Chapter 1) and the 'one health' approach (Chapter 8) are touching upon truly global agendas, and will need, as new visions for greater sustainability, to convince other international players.

### *Transforming the system, not just adjusting it*

Several chapters also show that, in some domains, the EU risks proposing adjustments that only intervene at the margins instead of creating real transformation. Biodiversity and agriculture (Chapter 5) is probably the key illustration of such a limit. The EU's focus on sustainable farming practices is an essential step in promoting biodiversity. Chapter 5 emphasizes the importance of integrating environmental goals into the Common Agricultural Policy to ensure long-term environmental resilience. Rewilding initiatives offer promising solutions to restore ecosystems damaged by intensive agriculture. But large-scale implementation of both solutions (sustainable farming and rewilding) will require overcoming institutional barriers and resistance from farming communities and industries that may see them as incompatible with their interests.

A second landmark example is the imperative to focus on circular economy instead of fighting overconsumption for example. The EU's transition toward a circular economy is reflected in its evolving waste management framework, particularly the Sustainable Batteries Regulation (Chapter 7). This regulation aims to reduce the environmental impacts of batteries throughout their lifecycle, underscoring the EU's drive to minimize waste and improve recycling processes. But such measures do not tackle the root cause of waste production: overconsumption. As recalled by youth (Chapter 12), Indigenous Peoples (Chapter 11) or communities (Chapters 10 and 4), it is also the very fundamental logic of the market that could be questioned.

### *Simplifying processes*

Many chapters explain how complicated the accumulation of different legislation and policies can be in the EU. For youth to participate in environmental governance (Chapter 12), one challenge relies on the many different mechanisms that have been developed, with none being really efficient. And yet, the involvement of youth in environmental policy will be crucial to sustain momentum and ensure that the EU's environmental agenda remains forward-thinking and inclusive. In the same line, despite progress, the harmonization of waste management policies across Member States continues to pose significant challenges. The Sustainable Batteries Regulation (Chapter 7) exemplifies the difficulty of ensuring compliance and fostering technological innovation in recycling processes. Among others, 'one health' policies, as explored in Chapter 8, represent an integrated approach that connects human, animal, and environmental health, positioning the EU as a leader in holistic environmental governance. Though increasingly popular, it is still in the early stages of full implementation and face barriers in terms of coordination across sectors and Member States.

## Future research

Overall, the EU's environmental governance framework has made significant progress in addressing global environmental challenges, from climate policy and waste management to marine biodiversity. However, the ongoing challenges—the harmonization of regulations across Member States and beyond the Member States, the balancing of economic growth with sustainability for a real transformation, and the coherence and integration of emerging environmental concerns—require continuous adaptation and innovative solutions. As the global environmental crisis deepens, the EU's ability to reconcile competing interests, adapt to new challenges, and collaborate both within and beyond its borders will be crucial in shaping a sustainable future for Europe and the world.

In addition to furthering the analysis contained in the different chapters of this volume, new topics could be further researched: in 2025, States are expected to adopt a new plastic treaty at the global level, and following such negotiations would be important. Most of all, as we conclude this second edition, the recent re-election of Donald Trump as American President and the adoption of his Presidential Action "putting America first in international environmental agreements"[1] means the United States is withdrawing again from the Paris Agreement under the United Nations Framework Convention on Climate Change. It also refuses any international commitment, or financial contribution, that could go against the economic interests of the American people. This represents a major shock for multilateralism more broadly, and environmental multilateralism more specifically. With another recent declaration indicating that the United States claims the Greenland territory in the Arctic,[2] it is becoming clear that the EU will face increasing challenges in promoting an environmental agenda that is viable worldwide.

## Notes

1 https://www.whitehouse.gov/presidential-actions/2025/01/putting-america-first-in-international-environmental-agreements/.
2 https://www.bbc.com/news/articles/c74x4m71pmjo.

# Index

For Product Safety Concerns and Information please contact our EU representative GPSR@taylorandfrancis.com
Taylor & Francis Verlag GmbH, Kaufingerstraße 24, 80331 München, Germany

www.ingramcontent.com/pod-product-compliance
Lightning Source LLC
LaVergne TN
LVHW010651300926
849161LV00005B/1100
* 9 7 8 1 0 3 2 7 6 6 0 1 0 *